Lecture Notes in Computer Science 9347

Advanced Research in Computing and Software Science

Subline of Lecture Notes in Computer Science

More information about this series at http://www.springer.com/series/7409

Martin Hoefer (Ed.)

Algorithmic Game Theory

8th International Symposium, SAGT 2015 Saarbrücken, Germany, September 28–30, 2015 Proceedings

Springer

Editor
Martin Hoefer
Max-Planck-Institut für Informatik
Saarbrücken
Germany

and

Cluster of Excellence M2CI
Universität des Saarlandes
Saarbrücken
Germany

ISSN 0302-9743 ISSN 1611-3349 (electronic)
Lecture Notes in Computer Science
ISBN 978-3-662-48432-6 ISBN 978-3-662-48433-3 (eBook)
DOI 10.1007/978-3-662-48433-3

Library of Congress Control Number: 2015949442

LNCS Sublibrary: SL3 – Information Systems and Applications, incl. Internet/Web, and HCI

Springer Heidelberg New York Dordrecht London

Printed on acid-free paper

Springer-Verlag GmbH Berlin Heidelberg is part of Springer Science+Business Media
(www.springer.com)

Preface

This volume contains the proceedings of the 8th International Symposium on Algorithmic Game Theory (SAGT), held in Saarbrücken, Germany, in September 2015.

The program of SAGT 2015 consisted of three invited lectures and 29 presentations of refereed submissions. After a careful reviewing process, the Program Committee selected 23 out of 63 submissions for regular presentation and invited 6 submissions to be presented in the form of a brief announcement at the conference. The committee feels that these brief announcements added inspiration and novelty to the program.

The accepted submissions were invited to these proceedings. They cover various important aspects of algorithmic game theory such as matching under preferences, cost sharing, mechanism design, social choice, auctions, networking, routing and fairness, as well as equilibrium computation.

We would like to thank all authors who submitted their research work and all Program Committee members and external reviewers for their effort in selecting the program for SAGT 2015.

July 2015 Martin Hoefer

Preface

Organization

Program Committee

Elliot Anshelevich	Rensselaer Polytechnic Institute, USA
Felix Brandt	TU Munich, Germany
Peter Bro Miltersen	Aarhus University, Denmark
Edith Elkind	University of Oxford, UK
Angelo Fanelli	CNRS, France
Felix Fischer	University of Cambridge, UK
Dimitris Fotakis	NTU Athens, Greece
Nick Gravin	Microsoft Research, USA
Monika Henzinger	University of Vienna, Austria
Martin Hoefer (*Chair*)	MPI Informatik and Saarland University, Germany
Max Klimm	TU Berlin, Germany
Piotr Krysta	University of Liverpool, UK
Jochen Könemann	University of Waterloo, Canada
Yishay Mansour	Tel Aviv University, Israel
Vahab Mirrokni	Google Research, USA
Herve Moulin	University of Glasgow, UK
Rahul Savani	University of Liverpool, UK
Guido Schäfer	CWI and VU Amsterdam, The Netherlands
Alexander Skopalik	Paderborn University, Germany
Vijay Vazirani	Georgia Tech, USA

Steering Committee

Elias Koutsoupias	University of Oxford, UK
Marios Mavronicolas	University of Cyprus, Cyprus
Dov Monderer	Technion, Israel
Burkhard Monien	Paderborn University, Germany
Christos Papadimitriou	UC Berkeley, USA
Giuseppe Persiano	University of Salerno, Italy
Paul Spirakis (*Chair*)	University of Liverpool, UK, and University of Patras, Greece

Organizing Team

Christina Fries	MPI Informatik, Germany
Martin Hoefer (*Chair*)	MPI Informatik and Saarland University, Germany

Bojana Kodric	MPI Informatik and Saarland University, Germany
Paresh Nakhe	MPI Informatik, Germany
Sabine Nermerich	Saarland University, Germany
Roxane Wetzel	MPI Informatik, Germany

Additional Reviewers

Rediet Abebe
Carlos Alos-Ferrer
Yoram Bachrach
Siddharth Barman
Dorothea Baumeister
Umang Bhaskar
Vittorio Bilò
Po-An Chen
Christine Cheng
Yun Kuen Cheung
Andreas Cord-Landwehr
Ágnes Cseh
Bart de Keijzer
Argyrios Deligkas
Nikhil Devanur
Maximilian Drees
Joanna Drummond
Paul Dütting
Amir Epstein
Hossein Esfandiari
Piotr Faliszewski
Linda Farczadi
John Fearnley
Matthias Feldotto
Aris Filos-Ratsikas
Tamas Fleiner
Lance Fortnow
Rupert Freeman
Ophir Friedler
Jugal Garg
Ofir Geri
Vasilis Gkatzelis
Gagan Goel
Negin Golrezaei
Yannai Gonczarowski
Tobias Harks
Paul Harrenstein
Johannes Hofbauer
Zsuzsanna Jankó
Ruben Juarez
Sangram Kadam
Yash Kanoria
Ian Kash
Telikepalli Kavitha
Bojana Kodric
Konstantinos Kollias
Scott Kominers
Maria Kyropoulou
Michel Lemaître
Pascal Lenzner
Yuqian Li
Thanasis Lianeas
Brendan Lucier
Thodoris Lykouris
Evangelos Markakis
Aranyak Mehta
Ruta Mehta
Igal Milchtaich
Paresh Nakhe
Renato Paes Leme
Katarzyna Paluch
Dominik Peters
Georgios Piliouras
John Postl
Morvarid Rahmani
Mona Rahn
Tim Roughgarden
Lena Schend
Kevin Schewior
Jan Christoph Schlegel
Daniel Schmand
Saeed Seddighin
Lior Seeman
Shreyas Sekar
Nisarg Shah
Martin Starnberger
Vasilis Syrgkanis
Bo Tang
Orestis Telelis
Clemens Thielen
Dimitris Tsipras
S. Matthew Weinberg
Sadra Yazdanbod
Li Zhang
Aviv Zohar

Contents

Auctions

Networking

Routing and Fairness

Equilibrium Computation

Matching Under Preferences

Stable Matchings with Ties, Master Preference Lists, and Matroid Constraints

Naoyuki Kamiyama[1,2](✉)

[1] Institute of Mathematics for Industry, Kyushu University, Fukuoka, Japan
kamiyama@imi.kyushu-u.ac.jp
[2] JST, PRESTO, Saitama, Japan

Abstract. In this paper, we consider a matroid generalization of the hospitals/residents problem with ties and master lists. In this model, the capacity constraints for hospitals are generalized to matroid constraints. By generalizing the algorithms of O'Malley for the hospitals/residents problem with ties and master lists, we give polynomial-time algorithms for deciding whether there exist a super-stable matching and a strongly stable matching in our model, and finding such matchings if they exist.

1 Introduction

The stable matching problem introduced by Gale and Shapley [1] is one of the most popular mathematical models of a matching problem in which agents have preferences. It is known [1] that if there exists no tie in preference lists, then there always exists a stable matching and we can find one in polynomial time. However, if there exist ties in preference lists, then the situation dramatically changes. In the stable matching problem with ties, three stability concepts were proposed by Irving [2]. The first one is called the weak stability. This stability concept guarantees that there exists no unmatched pair each of whom prefers the other to the current partner. It is known [2] that there always exists a weakly stable matching and we can find one in polynomial time by slightly modifying the algorithm of [1]. This algorithm can be generalized to the many-to-one setting. The second one is called the strong stability. This stability concept guarantees that there exists no unmatched pair such that (i) there exists an agent a in this pair that prefers the other to the current partner, and (ii) the agent in this pair other than a prefers a to the current partner, or is indifferent between a and the current partner. The last one is called the super-stability. This stability concept guarantees that there exists no unmatched pair each of whom prefers the other to the current partner, or is indifferent between the other and the current partner.

One of the most notable differences between the last two concepts and the stability concept in the stable matching problem without ties is that there may exist no stable matching [2]. From the algorithmic viewpoint, it is important to reveal whether we can check the existence of matchings satisfying such stability conditions in polynomial time. For the one-to-one setting (i.e., the stable matching problem with ties), Irving [2] proposed polynomial-time algorithms

M. Hoefer (Ed.): SAGT 2015, LNCS 9347, pp. 3–14, 2015.
DOI: 10.1007/978-3-662-48433-3_1

for finding a super-stable matching and a strongly stable matching (see also [3]). For the many-to-one setting (i.e., the hospitals/residents problem with ties), Irving, Manlove, and Scott [4] proposed a polynomial-time algorithm for finding a super-stable matching, and Irving, Manlove, and Scott [5] a polynomial-time algorithm for finding a strongly stable matching. It should be noted that Kavitha, Mehlhorn, Michail, and Paluch [6] proposed a faster algorithm for this setting.

In this paper, we consider a matroid generalization of the hospitals/residents problem with ties. In our model, the capacity constraints for hospitals are generalized to matroid constraints. In contrast to the stable matching problem without ties (and the weak stability) [7], to the best of our knowledge, it is open whether we can extend the results about the super-stability and the strong stability to the matroid setting. As a step toward the settlement of this question, we focus on the situation in which we are given a master list and the preference list of each hospital over residents is derived from this master list. For the stable matching problem with ties and master lists, Irving, Manlove, and Scott [8] gave simple polynomial-time algorithms for finding a super-stable matching and a strongly stable matching. Furthermore, O'Malley [9] gave polynomial-time algorithms for finding a super-stable matching and a strongly stable matching in the hospitals/residents problem with ties and master lists. In this paper, by generalizing the algorithms of [9], we give polynomial-time algorithms for finding a super-stable matching and a strongly stable matching in a matroid generalization of the hospitals/residents problem with ties and master lists, i.e., a partial positive answer for the above question.

2 Preliminaries

For a set X and an element x, define $X + x := X \cup \{x\}$ and $X - x := X \setminus \{x\}$. A pair $\mathbf{M} = (U, \mathcal{I})$ is called a *matroid*, if U is a finite set and $\mathcal{I}$ is a family of subsets of U satisfying the following conditions. (I0) $\emptyset \in \mathcal{I}$. (I1) If $I \in \mathcal{I}$ and $J \subseteq I$, then $J \in \mathcal{I}$. (I2) If $I, J \in \mathcal{I}$ and $|I| < |J|$, then there exists an element u in $J \setminus I$ such that $I + u \in \mathcal{I}$.

In this paper, we are given a simple (not necessarily complete) bipartite graph $G = (V, E)$ such that its vertex set V is partitioned into disjoint subsets R and H, and an edge in E connects a vertex in R and a vertex in H. We call a vertex in R (resp., H) a *resident* (resp., a *hospital*). For a vertex v in V and a subset F of E, we denote by $F(v)$ the set of edges in F incident to v. For a resident r in R and a hospital h in H, if there exists an edge in E connecting r and h, then we denote by (r, h) this edge. Furthermore, we are given a matroid $\mathbf{N} = (E, \mathcal{F})$ such that $\{e\} \in \mathcal{F}$ for any edge e in E. The matroid $\mathbf{N}$ represents constraints on assignment of residents to hospitals (it should be noted that in applications, $\mathbf{N}$ might be a direct sum of different matroids for hospitals).

For a resident r in R, we are given a reflexive and transitive binary relation $\succsim_r$ on $E(r) \cup \{\emptyset\}$ such that at least one of $e \succsim_r f$ and $f \succsim_r e$ holds for any edges e, f in $E(r)$, and $e \succsim_r \emptyset$ and $\emptyset \not\succsim_r e$ for any edge e in $E(r)$. For a resident r in R, the binary relation $\succsim_r$ represents the preference list of r. Namely, for a resident r in R and edges e, f in $E(r)$, if $e \succsim_r f$ holds, then r prefers e to f, or is indifferent between e and f. For a resident r in R, we use the notation $e \succ_r f$ (resp., $e \sim_r f$), if $e \succsim_r f$ and $f \not\succsim_r e$ (resp., $e \succsim_r f$ and $f \succsim_r e$). Furthermore, we are given a reflexive and transitive binary relation $\succsim_H$ on R such that at least one of $r \succsim_H s$ and $s \succsim_H r$ holds for any residents r, s in R. The binary relation $\succsim_H$ represents the master preference list of hospitals in H. We define $\succ_H$ and $\sim_H$ in the same way as $\succ_r$ and $\sim_r$ for a resident r in R.

A subset M of E is called a *matching* in G, if $|M(r)| \leq 1$ for any resident r in R, and $M \in \mathcal{F}$. For a matching M in G and a resident r in R such that $M(r) \neq \emptyset$, we do not distinguish between $M(r)$ and its element. For a matching M in G and an edge $e = (r, h)$ in $E \setminus M$, we say that r *weakly* (resp., *strictly*) *prefers* e on M, if $e \succsim_r M(r)$ (resp., $e \succ_r M(r)$). For a matching M in G and an edge $e = (r, h)$ in $E \setminus M$, we say that H *weakly* (resp., *strictly*) *prefers* e on M, if at least one of the following two conditions is satisfied. (P1) $M + e \in \mathcal{F}$. (P2) There exists an edge $f = (s, p)$ in M such that $M + e - f \in \mathcal{F}$ and $r \succsim_H s$ (resp., $r \succ_H s$).

A matching M in G is said to be *super-stable*, if there exists no edge (r, h) in $E \setminus M$ such that r and H weakly prefer (r, h) on M. A matching M in G is said to be *strongly stable*, if there exists no edge (r, h) in $E \setminus M$ such that r and H weakly prefer (r, h) on M, and at least one of r and H strictly prefers (r, h) on M. Our problem is to decide whether there exists a matching satisfying the above conditions, and find such a matching if one exists.

Here we give a concrete setting of our abstract model. Assume that H is partitioned into non-empty subsets $H_1, H_2, \ldots, H_k$, and we are given positive integers $d_1, d_2, \ldots, d_k$ and c_h for a hospital h in H. Then, define $\mathcal{F}$ as the family of subsets F of E such that $|F(h)| \leq c_h$ for any hospital h in H, and $\sum_{h \in H_i} |F(h)| \leq d_i$ for any $i = 1, 2, \ldots, k$. It is not difficult to see that $\mathbf{N} = (E, \mathcal{F})$ is a matroid. When we define $\mathbf{N}$ in this way, a subset M of E is a matching in G if and only if $|M(r)| \leq 1$ for any resident r in R, $|M(h)| \leq c_h$ for any hospital h in H, and $\sum_{h \in H_i} |M(h)| \leq d_i$ for any $i = 1, 2, \ldots, k$. These constraints are exactly the capacity constraints in the student-project allocation problem [10].

Basics of Matroids. Let $\mathbf{M} = (U, \mathcal{I})$ be a matroid. A subset of U belonging to $\mathcal{I}$ is called an *independent set* of $\mathbf{M}$. A subset C of U is called a *circuit* of $\mathbf{M}$, if C is not an independent set of $\mathbf{M}$, but any proper subset of C is an independent set of $\mathbf{M}$. Assume that we are given an independent set I of $\mathbf{M}$ and an element u in $U \setminus I$ such that $I + u \notin \mathcal{I}$. It is known [11, Proposition 1.1.6] that $I + u$ contains the unique circuit of $\mathbf{M}$, and u belongs to this circuit. This circuit is called the *fundamental circuit* of u, and denoted by $\mathsf{C}_{\mathbf{M}}(u, I)$. Define $\mathsf{C}^-_{\mathbf{M}}(u, I)$ as $\mathsf{C}_{\mathbf{M}}(u, I) - u$. It is known [11, p. 20, Exercise 5] that $\mathsf{C}_{\mathbf{M}}(u, I)$ is the set of elements w in $I + u$ such that $I + u - w \in \mathcal{I}$. Thus, for a matching M in G and an edge e in $E \setminus M$ such that $M + e \notin \mathcal{F}$, the condition (P2) can be restated as follows. There exists an edge $f = (s, p)$ in $\mathsf{C}^-_{\mathbf{N}}(e, M)$ such that $r \succsim_H s$ (resp., $r \succ_H s$). A maximal independent set of $\mathbf{M}$ is called a *base* of $\mathbf{M}$. The condition (I2) implies that any base of $\mathbf{M}$ has the same size. For a subset X of U, we define $\mathcal{I}|X := \{I \subseteq X \mid I \in \mathcal{I}\}$, and $\mathbf{M}|X := (X, \mathcal{I}|X)$. It is known [11, p. 20] that for any subset X of U, $\mathbf{M}|X$ is a matroid. For a subset X of U, we define $\mathsf{r}_{\mathbf{M}}(X)$ as the size of a base of $\mathbf{M}|X$. It is not difficult to see that for any independent set I of $\mathbf{M}$, I is a base of $\mathbf{M}$ if and only if $|I| = \mathsf{r}_{\mathbf{M}}(U)$. For disjoint subsets X, J of U, we define $\mathsf{p}(J; X) := \mathsf{r}_{\mathbf{M}}(J \cup X) - \mathsf{r}_{\mathbf{M}}(X)$. For a subset X of U, we define $\mathcal{I}/X := \{I \subseteq U \setminus X \mid \mathsf{p}(I; X) = |I|\}$, and $\mathbf{M}/X := (U \setminus X, \mathcal{I}/X)$. It is known [11, Proposition 3.1.6] that for any subset X of U, $\mathbf{M}/X$ is a matroid.

Theorem 1 (See [11, p. 15, Exercise 14]). *Assume that we are given a matroid* $\mathbf{M}$ *and circuits* C_1, C_2 *of* $\mathbf{M}$ *such that* $C_1 \cap C_2 \neq \emptyset$ *and* $C_1 \setminus C_2 \neq \emptyset$*. Then, for any element* u *in* $C_1 \cap C_2$ *and any element* w *in* $C_1 \setminus C_2$*, there exists a circuit* C *of* $\mathbf{M}$ *such that* $w \in C$ *and* C *is a subset of* $(C_1 \cup C_2) - u$.

The following lemma easily follows from Theorem 1.

Lemma 1. *Assume that we are given a matroid* $\mathbf{M} = (U, \mathcal{I})$ *and independent sets* I, J *of* $\mathbf{M}$ *such that* $I \subseteq J$*. Then, for any element* u *in* $U \setminus J$*, if* $I + u \notin \mathcal{I}$*, then* $J + u \notin \mathcal{I}$ *and* $\mathsf{C}_{\mathbf{M}}(u, I) = \mathsf{C}_{\mathbf{M}}(u, J)$.

Lemma 2. *Let* $\mathbf{M} = (U, \mathcal{I})$ *be a matroid, and let* $C, C_1, C_2, \ldots, C_k$ *be circuits of* $\mathbf{M}$*. Assume that we are given distinct elements* $u_1, u_2, \ldots, u_k$ *in* U *such that* $u_i \in C \cap C_i$ *and* $u_i \notin C_j$ *for any* $i, j = 1, 2, \ldots, k$ *with* $i \neq j$*, and there exists an element* w *in* U *such that* $w \in C \setminus C_i$ *for any* $i = 1, 2, \ldots, k$*. Then, there exists a circuit* C' *of* $\mathbf{M}$ *such that* $C' \subseteq (C \cup C_1 \cup C_2 \cup \cdots \cup C_k) \setminus \{u_1, u_2, \ldots, u_k\}$.

Proof. We consider the following procedure.

Step 1. Set $t := 1$ and $K_0 := C$.
Step 2. If $t \leq k$, then do the following steps. Otherwise, go to **Step 3**.
- **(2-a)** If $u_t \notin K_{t-1}$, then set $K_t := K_{t-1}$ and go to **Step (2-c)**.
- **(2-b)** Find a circuit K_t of $\mathbf{M}$ such that $w \in K_t$ and $K_t \subseteq (K_{t-1} \cup C_t) - u_t$.
- **(2-c)** Update $t := t + 1$, and go back to the beginning of **Step 2**.

Step 3. Output K_k and halt.

In **Step (2-b)**, Theorem 1 implies that there exists a circuit K_t of $\mathbf{M}$ satisfying the above properties. It is not difficult to see that $K_k \subseteq C \cup C_1 \cup C_2 \cup \cdots \cup C_k$ and $u_i \notin K_k$ for any $i = 1, 2, \ldots, k$. This completes the proof. □

Since the following lemma easily follows from known facts, we omit the proof.

Lemma 3. *Let* $\mathbf{M} = (U, \mathcal{I})$ *be a matroid, and let* X *be a subset of* U*.* (1) *Let* B *be an arbitrary base of* $\mathbf{M}|X$*. For any subset* I *of* $U \setminus X$*,* I *is an independent set of* $\mathbf{M}/X$ *if and only if* $I \cup B$ *is an independent set of* $\mathbf{M}$*.* (2) *For any base* B_1 *of* $\mathbf{M}|X$ *and any base* B_2 *of* $\mathbf{M}/X$*,* $B_1 \cup B_2$ *is a base of* $\mathbf{M}$*.* (3) *For any subset* I *of* U *such that* $I \cap X$ *is a base of* $\mathbf{M}|X$*, if* $I \setminus X$ *is not an independent set of* $\mathbf{M}/X$*, then* I *is not an independent set of* $\mathbf{M}$.

Let $\mathbf{M}_1 = (U, \mathcal{I}_1)$ and $\mathbf{M}_2 = (U, \mathcal{I}_2)$ be matroids on the same ground set. A subset I of U is called a *common independent set* of $\mathbf{M}_1$ and $\mathbf{M}_2$, if I belongs to $\mathcal{I}_1 \cap \mathcal{I}_2$. It is known that we can find a maximum-size common independent set of $\mathbf{M}_1$ and $\mathbf{M}_2$ in time bounded by a polynomial in $|U|$ and EO, where EO is the time required to decide whether X is an independent set of $\mathbf{M}_i$ for a subset X of U and $i = 1, 2$. If we use the algorithm proposed by Cunningham [12], then we can find a maximum-size common independent set in $O(|U|^{2.5}\mathsf{EO})$ time.

3 Super-Stable Matchings

In this section, we consider the problem of deciding whether there exists a super-stable matching in G, and finding such a matching if one exists. We first partition R into non-empty disjoint subsets $R_1, R_2, \ldots, R_n$ such that (i) $r \sim_H s$ for any $i = 1, 2, \ldots, n$ and any residents r, s in R_i, and (ii) $r \succ_H s$ for any $i, j = 1, 2, \ldots, n$ such that $i < j$ and any residents r in R_i and s in R_j. For $i, j = 1, 2, \ldots, n$ such that $i \leq j$, we define $R_{i,j} := R_i \cup R_{i+1} \cup \cdots \cup R_j$. Define $R_{n+1,n} = \emptyset$.

Our algorithm `SuperM` is described as follows.

Step 1. Set $i := 1$, $M_0 := \emptyset$, and $D_0 := \emptyset$.
Step 2. If $i \leq n$, then do the following steps. Otherwise, go to **Step 3.**
- **(2-a)** For a resident r in R_i, set T_r to be the set of edges e in $E(r) \setminus D_{i-1}$ such that $e \succsim_r f$ for any edge f in $E(r) \setminus D_{i-1}$.
- **(2-b)** If there exists a resident r in R_i such that $|T_r| > 1$, then output **null** and halt (i.e., there exists no super-stable matching in G).
- **(2-c)** Set $E_i := \cup_{r \in R_i} T_r$. If $M_{i-1} \cup E_i \notin \mathcal{F}$, then output **null** and halt.
- **(2-d)** Set $M_i := M_{i-1} \cup E_i$.
- **(2-e)** Set L_i to be the set of edges (r, h) in $E \setminus D_{i-1}$ such that $r \in R_{i+1,n}$ and $M_i + (r, h) \notin \mathcal{F}$. Furthermore, set $D_i := D_{i-1} \cup L_i$.
- **(2-f)** Update $i := i + 1$, and go back to the beginning of **Step 2.**

Step 3. Output M_n and halt (i.e., M_n is a super-stable matching in G).

In the sequel, we prove the correctness of the algorithm `SuperM`.

Lemma 4. *Assume that the algorithm* `SuperM` *halts when $i = \delta$. Then, there exists no super-stable matching N in G such that $N \cap D_{\delta-1} \neq \emptyset$.*

Proof. An edge e in $D_{\delta-1}$ is said to be *bad*, if there exists a super-stable matching N in G such that $e \in N$. Namely, our goal is to prove that there exists no bad edge in $D_{\delta-1}$. We prove this lemma by contradiction. Assume that there exists a bad edge in $D_{\delta-1}$. Define Δ as the set of integers ℓ in $\{1, 2, \ldots, \delta - 1\}$ such that there exists a bad edge in L_ℓ, and j as the minimum integer in Δ. Furthermore, let $e = (r, h)$ and N be a bad edge in L_j and a super-stable matching in G such that $e \in N$, respectively. Since $e \in L_j$, $M_j + e \notin \mathcal{F}$. Define $C := \mathsf{C_N}(e, M_j)$. If $C \subseteq N$, then this contradicts the fact that $N \in \mathcal{F}$. Thus, $C \setminus N \neq \emptyset$. In addition, for any edge (s, p) in $C \setminus N$, since $r \in R_{j+1,n}$ and $s \in R_{1,j}$, we have $s \succ_H r$.

We first consider the case where there exists an edge $f = (s, p)$ in $C \setminus N$ such that $N(s) \succ_s f$. Since $e \in N$, $e \neq f$. This implies that $f \in M_j$, i.e., $f \in T_s$. Thus, the definition of the algorithm `SuperM` implies that $j \neq 1$ and $N(s) \in L_k$ for an integer k in $\{1, 2, \ldots, j-1\}$. Since $L_k \subseteq D_{\delta-1}$ and N is super-stable, $N(s)$ is bad. This contradicts the minimality of j.

Next we consider the case where $f \succsim_s N(s)$ for any edge $f = (s, p)$ in $C \setminus N$. If there exists an edge f in $C \setminus N$ such that $N + f \in \mathcal{F}$, then this contradicts the fact that N is super-stable. Thus, we can assume that $N + f \notin \mathcal{F}$ for any edge f in $C \setminus N$. For an edge f in $C \setminus N$, we define $C_f := \mathsf{C_N}(f, N)$. Since N is super-stable, $t \succ_H s$ for any edge $f = (s, p)$ in $C \setminus N$ and any edge $g = (t, q)$ in

$C_f - f$. For any edge $f = (s, p)$ in $C \setminus N$, since $s \succ_H r$, this implies that $e \notin C_f$. In addition, $f \in C \cap C_f$ for any edge f in $C \setminus N$. Thus, Lemma 2 implies that there exists a circuit C' of $\mathbf{N}$ such that $C' \subseteq (C \cup C^*) \setminus (C \setminus N)$, where C^* is $\cup_{f \in C \setminus N} C_f$. Thus, since $C_f - f$ is a subset of N for any edge f in $C \setminus N$, C' is a subset of N. This contradicts the fact that $N \in \mathcal{F}$, and completes the proof. □

Lemma 5. *Assume that the algorithm* SuperM *halts when $i = \delta$ ($\leq n$), and we are given a super-stable matching N in G. Then, for any resident r in $R_{1,\delta}$ such that $T_r = \emptyset$ (resp., $T_r \neq \emptyset$), we have $N(r) = \emptyset$ (resp., $N(r) \in T_r$).*

Proof. For any resident r in $R_{1,\delta}$ such that $T_r = \emptyset$, we have $E(r) \subseteq D_{\delta-1}$. Thus, Lemma 4 implies that $N(r) = \emptyset$ for any resident r in $R_{1,\delta}$ such that $T_r = \emptyset$. Let Π be the set of residents r in $R_{1,\delta}$ such that $T_r \neq \emptyset$ and $N(r) \notin T_r$. If $\Pi = \emptyset$, then the proof is done. Thus, we assume that $\Pi \neq \emptyset$.

We first prove that there exists no resident r in Π such that $N(r) \succsim_r e$ for an edge e in T_r (notice that $f \sim_r g$ for any edges f, g in T_r). If there exists a resident r in Π such that $N(r) \succsim_r e$ for an edge e in T_r, then the definition of the algorithm SuperM implies that $N(r) \in D_{\delta-1}$, which contradicts Lemma 4.

Let Δ be the set of integers ℓ in $\{1, 2, \ldots, \delta\}$ such that $\Pi \cap R_\ell \neq \emptyset$, and let j and r be the minimum integer in Δ and a resident in $\Pi \cap R_j$, respectively. Let e be an edge in T_r. Then, since $N(r) \notin T_r$ and $e \in T_r$, we have $e \notin N$. Since N is super-stable and $e \succ_r N(r)$, we have $N + e \notin \mathcal{F}$ and $s \succ_H r$ for any edge (s, p) in $\mathsf{C}^-_{\mathbf{N}}(e, N)$. Since $\{e\} \in \mathcal{F}$ (i.e., $\mathsf{C}^-_{\mathbf{N}}(e, N) \neq \emptyset$), this implies that $j \neq 1$. Furthermore, $e \in T_r$ implies $e \notin D_{j-1}$. Thus, $e \notin D_{j-2}$ and $e \notin L_{j-1}$. These imply that $M_{j-1} + e \in \mathcal{F}$. In the sequel, we prove that $\mathsf{C}_{\mathbf{N}}(e, N) \subseteq M_{j-1} + e$, which contradicts the fact that $M_{j-1} + e \in \mathcal{F}$ and completes the proof.

Let $f = (s, p)$ be an edge in $\mathsf{C}^-_{\mathbf{N}}(e, N)$. Since $s \succ_H r$, $s \in R_k$ for an integer k in $\{1, 2, \ldots, j-1\}$. This and the definition of the algorithm SuperM imply that $|T_s| \leq 1$. If $T_s = \emptyset$, then in the same way as above, we can prove that $N(s) = \emptyset$, which contradicts the fact that $f = N(s)$. Thus, $|T_s| = 1$. In this case, the unique edge in T_s belongs to M_{j-1}. Since $s \in R_k$ and $k \leq j-1$, the minimality of j implies that $f = N(s) \in T_s$. Thus, $f \in M_{j-1}$. This completes the proof. □

Lemma 6. *If the algorithm* SuperM *outputs M_n, then M_n is a super-stable matching in G.*

Proof. It follows from the definition of the algorithm SuperM that $M_n \in \mathcal{F}$ and $|M_n(r)| \leq 1$ for any resident r in R. These imply that M_n is a matching in G. What remains is to prove that M_n is super-stable. Let $e = (r, h)$ be an edge in $E \setminus M_n$. We prove that at least one of r and H does not weakly prefer e on M_n. If $M_n(r) \succ_r e$, then the proof is done. Thus, we can assume that $e \succsim_r M_n(r)$. Assume that $r \in R_j$ for an integer j in $\{1, 2, \ldots, n\}$. If $M_n(r) = \emptyset$, then $T_r = \emptyset$, and thus $j \neq 1$ and $e \in L_k$ for an integer k in $\{1, 2 \ldots, j-1\}$. If $M_n(r) \neq \emptyset$, then $T_r = \{M_n(r)\}$. Since $e \neq M_n(r)$, $e \notin T_r$. Thus, since $e \succsim_r M_n(r)$, $j \neq 1$ and $e \in L_k$ for an integer k in $\{1, 2 \ldots, j-1\}$. In both cases, $M_k + e \notin \mathcal{F}$ and $s \succ_H r$ for any edge (s, p) in $\mathsf{C}^-_{\mathbf{N}}(e, M_k)$. Since $M_k \subseteq M_n$, Lemma 1 implies that $M_n + e \notin \mathcal{F}$ and $\mathsf{C}_{\mathbf{N}}(e, M_k) = \mathsf{C}_{\mathbf{N}}(e, M_n)$. Thus, $s \succ_H r$ for any edge $f = (s, p)$ in $\mathsf{C}^-_{\mathbf{N}}(e, M_n)$. This completes the proof. □

Lemma 7. *If the algorithm* `SuperM` *outputs* **null**, *then there exists no super-stable matching in* G.

Proof. We assume that the algorithm `SuperM` outputs **null** when $i = \delta$ ($\leq n$). We prove this lemma by contradiction. Assume that there exists a super-stable matching N in G.

We first consider the case where in **Step (2-b)** the algorithm `SuperM` outputs **null**. In this case, there exists a resident r in R_δ such that $|T_r| > 1$. Lemma 5 implies that $N(r) \in T_r$. Let e be an edge in T_r such that $e \neq N(r)$. Then, since e and $N(r)$ belong to T_r, we have $e \sim_r N(r)$. Thus, what remains is to prove that $N + e \in \mathcal{F}$ and/or there exists an edge $f = (s, p)$ in N such that $N + e - f \in \mathcal{F}$ and $r \succsim_H s$. This contradicts the fact that N is super-stable. If $N + e \in \mathcal{F}$, then the proof is done. Thus, we assume that $N + e \notin \mathcal{F}$ and $s \succ_H r$ (i.e., s belongs to $R_{1,\delta-1}$) for any edge (s, p) in $\mathsf{C}_{\mathbf{N}}(e, N)$. Since $e \in T_r$, $M_{\delta-1} + e \in \mathcal{F}$. Lemma 5 and the definition of the algorithm `SuperM` imply that $N(s) = M_{\delta-1}(s)$ for any resident s in $R_{1,\delta-1}$. These imply that $\mathsf{C}_{\mathbf{N}}(e, N)$ is a subset of $M_{\delta-1} + e$, which contradicts the fact that $M_{\delta-1} + e \in \mathcal{F}$.

Next we consider the case where in **Step (2-c)** the algorithm `SuperM` outputs **null**, i.e., $M_{\delta-1} \cup E_\delta \notin \mathcal{F}$. Lemma 5 and the definition of the algorithm `SuperM` imply that $M_{\delta-1}(r) = N(r)$ for any resident r in $R_{1,\delta-1}$, and $E_\delta(r) = T_r = N(r)$ for any resident r in R_δ. These imply that $M_{\delta-1} \cup E_\delta$ is a subset of N, which contradicts the fact that $N \in \mathcal{F}$. □

Lemmas 6 and 7 imply the following theorem.

Theorem 2. *The algorithm* `SuperM` *can decide whether there exists a super-stable matching in* G, *and find a super-stable matching if one exists.*

Here we consider the time complexity of the algorithm `SuperM`. Define $m := |E|$ and $\lambda := \max\{|R|, |H|\}$, and we denote by EO the time required to decide whether $F \in \mathcal{F}$ for a subset F of E. In addition, we assume that we can decide in $O(1)$ time whether $e \succsim_r f$ (resp., $r \succsim_H s$) for a resident r in R and edges e, f in $E(r)$ (resp., for residents r, s in R). For simplicity, we assume that EO is $\Omega(m)$ and $m \geq \lambda$. Then, we can construct the subsets $R_1, R_2, \ldots, R_n$ in $O(\lambda^2)$ time. The number of iterations of **Step 2** is at most λ, and **Step (2-e)** is the bottleneck of **Step 2**. The time complexity of this step is $O(m\mathsf{EO})$. Thus, the time complexity of the algorithm `SuperM` is $O(\lambda m \mathsf{EO})$.

4 Strongly Stable Matchings

In this section, we give an algorithm for the problem of deciding whether there exists a strongly stable matching in G, and finding such a matching if one exists. We define the subsets $R_1, R_2, \ldots, R_n$ of R and the notation $R_{i,j}$ in the same way as in Sect. 3.

Our algorithm `StrongM` is described as follows.

Step 1. Set $i := 1$, $M_0 := \emptyset$, $D_0 := \emptyset$, and $P_0 := \emptyset$.

Step 2. If $i \leq n$, then do the following steps. Otherwise, go to **Step 3**.

(2-a) For a resident r in R_i, set T_r to be the set of edges e in $E(r) \setminus D_{i-1}$ such that $e \succsim_r f$ for any edge f in $E(r) \setminus D_{i-1}$.

(2-b) Set $E_i := \bigcup_{r \in R_i} T_r$, $P_i := \bigcup_{\ell=1,2,\ldots,i} E_\ell$, $\mathbf{N}_i := (\mathbf{N}|P_i)/P_{i-1}$, and $R_i^* := \{r \in R_i \mid T_r \neq \emptyset\}$.

(2-c) If $r_{\mathbf{N}_i}(E_i) > |R_i^*|$, then output **null** and halt.

(2-d) Set $\mathcal{U}_i$ to be the family of subsets F of E_i such that $|F(r)| \leq 1$ for any resident r in R_i, and set $\mathbf{A}_i := (E_i, \mathcal{U}_i)$ (notice that $\mathbf{A}_i$ is a matroid). Then, find a maximum-size common independent set F_i of $\mathbf{A}_i$ and $\mathbf{N}_i$.

(2-e) If $|F_i| < |R_i^*|$, then output **null** and halt. Otherwise, set $M_i := M_{i-1} \cup F_i$.

(2-f) Set L_i to be the set of edges (r, h) in $E \setminus D_{i-1}$ such that $r \in R_{i+1,n}$ and $M_i + (r, h) \notin \mathcal{F}$. Furthermore, set $D_i := D_{i-1} \cup L_i$.

(2-g) Update $i := i + 1$, and go back to the beginning of **Step 2**.

Step 3. Output M_n and halt.

In the sequel, we prove the correctness of the algorithm `StrongM`.

Lemma 8. *Assume that the algorithm* `StrongM` *halts when* $i = \delta$. *Then,*

1. F_ℓ *is a base of* $\mathbf{N}_\ell$ *for any* $\ell = 1, 2, \ldots, \delta - 1$, *and*
2. M_ℓ *is a base of* $\mathbf{N}|P_\ell$ *for any* $\ell = 1, 2, \ldots, \delta - 1$.

Proof. We first consider the statement (1). Let us fix an integer ℓ in $\{1, 2, \ldots, \delta - 1\}$. Since the algorithm `StrongM` does not output **null** when $i = \ell$, we have $r_{\mathbf{N}_\ell}(E_\ell) \leq |R_\ell^*|$ and $|F_\ell| \geq |R_\ell^*|$. These imply that $r_{\mathbf{N}_\ell}(E_\ell) \leq |F_\ell|$. Furthermore, since F_ℓ is an independent set of $\mathbf{N}_\ell$, we have $|F_\ell| \leq r_{\mathbf{N}_\ell}(E_\ell)$. These imply that $|F_\ell| = r_{\mathbf{N}_\ell}(E_\ell)$, i.e., F_ℓ is a base of $\mathbf{N}_\ell$.

Next we prove the statement (2) by induction on ℓ. It is not difficult to see that the statement (2) for $\ell = 1$ is equivalent to the statement (1) for $\ell = 1$. Let ξ be an integer in $\{1, 2, \ldots, \delta - 2\}$. Assume that the statement (2) holds in the case of $\ell = \xi$, and we consider the case of $\ell = \xi + 1$. Since $\mathbf{N}|P_\xi = (\mathbf{N}|P_{\xi+1})|P_\xi$ and $F_{\xi+1}$ is a base of $\mathbf{N}_{\xi+1} = (\mathbf{N}|P_{\xi+1})/P_\xi$, it follows from the induction hypothesis and Lemma 3(2) that $M_{\xi+1}$ is a base of $\mathbf{N}|P_{\xi+1}$. This completes the proof. □

Lemma 9. *Assume that the algorithm* `StrongM` *halts when* $i = \delta$. *Then, there exists no strongly stable matching* N *in* G *such that* $N \cap D_{\delta-1} \neq \emptyset$.

Proof. We say that an edge e in $D_{\delta-1}$ is *bad*, if there exists a strongly stable matching N in G such that $e \in N$. Namely, our goal is to prove that there exists no bad edge in $D_{\delta-1}$. We prove this lemma by contradiction. Assume that there exists a bad edge in $D_{\delta-1}$. Define Δ as the set of integers ℓ in $\{1, 2, \ldots, \delta - 1\}$ such that there exists a bad edge in L_ℓ, and j as the minimum integer in Δ. Furthermore, let $e = (r, h)$ and N be a bad edge in L_j and a strongly stable matching in G such that $e \in N$, respectively. Since $e \in L_j$, $M_j + e \notin \mathcal{F}$. Define $C := \mathsf{C}_{\mathbf{N}}(e, M_j)$ (notice that $M_j \in \mathcal{F}$ follows from Lemma 8(2)). If C is a subset of N, then this contradicts the fact that $N \in \mathcal{F}$. Thus, $C \setminus N$ is not empty. For any edge (s,p) in $C \setminus N$, since $r \in R_{j+1,n}$ and $s \in R_{1,j}$, we have $s \succ_H r$.

We first consider the case where there exists an edge $f = (s,p)$ in $C \setminus N$ such that $N(s) \succ_s f$. Since $e \in N$, we have $e \neq f$. This implies that $f \in M_j$, i.e., $f \in T_s$. Thus, the definition of the algorithm StrongM implies that $j \neq 1$ and $N(s) \in L_k$ for an integer k in $\{1, 2, \ldots, j-1\}$. Since $L_k \subseteq D_{\delta-1}$ and N is strongly stable, $N(s)$ is bad, which contradicts the minimality of j.

Next we consider the case where $f \succsim_s N(s)$ for any edge $f = (s,p)$ in $C \setminus N$. If there exists an edge f in $C \setminus N$ such that $N + f \in \mathcal{F}$, then this contradicts the fact that N is strongly stable. Thus, we assume that $N + f \notin \mathcal{F}$ for any edge f in $C \setminus N$. For an edge f in $C \setminus N$, we define $C_f := \mathsf{C}_{\mathbf{N}}(f, N)$. Since N is strongly stable, $t \succsim_H s$ for any edge $f = (s,p)$ in $C \setminus N$ and any edge $g = (t,q)$ in $C_f - f$. For any edge $f = (s,p)$ in $C \setminus N$, since $s \succ_H r$, this implies that $e \notin C_f$. In addition, $f \in C \cap C_f$ for any edge f in $C \setminus N$. Thus, Lemma 2 implies that there exists a circuit C' of $\mathbf{N}$ such that $C' \subseteq N$. This contradicts the fact that $N \in \mathcal{F}$. □

Lemma 10. *Assume that the algorithm* StrongM *halts when $i = \delta$ ($\leq n$), and we are given a strongly stable matching N in G. Then, for any resident r in $R_{1,\delta}$ such that $T_r = \emptyset$ (resp., $T_r \neq \emptyset$), we have $N(r) = \emptyset$ (resp., $N(r) \in T_r$).*

Proof. For any resident r in $R_{1,\delta}$ such that $T_r = \emptyset$, we have $E(r) \subseteq D_{\delta-1}$. Thus, Lemma 9 implies that $N(r) = \emptyset$ for any resident r in $R_{1,\delta}$ such that $T_r = \emptyset$. Let Π be the set of residents r in $R_{1,\delta}$ such that $T_r \neq \emptyset$ and $N(r) \notin T_r$. If $\Pi = \emptyset$, then the proof is done. Thus, we assume that $\Pi \neq \emptyset$.

We first prove that there exists no resident r in Π such that $N(r) \succsim_r e$ for an edge e in T_r. If there exists a resident r in Π such that $N(r) \succsim_r e$ for an edge e in T_r, then the definition of the algorithm StrongM implies that $N(r) \in D_{\delta-1}$, which contradicts Lemma 9.

Let Δ be the set of integers ℓ in $\{1, 2, \ldots, \delta\}$ such that $\Pi \cap R_\ell \neq \emptyset$, and let j and r be the minimum integer in Δ and a resident in $\Pi \cap R_j$, respectively. Let e be an edge in T_r. Then, since $N(r) \notin T_r$ and $e \in T_r$, we have $e \notin N$. Since N is strongly stable, $N + e \notin \mathcal{F}$ and $s \succ_H r$ for any edge (s,p) in $\mathsf{C}^{-}_{\mathbf{N}}(e, N)$. This implies that $j \neq 1$. Define $C := \mathsf{C}_{\mathbf{N}}(e, N)$. Furthermore, since $e \in T_r$, we have $M_{j-1} + e \in \mathcal{F}$. Thus, C is not a subset of $M_{j-1} + e$.

Here we prove that for any edge f in $C \setminus (M_{j-1} + e)$, we have $M_{j-1} + f \notin \mathcal{F}$. Before proving this, we prove that we can complete the proof of this lemma by using this. For an edge f in $C \setminus (M_{j-1} + e)$, we define $C_f := \mathsf{C}_{\mathbf{N}}(f, M_{j-1})$. For any edge f in $C \setminus (M_{j-1} + e)$, since $e \notin M_{j-1}$ and $f \neq e$, we have $e \notin C_f$. In addition, $f \in C \cap C_f$ for any edge f in $C \setminus (M_{j-1} + e)$. Thus, Lemma 2 implies that there exists a circuit C' of $\mathbf{N}$ that $C' \subseteq M_{j-1} + e$, which contradicts the fact that $M_{j-1} + e \in \mathcal{F}$. This completes the proof.

What remains is to prove that for any edge f in $C \setminus (M_{j-1} + e)$, we have $M_{j-1} + f \notin \mathcal{F}$. Let us fix an edge $f = (s,p)$ in $C \setminus (M_{j-1} + e)$. Since $s \succ_H r$, $s \in R_k$ for an integer k in $\{1, 2, \ldots, j-1\}$. Since $f = N(s)$, we can prove that $T_s \neq \emptyset$ in the same way as above. Since $s \in R_k$ and $k \leq j-1$, the minimality of j implies that $f \in T_s$, i.e., $f \in E_k$. Since $f \notin M_{j-1} + e$ implies that $f \notin F_k$

and Lemma 8(1) implies that F_k is a base of $\mathbf{N}_k$, $F_k + f$ is not an independent set of $\mathbf{N}_k = (\mathbf{N}|P_k)/P_{k-1}$. Thus, since Lemma 8(2) implies that M_{k-1} is a base of $(\mathbf{N}|P_k)|P_{k-1}$, Lemma 3(3) implies that $M_k + f \notin \mathcal{F}$. Thus, since $M_k \subseteq M_{j-1}$, $M_{j-1} + f \notin \mathcal{F}$. This completes the proof. □

Lemma 11. *Assume that the algorithm* `StrongM` *halts when* $i = \delta$ $(\leq n)$, *and we are given a strongly stable matching* N *in* G. *Then, for any* $\ell = 1, 2, \ldots, \delta$, *the following statements hold.*

1. *If* $\ell > 1$, *then* $N \cap P_{\ell-1}$ *is a base of* $\mathbf{N}|P_{\ell-1}$.
2. $N \cap E_\ell$ *is a base of* $\mathbf{N}_\ell$.

Proof. We prove this lemma by induction on ℓ. We first consider the case of $\ell = 1$, and we prove the statement (2) by contradiction. Assume that $N \cap E_1$ is not a base of $\mathbf{N}_1 = \mathbf{N}|E_1$. Since $N \in \mathcal{F}$, $N \cap E_1$ is an independent set of $\mathbf{N}_1$. Thus, there exists an edge $e = (r, h)$ in $E_1 \setminus N$ such that $(N \cap E_1) + e \in \mathcal{F}$. Since $e \in E_1$ (i.e., $e \in T_r$), Lemma 10 implies that $e \sim_r N(r)$. If $N + e \in \mathcal{F}$, then this contradicts the fact that N is strongly stable. Thus, we assume that $N + e \notin \mathcal{F}$. Since $(N \cap E_1) + e \in \mathcal{F}$, we have $\mathsf{C}_{\mathbf{N}}(e, N) \not\subseteq (N \cap E_1) + e$. Let $f = (s, p)$ be an edge in $\mathsf{C}_{\mathbf{N}}(e, N) \setminus ((N \cap E_1) + e)$. Since $f = N(s)$, if $s \in R_1$, then Lemma 10 implies that $f \in T_s$, and thus $f \in E_1$. This contradicts the fact that $f \notin E_1$. Thus, $s \in R_{2,n}$ and $r \succ_H s$. This contradicts the fact that N is strongly stable.

Let ξ be an integer in $\{1, 2, \ldots, \delta - 1\}$. Assume that this lemma holds in the case of $\ell = \xi$, and we consider the case of $\ell = \xi + 1$. We first prove the statement (1). If $\xi = 1$, then the statement (1) in the case of $\ell = \xi + 1$ is equivalent to the statement (2) in the case of $\ell = 1$. Thus, we assume that $\xi \geq 2$. The induction hypothesis implies that $N \cap P_{\xi-1}$ is a base of $(\mathbf{N}|P_\xi)|P_{\xi-1}$ and $N \cap E_\xi$ is a base of $(\mathbf{N}|P_\xi)/P_{\xi-1}$. Thus, Lemma 3(2) implies that $N \cap P_\xi$ is a base of $\mathbf{N}|P_\xi$. This completes the proof.

Next we prove the statement (2) by contradiction. Assume that $N \cap E_{\xi+1}$ is not a base of $\mathbf{N}_{\xi+1}$. Since $N \in \mathcal{F}$, $N \cap P_{\xi+1}$ is an independent set of $\mathbf{N}|P_{\xi+1}$. In addition, the statement (1) in the case of $\ell = \xi + 1$ implies that $N \cap P_\xi$ is a base of $\mathbf{N}|P_\xi = (\mathbf{N}|P_{\xi+1})|P_\xi$. These facts and Lemma 3(1) imply that $N \cap E_{\xi+1}$ is an independent set of $\mathbf{N}_{\xi+1} = (\mathbf{N}|P_{\xi+1})/P_\xi$. Thus, there exists an edge $e = (r, h)$ in $E_{\xi+1} \setminus N$ such that $(N \cap E_{\xi+1}) + e$ is an independent set of $\mathbf{N}_{\xi+1}$. Then, since $N \cap P_\xi$ is a base of $(\mathbf{N}|P_{\xi+1})|P_\xi$, Lemma 3(1) implies that $(N \cap P_{\xi+1}) + e \in \mathcal{F}$. Since $e \in E_{\xi+1}$ (i.e., $e \in T_r$), Lemma 10 implies that $e \sim_r N(r)$. If $N + e \in \mathcal{F}$, then this contradicts the fact that N is strongly stable. Thus, we consider the case where $N + e \notin \mathcal{F}$. Since $(N \cap P_{\xi+1}) + e \in \mathcal{F}$, $\mathsf{C}_{\mathbf{N}}(e, N)$ is not a subset of $(N \cap P_{\xi+1}) + e$. Let $f = (s, p)$ be an edge in $\mathsf{C}_{\mathbf{N}}(e, N) \setminus ((N \cap P_{\xi+1}) + e)$. Since $f = N(s)$, if $s \in R_{1,\xi+1}$, then Lemma 10 implies that $f \in T_s$, and thus $f \in P_{\xi+1}$. This contradicts the fact that $f \notin P_{\xi+1}$. Thus, we have $s \in R_{\xi+2,n}$ and $r \succ_H s$. This contradicts the fact that N is strongly stable. □

Lemma 12. *If the algorithm* `StrongM` *outputs* M_n, *then* $T_r = \emptyset$ *for any resident* r *in* R *such that* $M_n(r) = \emptyset$.

Proof. Let j be an integer in $\{1, 2, \ldots, n\}$. It follows from the definition of the algorithm StrongM that $r_{\mathbf{N}_j}(E_j) \le |R_j^*| \le |F_j|$. Furthermore, since F_j is an independent set of $\mathbf{N}_j$, we have $|F_j| \le r_{\mathbf{N}_j}(E_j)$. Thus, $|F_j| = |R_j^*|$. Since F_j is a subset of $\cup_{r \in R_j} T_r$ and $|F_j(r)| \le 1$ for any resident r in R_j, $|F_j(r)| = 1$ for any resident r in R_j^*. Thus, for any resident r in R_j such that $M_n(r) = \emptyset$, since $F_j(r) = M_n(r)$, we have $r \notin R_j^*$, i.e., $T_r = \emptyset$. This completes the proof. □

Lemma 13. *If the algorithm* StrongM *outputs* M_n, *then* M_n *is a strongly stable matching in* G.

Proof. The definition of the algorithm StrongM implies that $|M_n(r)| \le 1$ for any resident r in R. In addition, Lemma 8(2) implies that $M_n \in \mathcal{F}$. These facts imply that M_n is a matching in G. Thus, what remains is to prove that M_n is strongly stable. Let $e = (r, h)$ be an edge in $E \setminus M_n$. If $M_n(r) \succ_r e$, then r does not weakly prefer e on M_n. Thus, we do not need to consider this case. Assume that $e \succsim_r M_n(r)$ and $r \in R_j$ for an integer j in $\{1, 2, \ldots, n\}$.

We first assume that $e \succ_r M_n(r)$. If $M_n(r) = \emptyset$, then Lemma 12 implies that $T_r = \emptyset$. Thus, $j \ne 1$ and $e \in L_k$ for an integer k in $\{1, 2 \ldots, j-1\}$. If $M_n(r) \ne \emptyset$, then the definition of the algorithm StrongM implies that $M_n(r) \in T_r$. Thus, $j \ne 1$ and $e \in L_k$ for an integer k in $\{1, 2 \ldots, j-1\}$. In both cases, $M_k + e \notin \mathcal{F}$ and $s \succ_H r$ for any edge (s, p) in $\mathsf{C}_{\mathbf{N}}^-(e, M_k)$. Since $M_k \subseteq M_n$, Lemma 1 implies that $M_n + e \notin \mathcal{F}$ and $\mathsf{C}_{\mathbf{N}}(e, M_k) = \mathsf{C}_{\mathbf{N}}(e, M_n)$. Therefore, $s \succ_H r$ for any edge $f = (s, p)$ in $\mathsf{C}_{\mathbf{N}}^-(e, M_n)$. This implies that H does not weakly prefer e on M_n.

Next we assume that $e \sim_r M_n(r)$. This implies that $M_n(r) \ne \emptyset$ and $M_n(r) \in T_r$. We first consider the case where $e \notin T_r$. In this case, the definition of the algorithm StrongM implies that $j \ne 1$ and $e \in L_k$ for an integer k in $\{1, 2, \ldots, j-1\}$. Thus, we can treat this case in the same way as the case where $e \succ_r M_n(r)$. Next we consider the case where $e \in T_r$. In this case, since $e \notin M_n$, $e \in E_j \setminus F_j$. Since Lemma 8(1) implies that F_j is a base of $\mathbf{N}_j$, $F_j + e$ is not an independent set of $(\mathbf{N}|P_j)/P_{j-1}$. Furthermore, Lemma 8(2) implies that M_{j-1} is a base of $(\mathbf{N}|P_j)|P_{j-1}$. Thus, it follows from Lemma 3(3) that $M_j + e \notin \mathcal{F}$. Since $M_j \subseteq P_j$, $s \succsim_H r$ for any edge (s, p) in $\mathsf{C}_{\mathbf{N}}^-(e, M_j)$. Furthermore, since $M_j \subseteq M_n$, Lemma 1 implies that $M_n + e \notin \mathcal{F}$ and $\mathsf{C}_{\mathbf{N}}(e, M_j) = \mathsf{C}_{\mathbf{N}}(e, M_n)$. Thus, $s \succsim_H r$ for any edge $f = (s, p)$ in $\mathsf{C}_{\mathbf{N}}^-(e, M_n)$. This completes the proof. □

Lemma 14. *If the algorithm* StrongM *outputs* **null**, *then there exists no strongly stable matching in* G.

Proof. We assume that the algorithm StrongM outputs **null** when $i = \delta$ $(\le n)$. We prove this lemma by contradiction. Assume that there exists a strongly stable matching N in G.

We first consider the case where in **Step (2-c)** the algorithm StrongM outputs **null**, i.e., $r_{\mathbf{N}_\delta}(E_\delta) > |R_\delta^*|$. Since Lemma 11(2) implies that $N \cap E_\delta$ is a base of $\mathbf{N}_\delta$, we have $|N \cap E_\delta| = r_{\mathbf{N}_\delta}(E_\delta)$. Thus, $|N \cap E_\delta| > |R_\delta^*|$. However, since N is a matching in G (i.e., $|N(r)| \le 1$ for any resident r in R_δ) and E_δ is $\cup_{r \in R_\delta^*} T_r$, $|N \cap E_\delta| \le |R_\delta^*|$. This contradicts the fact that $|N \cap E_\delta| > |R_\delta^*|$.

Next we consider the case where in **Step (2-e)** the algorithm `StrongM` outputs **null**, i.e., we have $|F_\delta| < |R^*_\delta|$. Since N is a matching in G, $|N(r)| \leq 1$ for any resident r in R_δ. Furthermore, Lemma 11(2) implies that $N \cap E_\delta$ is an independent set of $\mathbf{N}_\delta$. Thus, $N \cap E_\delta$ is a common independent set of $\mathbf{A}_\delta$ and $\mathbf{N}_\delta$. Since F_δ is a maximum-size common independent set of $\mathbf{A}_\delta$ and $\mathbf{N}_\delta$, we have $|N \cap E_\delta| \leq |F_\delta| < |R^*_\delta|$. This implies that there exists a resident r in R^*_δ such that $N(r) \notin E_\delta$. However, since $T_r \subseteq E_\delta$, this contradicts Lemma 10. □

Lemmas 13 and 14 imply the following theorem.

Theorem 3. *The algorithm* `StrongM` *can decide whether there exists a strongly stable matching in* G*, and find a strongly stable matching if one exists.*

Here we consider the time complexity of the algorithm `StrongM` under the same assumption in Sect. 3. It is not difficult to see that the total time complexity of **Step 2** except **Step (2-d)** is $O(\lambda m \mathsf{EO})$. We consider the total time complexity of **Step (2-d)**. Let ℓ be an integer in $\{1, 2, \ldots, n\}$. Since Lemma 8(2) implies that $M_{\ell-1}$ is a base of $\mathbf{N}|P_{\ell-1}$, Lemma 3(1) implies that we can decide in $O(\mathsf{EO})$ time whether F is an independent set of $\mathbf{N}_\ell$ for a subset F of E_ℓ. Thus, the time complexity of **Step (2-d)** is $O(|E_\ell|^{2.5}\mathsf{EO})$ for $i = \ell$. Since $\sum_{\ell=1}^{n} |E_\ell|$ is $O(m)$, the total time complexity of this step is $O(m^{2.5}\mathsf{EO})$, and thus the time complexity of the algorithm `StrongM` is $O(m^{2.5}\mathsf{EO})$.

References

1. Gale, D., Shapley, L.S.: College admissions and the stability of marriage. Am. Math. Monthly **69**(1), 9–15 (1962)
2. Irving, R.W.: Stable marriage and indifference. Discrete Appl. Math. **48**(3), 261–272 (1994)
3. Manlove, D.F.: Stable marriage with ties and unacceptable partners. Technical report TR-1999-29, The University of Glasgow, Department of Computing Science (1999)
4. Irving, R.W., Manlove, D.F., Scott, S.: The hospitals/residents problem with ties. In: Halldórsson, M.M. (ed.) SWAT 2000. LNCS, vol. 1851, pp. 259–271. Springer, Heidelberg (2000)
5. Irving, R.W., Manlove, D.F., Scott, S.: Strong stability in the hospitals/residents problem. STACS 2003. LNCS, vol. 2607, pp. 439–450. Springer, Heidelberg (2003)
6. Kavitha, T., Mehlhorn, K., Michail, D., Paluch, K.: Strongly stable matchings in time $O(nm)$ and extension to the hospitals-residents problem. ACM Trans. Algorithms **3**(2), 1–18 (2007)
7. Fleiner, T.: A fixed-point approach to stable matchings and some applications. Math. Oper. Res. **28**(1), 103–126 (2003)
8. Irving, R.W., Manlove, D.F., Scott, S.: The stable marriage problem with master preference lists. Discrete Appl. Math. **156**(15), 2959–2977 (2008)
9. O'Malley, G.: Algorithmic aspects of stable matching problems. Ph.D. thesis, The University of Glasgow (2007)
10. Abraham, D.J., Irving, R.W., Manlove, D.F.: Two algorithms for the student-project allocation problem. J. Discrete Algorithms **5**(1), 73–90 (2007)
11. Oxley, J.G.: Matroid Theory, 2nd edn. Oxford University Press, New York (2011)
12. Cunningham, W.H.: Improved bounds for matroid partition and intersection algorithms. SIAM J. Comput. **15**(4), 948–957 (1986)

Stable Marriage and Roommates Problems with Restricted Edges: Complexity and Approximability

Ágnes Cseh[1](✉) and David F. Manlove[2]

[1] Institute for Mathematics, TU Berlin, Berlin, Germany
cseh@math.tu-berlin.de
[2] School of Computing Science, University of Glasgow, Glasgow, UK
David.Manlove@glasgow.ac.uk

Abstract. In the stable marriage and roommates problems, a set of agents is given, each of them having a strictly ordered preference list over some or all of the other agents. A matching is a set of disjoint pairs of mutually acceptable agents. If any two agents mutually prefer each other to their partner, then they block the matching, otherwise, the matching is said to be stable. We investigate the complexity of finding a solution satisfying additional constraints on restricted pairs of agents. Restricted pairs can be either *forced* or *forbidden*. A stable solution must contain all of the forced pairs, while it must contain none of the forbidden pairs.

Dias et al. [5] gave a polynomial-time algorithm to decide whether such a solution exists in the presence of restricted edges. If the answer is no, one might look for a solution close to optimal. Since optimality in this context means that the matching is stable and satisfies all constraints on restricted pairs, there are two ways of relaxing the constraints by permitting a solution to: (1) be blocked by as few as possible pairs, or (2) violate as few as possible constraints on restricted pairs.

Our main theorems prove that for the (bipartite) stable marriage problem, case (1) leads to NP-hardness and inapproximability results, whilst case (2) can be solved in polynomial time. For non-bipartite stable roommates instances, case (2) yields an NP-hard but (under some cardinality assumptions) 2-approximable problem. In the case of NP-hard problems, we also discuss polynomially solvable special cases, arising from restrictions on the lengths of the preference lists, or upper bounds on the numbers of restricted pairs.

1 Introduction

In the classical *stable marriage problem* (SM) [10], a bipartite graph is given, where one color class symbolises a set of men U and the other color class stands

Á. Cseh—Supported by COST Action IC1205 on Computational Social Choice and by the Deutsche Telekom Stiftung. Part of this work was carried out whilst visiting the University of Glasgow.
D.F. Manlove—Supported by grant EP/K010042/1 from the Engineering and Physical Sciences Research Council.

M. Hoefer (Ed.): SAGT 2015, LNCS 9347, pp. 15–26, 2015.
DOI: 10.1007/978-3-662-48433-3_2

for a set of women W. Man u and woman w are connected by edge uw if they find one another mutually acceptable. Each participant provides a strictly ordered preference list of the acceptable agents of the opposite gender. An edge uw *blocks* matching M if it is not in M, but each of u and w is either unmatched or prefers the other to their partner. A *stable matching* is a matching not blocked by any edge. From the seminal paper of Gale and Shapley [10], we know that the existence of such a stable solution is guaranteed and one can be found in linear time. Moreover, the solutions form a distributive lattice [20]. The two extreme points of this lattice are called the *man-* and *woman-optimal stable matchings* [10]. These assign each man/woman their best partner reachable in any stable matching. Another interesting and useful property of stable solutions is the so-called Rural Hospitals Theorem. Part of this theorem states that if an agent is unmatched in one stable matching, then all stable solutions leave him unmatched [11].

One of the most widely studied extensions of SM is the *stable roommates problem* (SR) [10,14], defined on general graphs instead of bipartite graphs. The notion of a blocking edge is as defined above (except that it can now involve any two agents in general), but several results do not carry over to this setting. For instance, the existence of a stable solution is not guaranteed any more. On the other hand, there is a linear-time algorithm to find a stable matching or report that none exists [14]. Moreover, the corresponding variant of the Rural Hospitals Theorem holds in the roommates case as well: the set of matched agents is the same for all stable solutions [12].

Both SM and SR are widely used in various applications. In markets where the goal is to maximise social welfare instead of profit, the notion of stability is especially suitable as an optimality criterion [22]. For SM, the oldest and most common area of applications is employer allocation markets [24]. On one side, job applicants are represented, while the job openings form the other side. Each application corresponds to an edge in the bipartite graph. The employers rank all applicants to a specific job offer and similarly, each applicant sets up a preference list of jobs. Given a proposed matching M of applicants to jobs, if an employer-applicant pair exists such that the position is not filled or a worse applicant is assigned to it, and the applicant received no contract or a worse contract, then this pair blocks M. In this case the employer and applicant find it mutually beneficial to enter into a contract outside of M, undermining its integrity. If no such blocking pair exists, then M is stable. Stability as an underlying concept is also used to allocate graduating medical students to hospitals in many countries [23]. SR on the other hand has applications in the area of P2P networks [9].

Forced and forbidden edges in SM and SR open the way to formulate various special requirements on the sought solution. Such edges now form part of the extended problem instance: if an edge is *forced*, it must belong to a constructed stable matching, whilst if an edge is *forbidden*, it must not. In certain market situations, a contract is for some reason particularly important, or to the contrary, not wished by the majority of the community or by the central authority in control. In such cases, forcing or forbidding the edge and then seeking a stable solution ensures that the wishes on these specific contracts are fulfilled while stability is guaranteed. Henceforth, the term *restricted edge* will be used to refer either to a forbidden edge or a forced edge. The remaining edges of the graph are referred as *unrestricted edges*.

Note that simply deleting forbidden edges or fixing forced edges and searching for a stable matching on the remaining instance does not solve the problem of finding a stable matching with restricted edges. Deleted edges (corresponding to forbidden edges, or those adjacent to forced edges) can block that matching. Therefore, to meet both requirements on restricted edges and stability, more sophisticated methods are needed.

The attention of the community was drawn very early to the characterization of stable matchings that must contain a prescribed set of edges. In the seminal book of Knuth [20], forced edges first appeared under the term *arranged marriages*. Knuth presented an algorithm that finds a stable matching with a given set of forced edges or reports that none exists. This method runs in $O(n^2)$ time, where n denotes the number of vertices in the graph. Gusfield and Irving [12] provided an algorithm based on rotations that terminates in $O(|Q|^2)$ time, following $O(n^4)$ pre-processing time, where Q is the set of forced edges. This latter method is favoured over Knuth's if multiple forced sets of small cardinality are proposed.

Forbidden edges appeared only in 2003 in the literature, and were first studied by Dias et al. [5]. In their paper, complete bipartite graphs were considered, but the methods can easily be extended to incomplete preference lists. Their main result was the following (in the following theorem, and henceforth, m is the total number of edges in the graph).

Theorem 1 (Dias et al. [5]). *The problem of finding a stable matching in a* SM *instance with forced and forbidden edges or reporting that none exists is solvable in $O(m)$ time.*

While Knuth's method relies on basic combinatorial properties of stable matchings, the other two algorithms make use of *rotations*. We refer the reader to [12] for background on these. The problem of finding a stable matching with forced and forbidden edges can easily be formulated as a weighted stable matching problem (that is, we seek a stable matching with minimum weight, where the weight of a matching M is the sum of the weights of the edges in M). Let us assign all forced edges weight 1, all forbidden edges weight -1, and all remaining edges weight 0. A stable matching satisfying all constraints on restricted edges exists if and only if there is a stable matching of weight $|Q|$ in the weighted instance, where Q is the set of forced edges. With the help of rotations, maximum weight stable matchings can be found in polynomial time [6,7,15,19].

Since finding a weight-maximal stable matching in SR instances is an NP-hard task [6], it follows that solving the problem with forced and forbidden edges requires different methods from the aforementioned weighted transformation. Fleiner et al. [8] showed that any SR instance with forbidden edges can be converted into another stable matching problem involving ties that can be solved in $O(m)$ time [16] and the transformation has the same time complexity as well. Forced edges can easily be eliminated by forbidding all edges adjacent to them, therefore we can state the following result.

Theorem 2 (Fleiner et al. [8]). *The problem of finding a stable matching in an* SR *instance with forced and forbidden edges or reporting that none exists is solvable in* $O(m)$ *time.*

As we have seen so far, answering the question as to whether a stable solution containing all forced and avoiding all forbidden edges exists can be solved efficiently in the case of both SM and SR. We thus concentrate on cases where the answer to this question is no. What kind of approximate solutions exist then and how can we find them?

Our Contribution. Since optimality is defined by two criteria, it is straightforward to define approximation from those two points of view. In case BP, all constraints on restricted edges must be satisfied, and we seek a matching with the minimum number of blocking edges. In case CV, we seek a stable matching that violates the fewest constraints on restricted edges. The optimization problems that arise from each of these cases are defined formally in Sect. 2.

In Sect. 3, we consider case BP: that is, all constraints on restricted edges must be fulfilled, while the number of blocking edges is minimised. We show that in the SM case, this problem is computationally hard and not approximable within $n^{1-\varepsilon}$ for any $\varepsilon > 0$, unless $\mathsf{P} = \mathsf{NP}$. We also discuss special cases for which this problem becomes tractable. This occurs if the maximum degree of the graph is at most 2 or if the number of blocking edges in the optimal solution is a constant. We point out a striking difference in the complexity of the two cases with only forbidden and only forced edges: the problem is polynomially solvable if the number of forbidden edges is a constant, but by contrast it is NP-hard even if the instance contains a single forced edge. We also prove that when the restricted edges are either all forced or all forbidden, the optimization problem remains NP-hard even on very sparse instances, where the maximum degree of a vertex is 3.

Case CV, where the number of violated constraints on restricted edges is minimised while stability is preserved, is studied in Sect. 4. It is a rather straightforward observation that in SM, the setting can be modelled and efficiently solved with the help of edge weights. Here we show that on non-bipartite graphs, the problem becomes NP-hard, but 2-approximable if the number of forced edges is sufficiently large or zero. As in case BP, we also discuss the complexity of degree-constrained restrictions and establish that the NP-hardness results remain intact even for graphs with degree at most 3, while the case with degree at most 2 is polynomially solvable.

A structured overview of our results is contained in Table 1.

2 Preliminaries and Techniques

In this section, we introduce the notation used in the remainder of the paper and also define the key problems that we investigate later. A *Stable Marriage*

Table 1. Summary of results

	Stable marriage	Stable roommates
case BP: min # blocking edges	NP-hard to approximate within $n^{1-\varepsilon}$	NP-hard to approximate within $n^{1-\varepsilon}$
case CV: min # violated restricted edge constraints	solvable in polynomial time	NP-hard; 2-approximable if $\|Q\|$ is large or 0

instance (SM) $\mathcal{I} = (G, O)$ consists of a bipartite graph $G = (U \cup W, E)$ with n vertices and m edges, and a set O: the set of strictly ordered, but not necessarily complete preference lists. These lists are provided on the set of adjacent vertices at each vertex. The *Stable Roommates Problem* (SR) differs from SM in one sense: the underlying graph G need not be bipartite. In both SM and SR, a matching in G is sought, assigning each agent to at most one partner. An edge $uw \in E \setminus M$ *blocks* matching M if u is unmatched or it prefers w to its partner in M and w is unmatched or it prefers u to its partner in M. A matching that is not blocked by any edge is called *stable*.

As already mentioned in the introduction, an SR instance need not admit a stable solution. The number of blocking edges is a characteristic property of every matching. The set of edges blocking M is denoted by $bp(M)$. A natural goal is to find a matching minimising $|bp(M)|$. For convenience, the minimum number of edges blocking any matching of an instance $\mathcal{I}$ is denoted by $bp(\mathcal{I})$. Following the consensus in the literature, matchings blocked by $bp(\mathcal{I})$ edges are called *almost stable matchings*. This approach has a broad literature: almost stable matchings have been investigated in SM [3,13,18] and SR [1,2] instances.

All problems investigated in this paper deal with at least one set of restricted edges. The set of forbidden edges is denoted by P, while Q stands for the set of forced edges. We assume throughout the paper that $P \cap Q = \emptyset$. A matching M satisfies all constraints on restricted edges if $M \cap P = \emptyset$ and $M \cap Q = Q$.

In Fig. 1, a sample SM instance on four men and four women can be seen. The preference ordering is shown above or below the vertices. For instance,

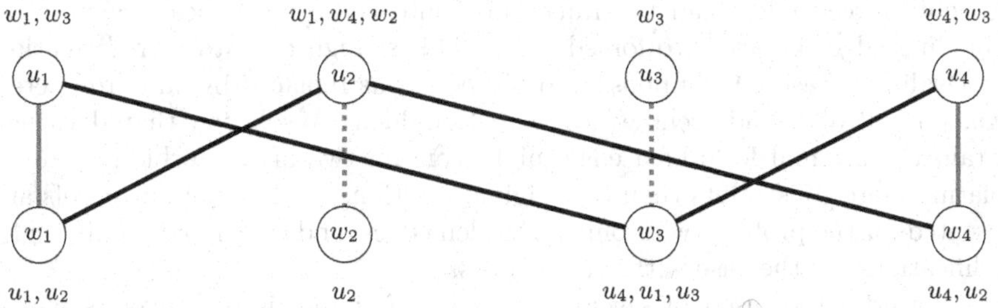

Fig. 1. A sample stable marriage instance with forbidden edges

vertex u_2 ranks w_1 best, then w_4, and w_2 last. The set of forbidden edges $P = \{u_2w_2, u_3w_3\}$ is marked by dotted gray edges. The unique stable matching $M = \{u_1w_1, u_2w_2, u_3w_3, u_4w_4\}$ contains both forbidden edges. Later on, we will return to this sample instance to demonstrate approximation concepts on it.

The first approximation concept (case BP described in Sect. 1) is to seek a matching M that satisfies all constraints on restricted edges, but among these matchings, it admits the minimum number of blocking edges. This leads to the following problem definition.

Problem 1. MIN BP SR RESTRICTED
Input: $\mathcal{I} = (G, O, P, Q)$; *an* SR *instance, a set of forbidden edges* P *and a set of forced edges* Q.
Output: A matching M *such that* $M \cap P = \emptyset$, $Q \subseteq M$ *and* $|bp(M)| \leq |bp(M')|$ *for every matching* M' *in* G *satisfying* $M' \cap P = \emptyset$, $Q \subseteq M'$.

Special attention is given to two special cases of MIN BP SR RESTRICTED: in MIN BP SR FORBIDDEN, $Q = \emptyset$, while in MIN BP SR FORCED, $P = \emptyset$. Note that an instance of MIN BP SR FORCED or MIN BP SR RESTRICTED can always be transformed into an instance of MIN BP SR FORBIDDEN by forbidding all edges that are adjacent to a forced edge. This transformation does not affect the number of blocking edges.

According to the other intuitive approximation concept (case CV described in Sect. 1), stability constraints need to be fulfilled, while some of the constraints on restricted edges are relaxed. The goal is to find a stable matching that violates as few constraints on restricted edges as possible.

Problem 2. SR MIN RESTRICTED VIOLATIONS
Input: $\mathcal{I} = (G, O, P, Q)$; *an* SR *instance, a set of forbidden edges* P *and a set of forced edges* Q.
Output: A stable matching M *such that* $|M \cap P| + |Q \setminus M| \leq |M' \cap P| + |Q \setminus M'|$ *for every stable matching* M' *in* G.

Just as in the previous approximation concept (referred to as case BP in Sect. 1), we separate the two subcases with only forbidden and only forced edges. If $Q = \emptyset$, SR MIN RESTRICTED VIOLATIONS is referred as SR MIN FORBIDDEN, while if $P = \emptyset$, the problem becomes SR MAX FORCED. In case BP, the subcase with only forced edges can be transformed into the other subcase, simply by forbidding edges adjacent to forced edges. This straightforward transformation is not valid for case CV. Suppose a forced edge was replaced by an unrestricted edge, but all of its adjacent edges were forbidden. A solution that does not contain the original forbidden edge might contain two of the forbidden edges, violating more constraints than the original solution. Yet most of our proofs are presented for the problem with only forbidden edges, and they require only slight modifications for the case with forced edges.

A powerful tool used in several proofs in our paper is to convert some of these problems into a weighted SM or SR problem, where the goal is to find a stable matching with the highest edge weight, taken over all stable matchings.

Irving et al. [15] were the first to show that the weighted SM can be solved in $O(n^4 \log n)$ time (where n is the number of vertices) if the weight function is monotone in the preference ordering, non-negative and integral. Feder [6,7] shows a method to drop the monotonicity requirement. He also presents the best known bound for the runtime of an algorithm for finding a minimum weight stable matching in SM: $O(n^2 \cdot \log(\frac{K}{n^2} + 2) \cdot \min\{n, \sqrt{K}\})$, where K is the weight of an optimal solution. Redesigning the weight function to avoid the monotonicity requirement using Feder's method can radically increase K. For weighted SR, finding an optimal matching is NP-hard, but 2-approximable, under the assumption of monotone, non-negative and integral weights [6]. These constraints restrict the practical use of Feder's results to a large extent. Fortunately, linear programming techniques allow the majority of the conditions to be dropped while retaining polynomial-time solvability. Weighted SM can be solved to optimality with arbitrary real-valued weight functions [19], and a 2-approximation for weighted SR can be found for every non-negative weight function [25].

In all discussed problems, n is the number of vertices and m is the number of edges in the graph underlying the particular problem instance. When considering the restriction of any of the above problems to the case of a bipartite graph SR is replaced by SM in the problem name. Finally, we note that all proofs can be found in the full version of the paper [4].

3 Almost Stable Matchings with Restricted Edges

In this section, constraints on restricted edges must be fulfilled strictly, while the number of blocking edges is minimised. Our results are presented in three subsections, and most of the results are given for MIN BP SM RESTRICTED. Firstly, in Sect. 3.1, basic complexity results are discussed. In particular, we prove that the studied problem MIN BP SM RESTRICTED is in general NP-hard and very difficult to approximate. Thus, restricted cases are analyzed in Sect. 3.2. First we assume that the number of forbidden, forced or blocking edges can be considered as a constant. Due to this assumption, two of the three problems that naturally follow from imposing these restrictions become tractable, but surprisingly, not all of them. Then, degree-constrained cases are discussed. We show that the NP-hardness result for MIN BP SM RESTRICTED holds even for instances where each preference list is of length at most 3, while on graphs with maximum degree 2, the problems become tractable. Finally, in Sect. 3.3 we mention the problem MIN BP SR RESTRICTED and briefly elaborate on how results established for the bipartite case carry over to the SR case.

3.1 General Complexity and Approximability Results

When minimising the number of blocking edges, one might think that removing the forbidden edges temporarily and then searching for a stable solution in the remaining instance leads to an optimal solution. Such a matching can only be blocked by forbidden edges, but as the upcoming example demonstrates, optimal solutions are sometimes blocked by unrestricted edges exclusively. In some instances, all almost stable solutions admit only non-forbidden blocking edges. Moreover, a man- or woman-optimal almost stable matching with forbidden edges does not always exist.

Let us recall the SM instance in Fig. 1. In the graph with edge set $E(G) \setminus P$, a unique stable matching exists: $M = \{u_1w_1, u_4w_4\}$. Matching M is blocked by both forbidden edges in the original instance. On the other hand, matching $M_1 = \{u_1w_1, u_2w_4, u_4w_3\}$ is blocked by exactly one edge: $bp(M_1) = u_4w_4$. Similarly, matching $M_2 = \{u_1w_3, u_2w_1, u_4w_4\}$ is blocked only by u_1w_1. Therefore, M_1 and M_2 are both almost stable matchings and $bp(\mathcal{I}) = 1$. One can easily check that M_1 and M_2 are the only matchings with the minimum number of blocking edges. They both are blocked only by unrestricted edges. Moreover, M_1 is better for u_1, w_1 and w_3, whereas M_2 is preferred by u_2, u_4 and w_4.

We now present two results demonstrating the NP-hardness and inapproximability of special cases of MIN BP SM RESTRICTED.

Theorem 3. MIN BP SM FORBIDDEN *and* MIN BP SM FORCED *are* NP-*hard.*

Theorem 4. *Each of* MIN BP SM FORBIDDEN *and* MIN BP SM FORCED *is not approximable within a factor of* $n^{1-\varepsilon}$, *for any* $\varepsilon > 0$, *unless* P = NP.

3.2 Bounded Parameters

Our results presented so far show that MIN BP SM RESTRICTED is computationally hard even if $P = \emptyset$ or $Q = \emptyset$. Yet if certain parameters of the instance or the solution can be considered as a constant, the problem can be solved in polynomial time. Theorem 5 firstly shows that this is true for MIN BP SM FORBIDDEN.

Theorem 5. MIN BP SM FORBIDDEN *is solvable in* $O(n^2 m^L)$ *time, where* $L = |P|$, *which is polynomial if* L *is a constant.*

In sharp contrast to the previous result on polynomial solvability when the number of forbidden edges is small, we state the following theorem for the MIN BP SM FORCED problem.

Theorem 6. MIN BP SM FORCED *is* NP-*hard even if* $|Q| = 1$.

On the other hand, a counterpart to Theorem 5 holds in the case of MIN BP SM RESTRICTED if the number of blocking pairs in an optimal solution is a constant.

Theorem 7. MIN BP SM RESTRICTED *is solvable in* $O(m^{L+1})$ *time, where* L *is the minimum number of edges blocking an optimal solution, which is polynomial if* L *is a constant.*

Next we study the case of degree-constrained graphs, because for most hard SM and SR problems, it is the most common special case to investigate [2,13,21]. Here, we show that MIN BP SM RESTRICTED remains computationally hard even for instances with preference lists of length at most 3. On the other hand, the problem can be solved by identifying forbidden subgraphs when the length of preference lists is bounded by 2.

Theorem 8. MIN BP SM FORBIDDEN *and* MIN BP SM FORCED *are* NP*-hard even if each agent's preference list consists of at most 3 elements.*

Theorem 9. MIN BP SM RESTRICTED *is solvable in* $O(n)$ *time if each preference list consists of at most 2 elements.*

Even with the previous two theorems, we have not quite drawn the line between tractable and hard cases in terms of vertex degrees. The complexity of MIN BP SM RESTRICTED remains open for the case when preference lists are of length at most 2 on one side of the bipartite graph and they are of unbounded length on the other side. However we believe that this problem is solvable in polynomial time.

Conjecture 1. MIN BP SM RESTRICTED *is solvable in polynomial time if each woman's preference list consists of at most 2 elements.*

3.3 Stable Roommates Problem

Having discussed several cases of SM, we turn our attention to non-bipartite instances. Since SM is a restriction of SR, all established results on the NP-hardness and inapproximability of MIN BP SM RESTRICTED carry over to the non-bipartite SR case. As a matter of fact, more is true, since MIN BP SR RESTRICTED is NP-hard and difficult to approximate even if $P = \emptyset$ and $Q = \emptyset$ [1]. We summarise these observations as follows.

Remark 1. By Theorems 3 and 4, MIN BP SR FORBIDDEN and MIN BP SR FORCED are NP-hard and not approximable within $n^{1-\varepsilon}$, for any $\varepsilon > 0$, unless $\mathsf{P} = \mathsf{NP}$. Moreover Theorems 8 and 6 imply that MIN BP SR FORBIDDEN and MIN BP SR FORCED are NP-hard even if all preference lists are of length at most 3 or, in the latter case, $|Q| = 1$. Finally MIN BP SR RESTRICTED is NP-hard and not approximable within $n^{\frac{1}{2}-\varepsilon}$, for any $\varepsilon > 0$, unless $\mathsf{P} = \mathsf{NP}$, even if $P = \emptyset$ and $Q = \emptyset$ [1].

As for the polynomially solvable cases, the proofs of Theorems 5, 7 and 9 carry over without applying any modifications, giving the following.

Remark 2. MIN BP SR FORBIDDEN is solvable in polynomial time if $|P|$ is a constant. MIN BP SR RESTRICTED is solvable in polynomial time if the minimal number of edges blocking an optimal solution is a constant.

4 Stable Matchings with the Minimum Number of Violated Constraints on Restricted Edges

In this section, we study the second intuitive approximation concept. The sought matching is stable and violates as few constraints on restricted edges aspossible.

We return to our example that already appeared in Fig. 1. As already mentioned earlier, the instance admits a single stable matching, namely $M = \{u_1w_1, u_2w_2, u_3w_3, u_4w_4\}$. Since M contains both forbidden edges, the minimum number of violated constraints on restricted edges is 2.

As mentioned in Sect. 1, a weighted stable matching instance models SM MIN RESTRICTED VIOLATIONS.

Theorem 10. SM MIN RESTRICTED VIOLATIONS *is solvable in polynomial time.*

In the SR context, finding a minimum weight stable matching is NP-hard [6], so the above technique for SM does not carry over to SR. Indeed special cases of SR MIN RESTRICTED VIOLATIONS are NP-hard, as the following result shows.

Theorem 11. SR MIN FORBIDDEN *and* SR MAX FORCED *are* NP*-hard.*

In our proof, we reduce the Minimum Vertex Cover problem to these two problems. Minimum Vertex Cover is NP-hard and cannot be approximated within a factor of $2-\varepsilon$ for any positive ε, unless the Unique Games Conjecture is false [17]. The reduction also answers basic questions about the approximability of these problems. Since any vertex cover on K vertices can be interpreted as a stable matching containing K forbidden edges in SR MIN FORBIDDEN and vice versa, the $(2-\varepsilon)$-inapproximability result carries over. The same holds for the number of violated forced edge constraints in SR MAX FORCED. On the positive side, we can close the gap with the best possible approximation ratio if $Q = \emptyset$ or $|Q|$ is sufficiently large. To derive this result, we use the 2-approximability of weighted SR for non-negative weight functions [25]. Due to the non-negativity constraint, the case of $0 < |Q| < |M|$ remains open.

Theorem 12. *If* $|Q| \geq |M|$ *for a stable matching* M*, then* SR MIN RESTRICTED VIOLATIONS *is 2-approximable in polynomial time.*

When studying SR MAX FORCED, we measured optimality by keeping track of the number of violated constraints. One might find it more intuitive instead to maximise $|Q \cap M|$, the number of forced edges in the stable matching. Our NP-hardness proof for SR MAX FORCED remains intact, but the approximability results need to be revisited. In fact, this modification of the measure changes the approximability of the problem as well:

Theorem 13. *For* SR MAX FORCED*, the maximum of* $|Q \cap M|$ *cannot be approximated within* $n^{\frac{1}{2}-\varepsilon}$ *for any* $\varepsilon > 0$*, unless* P = NP*.*

We now turn to the complexity of SR MIN RESTRICTED VIOLATIONS and its variants when the degree of the underlying graph is bounded or some parameter of the instance can be considered as a constant.

Theorem 14. SR MIN FORBIDDEN *and* SR MAX FORCED *are* NP*-hard even if every preference list is of length at most 3.*

Theorem 15. SR MIN RESTRICTED VIOLATIONS *is solvable in* $O(n)$ *time if every preference list is of length at most 2.*

Theorem 16. SR MIN RESTRICTED VIOLATIONS *is solvable in polynomial time if the number of restricted edges or the minimal number of violated constraints is constant.*

5 Conclusion

In this paper, we investigated the stable marriage and the stable roommates problems on graphs with forced and forbidden edges. Since a solution satisfying all constraints need not exist, two relaxed problems were defined. In MIN BP SM RESTRICTED, constraints on restricted edges are strict, while a matching with the minimum number of blocking edges is searched for. On the other hand, in SR MIN RESTRICTED VIOLATIONS, we seek stable solutions that violate as few constraints on restricted edges as possible. For both problems, we determined the complexity and studied several special cases.

One of the most striking open questions is the approximability of SR MIN RESTRICTED VIOLATIONS if $0 < |Q| < |M|$. Our other open question is formulated as Conjecture 1: the complexity of MIN BP SM RESTRICTED is not known if each woman's preference list consists of at most 2 elements. A more general direction of further research involves the SM MIN RESTRICTED VIOLATIONS problem. We have shown that it can be solved in polynomial time, due to algorithms for maximum weight stable marriage. The following question arises naturally: is there a faster method for SM MIN RESTRICTED VIOLATIONS that avoids reliance on Feder's algorithm or linear programming methods?

Acknowledgements. We would like to thank the anonymous reviewers for their valuable comments, which have helped to improve the presentation of this paper.

References

1. Ageev, A.A., Kononov, A.V.: Approximation algorithms for scheduling problems with exact delays. In: Erlebach, T., Kaklamanis, C. (eds.) WAOA 2006. LNCS, vol. 4368, pp. 1–14. Springer, Heidelberg (2007)
2. Biró, P., Manlove, D., McDermid, E.: "Almost stable" matchings in the roommates problem with bounded preference lists. Theor. Comput. Sci. **432**, 10–20 (2012)
3. Biró, P., Manlove, D., Mittal, S.: Size versus stability in the marriage problem. Theor. Comput. Sci. **411**, 1828–1841 (2010)
4. Cseh, A., Manlove, D.: Stable marriage and roommates problems with restricted edges: complexity and approximability. Technical Report 1412.0271, Computing Research Repository, Cornell University Library, 2015. http://arxiv.org/abs/1412.0271
5. Dias, V., da Fonseca, G., de Figueiredo, C., Szwarcfiter, J.: The stable marriage problem with restricted pairs. Theor. Comput. Sci. **306**(1–3), 391–405 (2003)

6. Feder, T.: A new fixed point approach for stable networks and stable marriages. J. Comput. Syst. Sci. **45**, 233–284 (1992)
7. Feder, T.: Network flow and 2-satisfiability. Algorithmica **11**(3), 291–319 (1994)
8. Fleiner, T., Irving, R., Manlove, D.: Efficient algorithms for generalised stable marriage and roommates problems. Theor. Comput. Sci. **381**(1–3), 162–176 (2007)
9. Gai, A.-T., Lebedev, D., Mathieu, F., de Montgolfier, F., Reynier, J., Viennot, L.: Acyclic preference systems in P2P networks. In: Kermarrec, A.-M., Bougé, L., Priol, T. (eds.) Euro-Par 2007. LNCS, vol. 4641, pp. 825–834. Springer, Heidelberg (2007)
10. Gale, D., Shapley, L.: College admissions and the stability of marriage. Am. Math. Monthly **69**, 9–15 (1962)
11. Gale, D., Sotomayor, M.: Some remarks on the stable matching problem. Discrete Appl. Math. **11**, 223–232 (1985)
12. Gusfield, D., Irving, R.: The Stable Marriage Problem: Structure and Algorithms. MIT Press, Cambridge (1989)
13. Hamada, K., Iwama, K., Miyazaki, S.: An improved approximation lower bound for finding almost stable maximum matchings. Inf. Process. Lett. **109**(18), 1036–1040 (2009)
14. Irving, R.: An efficient algorithm for the "stable roommates" problem. J. Algorithms **6**, 577–595 (1985)
15. Irving, R., Leather, P., Gusfield, D.: An efficient algorithm for the "optimal" stable marriage. J. ACM **34**(3), 532–543 (1987)
16. Irving, R., Manlove, D.: The stable roommates problem with ties. J. Algorithms **43**, 85–105 (2002)
17. Khot, S., Regev, O.: Vertex cover might be hard to approximate to within $2 - \varepsilon$. J. Comput. Syst. Sci. **74**(3), 335–349 (2008)
18. Khuller, S., Mitchell, S., Vazirani, V.: On-line algorithms for weighted bipartite matching and stable marriages. Theor. Comput. Sci. **127**, 255–267 (1994)
19. Király, T., Pap, J.: Total dual integrality of Rothblum's description of the stable-marriage polyhedron. Math. Oper. Res. **33**(2), 283–290 (2008)
20. Knuth, D.: Mariages Stables. (Les Presses de L'Université de Montréal, 1976). English translation in Stable Marriage and its Relation to Other Combinatorial Problems. CRM Proceedings and Lecture Notes, vol. 10. American Mathematical Society (1997)
21. O'Malley, G.: Algorithmic Aspects of Stable Matching Problems. Ph.D thesis, University of Glasgow, Department of Computing Science (2007)
22. Roth, A.: The evolution of the labor market for medical interns and residents: a case study in game theory. J. Political Econ. **92**(6), 991–1016 (1984)
23. Roth, A.: Deferred acceptance algorithms: history, theory, practice, and open questions. Int. J. Game Theor. **36**(3–4), 537–569 (2008)
24. Roth, A., Sotomayor, M.: Two-Sided Matching: A Study in Game-Theoretic Modeling and Analysis. Econometric Society Monographs, vol. 18. Cambridge University Press, Cambridge (1990)
25. Teo, C.-P., Sethuraman, J.: The geometry of fractional stable matchings and its applications. Math. Oper. Res. **23**(4), 874–891 (1998)

Pareto Optimal Matchings in Many-to-Many Markets with Ties

Katarína Cechlárová[1], Pavlos Eirinakis[2], Tamás Fleiner[3], Dimitrios Magos[4], David F. Manlove[5], Ioannis Mourtos[2], Eva Oceľáková[1], and Baharak Rastegari[5](✉)

[1] Institute of Mathematics, Faculty of Science, P.J. Šafárik University, Košice, Slovakia

[2] Department of Management Science and Technology, Athens University of Economics and Business, Athens, Greece

[3] Department of Computer Science and Information Theory, Budapest University of Technology and Economics and MTA-ELTE Egerváry Research Group, Budapest, Hungary

[4] Department of Informatics, Technological Educational Institute of Athens, Egaleo, Greece

[5] School of Computing Science, University of Glasgow, Glasgow, UK
baharak.rastegari@glasgow.ac.uk

Abstract. We consider Pareto-optimal matchings (POMs) in a many-to-many market of applicants and courses where applicants have preferences, which may include ties, over individual courses and lexicographic preferences over sets of courses. Since this is the most general setting examined so far in the literature, our work unifies and generalizes several known results. Specifically, we characterize POMs and introduce the *Generalized Serial Dictatorship Mechanism with Ties (GSDT)* that effectively handles ties via properties of network flows. We show that GSDT can generate all POMs using different priority orderings over the applicants, but it satisfies truthfulness only for certain such orderings. This shortcoming is not specific to our mechanism; we show that any mechanism generating all POMs in our setting is prone to strategic manipulation. This is in contrast to the one-to-one case (with or without ties), for which truthful mechanisms generating all POMs do exist.

Keywords: Pareto optimality · Many-to-many matching · Serial dictatorship · Truthfulness

---

This research has been co-financed by the European Union (European Social Fund - ESF) and Greek national funds under Thales grant MIS 380232 (Eirinakis, Magos, Mourtos), by grant EP/K01000X/1 from the Engineering and Physical Sciences Research Council (Manlove, Rastegari), grants VEGA 1/0344/14, 1/0142/15 from the Slovak Scientific grant agency VEGA (Cechlárová), student grant VVGS-PF-2014-463 (Oceľáková) and OTKA grant K108383 (Fleiner). The authors gratefully acknowledge the support of COST Action IC1205 Computational Social Choice.

M. Hoefer (Ed.): SAGT 2015, LNCS 9347, pp. 27–39, 2015.
DOI: 10.1007/978-3-662-48433-3_3

1 Introduction

We study a many-to-many matching market that involves two finite disjoint sets, a set of applicants and a set of courses. Each applicant finds a subset of courses acceptable and has a preference ordering, not necessarily strict, over these courses. Courses do not have preferences. Moreover, each applicant has a quota on the number of courses she can attend, while each course has a quota on the number of applicants it can admit.

A matching is a set of applicant-course pairs such that each applicant is paired only with acceptable courses and the quotas associated with the applicants and the courses are respected. The problem of finding an "optimal" matching given the above market is called the *Course Allocation problem* (CA). Although various optimality criteria exist, *Pareto optimality* (or *Pareto efficiency*) remains the most popular one (see, e.g., [1,2,9,17]). Pareto optimality is a fundamental concept that economists regard as a minimal requirement for a "reasonable" outcome of a mechanism. A matching is a Pareto optimal matching (POM) if there is no other matching in which no applicant is worse off and at least one applicant is better off. Our work examines Pareto optimal many-to-many matchings in the setting where applicants' preferences may include ties.

In the special case where each applicant and course has quota equal to one, our setting reduces to the extensively studied *House Allocation problem* (HA) [1,13], also known as the *Assignment problem* [5,11]. Computational aspects of HA have been examined thoroughly [2,16] and particularly for the case where applicants' preferences are strict. In [2] the authors provide a characterization of POMs in the case of strict preferences and utilize it in order to construct polynomial-time algorithms for checking whether a given matching is a POM and for finding a POM of maximum size. They also show that any POM in an instance of HA with strict preferences can be obtained through the well-known *Serial Dictatorship Mechanism (SDM)* [1]. SDM is a straightforward greedy algorithm that allocates houses sequentially according to some exogenous priority ordering of the applicants, giving each applicant her most-preferred vacant house.

Recently, the above results have been extended in two different directions. The first one [15] considers HA in settings where preferences may include ties. Prior to [15], few works in the literature had considered extensions of SDM to such settings. The difficulty regarding ties, observed already in [18], is that the assignments made in the individual steps of the SDM are not unique, and an unsuitable choice may result in an assignment that violates Pareto optimality. In [6] and [18] an implicit extension of SDM is provided (in the former case for *dichotomous preferences*, where an applicant's preference list comprises a single tie containing all acceptable houses), but without an explicit description of an algorithmic procedure. [15] describe a mechanism called the *Serial Dictatorship Mechanism with Ties (SDMT)* that combines SDM with the use of augmenting paths to ensure Pareto optimality. In an augmentation step, applicants already assigned a house may exchange it for another, equally preferred one, to enable another applicant to take a house that is most preferred given the assignments made so far. They also show that any POM in an instance of HA with ties can be obtained by an execution of SDMT and also describe the so-called *Random Serial Dictatorship Mechanism with Ties (RSDMT)* whose (expected) approximation ratio is $\frac{e}{e-1}$ with respect to the maximum-size POM.

The second direction [9] extends the results of [2] to the many-to-many setting (i.e., CA) with strict preferences, while also allowing for a structure of applicant-wise acceptable sets that is more general than the one implied by quotas; namely, [9] assumes that each applicant selects from a family of course subsets that is downward closed. This work provides a characterization of POMs assuming that the preferences of applicants over sets of courses are obtained from their (strict) preferences over individual courses in a lexicographic manner; using this characterization, it is shown that deciding whether a given matching is a POM can be accomplished in polynomial time. In addition, [9] generalizes SDM to provide the *Generalized Serial Dictatorship (GSD)* mechanism, which can be used to obtain any POM for CA under strict preferences. The main idea of GSD is to allow each applicant to choose not her most preferred set of courses at once but, instead, only one course at a time (i.e., the most preferred among non-full courses that can be added to the courses already chosen). This result is important as the version of SDM where an applicant chooses immediately her most preferred set of courses cannot obtain all POMs.

Our Contribution. In the current work, we combine the directions appearing in [15] and [9] to explore the many-to-many setting in which applicants have preferences, which may include ties, over individual courses. We extend these preferences to sets of courses lexicographically, since lexicographic set preferences naturally describe human behavior [12], they have already been considered in models of exchange of indivisible goods [9,10] and also possess theoretically interesting properties including responsiveness [14].

We provide a characterization of POMs in this setting and introduce the *Generalized Serial Dictatorship Mechanism with Ties (GSDT)* that generalizes both SDMT and GSD. SDM assumes a priority ordering over the applicants, according to which applicants are served one by one by the mechanism. Since in our setting applicants can be assigned more than one course, each applicant can return to the ordering several times (up to her quota), each time choosing just one course. The idea of using augmenting paths [15] has to be employed carefully to ensure that during course shuffling no applicant replaces a previously assigned course for a less preferred one. To achieve this, we utilize methods and properties of network flows. Although we prove that GSDT can generate all POMs using different priority orderings over applicants, we also observe that some of the priority orderings guarantee truthfulness whereas some others do not. That is, there may exist priority orderings for which some applicant benefits from misrepresenting her preferences. This is in contrast to SDM and SDMT in the one-to-one case in the sense that all executions of these mechanisms induce truthfulness. This shortcoming however is not specific to our mechanism, since we establish that any mechanism generating all POMs is prone to strategic manipulation by one or more applicants.

Remark. [4] presented a general mechanism for computing Pareto optimal outcomes in hedonic games which includes the many-to-many matching problem with ties. However their mechanism was not presented in a form that is specific to our setting and no explicit bound for the time complexity was given in [4].

Organization of the Paper. In Sect. 2 we define our notation and terminology. The characterization is provided in Sect. 3, while GSDT is presented in Sect. 4. A discussion on applicants' incentives in GSDT is provided in Sect. 5. Missing proofs can be found in the full version of this paper [8].

2 Preliminary Definitions of Notation and Terminology

Let $A = \{a'_1, a'_2, \cdots, a'_{n_1}\}$ be the set of applicants, $C = \{c_1, c_2, \cdots, c_{n_2}\}$ the set of courses and $[i]$ denote the set $\{1, 2, \ldots, i\}$. Each applicant a' has a quota $b(a')$ that denotes the maximum number of courses a' can accommodate into her schedule, and likewise each course c has a quota $q(c)$ that denotes the maximum number of applicants it can admit. Each applicant finds a subset of courses acceptable and has a transitive and complete preference ordering, not necessarily strict, over these courses. We write $c \succ_{a'} c'$ to denote that applicant a' *(strictly) prefers* course c to course c', and $c \simeq_{a'} c'$ to denote that a' is *indifferent between* c and c'. We write $c \succeq_{a'} c'$ to denote that a' either prefers c to c' or is indifferent between them, and say that a' *weakly prefers* c to c'.

Because of indifference, each applicant divides her acceptable courses into *indifference classes* such that she is indifferent between the courses in the same class and has a strict preference over courses in different classes. Let $C_t^{a'}$ denote the t'th indifference class, or *tie*, of applicant a' where $t \in [n_2]$. We assume that $C_t^{a'} = \emptyset$ implies $C_{t'}^{a'} = \emptyset$ for all $t' > t$. Let the preference list of any applicant a' be the tuple of sets $C_t^{a'}$, i.e., $P(a') = (C_1^{a'}, C_2^{a'}, \cdots, C_{n_2}^{a'})$; occasionally we consider $P(a')$ to be a set itself and write $c \in P(a')$ instead of $c \in C_t^{a'}$ for some t. We denote by $\mathcal{P}$ the joint preference profile of all applicants, and by $\mathcal{P}(-a')$ the joint profile of all applicants except a'. Under these definitions, an instance of CA is denoted by $I = (A, C, \mathcal{P}, b, q)$. Such an instance appears in Table 1.

A *(many-to-many) assignment* μ is a subset of $A \times C$. For $a' \in A$, $\mu(a') = \{c \in C : (a', c) \in \mu\}$ and for $c \in C$, $\mu(c) = \{a' \in A : (a', c) \in \mu\}$. An assignment μ is a *matching* if $\mu(a') \subseteq P(a')$, $|\mu(a')| \leq b(a')$ for each $a' \in A$ and $|\mu(c)| \leq q(c)$ for each $c \in C$. We say that a' is *exposed* if $|\mu(a')| < b(a')$, and is *full* otherwise. Analogous definitions of exposed and full hold for courses.

Table 1. An instance I of CA.

Applicant	Quota	Preference list	Course	Quota
a'_1	2	$(\{c_1, c_2\}, \{c_3\}, \emptyset)$	c_1	2
a'_2	3	$(\{c_2\}, \{c_1, c_3\}, \emptyset)$	c_2	1
a'_3	2	$(\{c_3\}, \{c_2\}, \{c_1\})$	c_3	1

For an applicant a' and a set of courses S, we define the *generalized characteristic vector* $\chi_{a'}(S)$ as the vector $(|S \cap C_1^{a'}|, |S \cap C_2^{a'}|, \ldots, |S \cap C_{n_2}^{a'}|)$. We assume that for any two sets of courses S and U, a' prefers S to U if and only if $\chi'_a(S) >_{lex} \chi'_a(U)$, i.e., if and only if there is an indifference class $C_t^{a'}$ such that $|S \cap C_t^{a'}| > |U \cap C_t^{a'}|$ and $|S \cap C_{t'}^{a'}| = |U \cap C_{t'}^{a'}|$ for all $t' < t$. If a' neither prefers S to U nor U to S, then she is indifferent between S and U. We write $S \succ'_a U$ if a' prefers S to U, $S \simeq'_a U$ if a' is indifferent between S and U, and $S \succeq_{a'} U$ if a' weakly prefers S to U.

A matching μ is a *Pareto optimal matching (POM)* if there is no other matching in which some applicant is better off and none is worse off. Formally, μ is Pareto optimal if there is no matching μ' such that $\mu'(a') \succeq_{a'} \mu(a')$ for all $a' \in A$, and $\mu'(a'') \succ_{a''} \mu(a'')$ for some $a'' \in A$. If such a μ' exists, we say that μ' *Pareto dominates* μ.

A *deterministic mechanism* ϕ maps an instance to a matching, i.e. $\phi : I \mapsto \mu$ where I is a CA instance and μ is a matching in I. A *randomized mechanism* ϕ maps an instance to a distribution over possible matchings. Applicants' preferences are private knowledge and an applicant may prefer not to reveal her preferences truthfully. A deterministic mechanism is *truthful* if all applicants always finds it best to declare their true preferences, no matter what other applicants declare. A randomized mechanism ϕ is *universally truthful* if it is a probability distribution over deterministic truthful mechanisms.

3 Characterizing Pareto Optimal Matchings

Manlove [16, Sect. 6.2.2.1] provided a characterization of Pareto optimal matchings in HA with preferences that may include indifference. He defined three different types of *coalitions* with respect to a given matching such that the existence of either means that a subset of applicants can trade among themselves (possibly using some exposed course) and ensure that, at the end, no one is worse off and at least one applicant is better off. He also showed that if no such coalition exists, then the matching is guaranteed to be Pareto optimal. We show that this characterization extends to the many-to-many setting, although the proof is more complex and involved than in the one-to-one setting.

In what follows we assume that in each sequence $\mathfrak{C}$ no applicant or course appears more than once.

An *alternating path coalition* w.r.t. μ comprises a sequence $\mathfrak{C} = \langle c_{j_0}, a'_{i_0}, c_{j_1}, a'_{i_1}, \ldots, c_{j_{r-1}} a'_{i_{r-1}}, c_{j_r} \rangle$ where $r \geq 1$, $c_{j_k} \in \mu(a'_{i_k})$ $(0 \leq k \leq r-1)$, $c_{j_k} \notin \mu(a'_{i_{k-1}})$ $(1 \leq k \leq r)$, a'_{i_0} is full, and c_{j_r} is an exposed course. Furthermore, a'_{i_0} prefers c_{j_1} to c_{j_0} and, if $r \geq 2$, a'_{i_k} weakly prefers $c_{j_{k+1}}$ to c_{j_k} $(1 \leq k \leq r-1)$.

An *augmenting path coalition* w.r.t. μ comprises a sequence $\mathfrak{C} = \langle a'_{i_0}, c_{j_1}, a'_{i_1}, \ldots, c_{j_{r-1}} a'_{i_{r-1}}, c_{j_r} \rangle$ where $r \geq 1$, $c_{j_k} \in \mu(a'_{i_k})$ $(1 \leq k \leq r-1)$, $c_{j_k} \notin \mu(a'_{i_{k-1}})$ $(1 \leq k \leq r)$, a'_{i_0} is an exposed applicant, and c_{j_r} is an exposed course. Furthermore, a'_{i_0} finds c_{j_1} acceptable and, if $r \geq 2$, a'_{i_k} weakly prefers $c_{j_{k+1}}$ to c_{j_k} $(1 \leq k \leq r-1)$.

A *cyclic coalition* w.r.t. μ comprises a sequence $\mathfrak{C} = \langle c_{j_0}, a'_{i_0}, c_{j_1}, a'_{i_1}, \ldots, c_{j_{r-1}}, a'_{i_{r-1}} \rangle$ where $r \geq 2$, $c_{j_k} \in \mu(a'_{i_k})$ $(0 \leq k \leq r-1)$, and $c_{j_k} \notin \mu(a'_{i_{k-1}})$ $(1 \leq k \leq r)$. Furthermore, a'_{i_0} prefers c_{j_1} to c_{j_0} and a'_{i_k} weakly prefers $c_{j_{k+1}}$ to c_{j_k} $(1 \leq k \leq r-1)$. (All subscripts are taken modulo r when reasoning about cyclic coalitions).

We define an *improving coalition* to be an alternating path coalition, an augmenting path coalition or a cyclic coalition. Given an improving coalition $\mathfrak{C}$, the matching

$$\mu^{\mathfrak{C}} = (\mu \setminus \{(a'_{i_k}, c_{j_k}) : \delta \leq k \leq r-1\}) \cup \{(a'_{i_k}, c_{j_{k+1}}) : 0 \leq k \leq r-1\}\} \quad (1)$$

is defined to be the matching obtained from μ by *satisfying* $\mathfrak{C}$ ($\delta = 1$ in the case that $\mathfrak{C}$ is an augmenting path coalition, otherwise $\delta = 0$).

The following theorem gives a necessary and sufficient condition for a matching to be Pareto optimal.

Theorem 1. *Given a CA instance I, a matching μ is a Pareto optimal matching in I if and only if μ admits no improving coalition.*

4 Constructing Pareto Optimal Matchings

We propose an algorithm for finding a POM in an instance of CA, which is in a certain sense a generalization of Serial Dictatorship thus named 'Generalized Serial Dictatorship with ties' (GSDT). The algorithm starts by setting the quotas of all applicants to 0 and those of courses are set at the original values given by q. At each stage i, the algorithm selects a single applicant whose original capacity has not been reached, and increases only her capacity by 1. The algorithm terminates after $B = \sum_{a \in A} b(a)$ stages, i.e., once the original capacities of all applicants have been reached. In that respect, the algorithm assumes a 'multisequence' $\Sigma = (a^1, a^2, \ldots, a^B)$ of applicants such that each applicant a appears $b(a)$ times in Σ; e.g., for the instance of Table 1 and the sequence $\Sigma = (a'_1, a'_1, a'_2, a'_2, a'_3, a'_2, a'_3)$, the vector of capacities evolves as follows:

$$(0,0,0), (1,0,0), (2,0,0), (2,1,0), (2,2,0), (2,2,1), (2,3,1), (2,3,2).$$

Let us denote the vector of applicants' capacities in stage i by b^i, i.e., b^0 is the all-zeroes vector and $b^B = b$. Clearly, each stage corresponds to an instance I^i similar to the original instance except for the capacities vector b^i. At each stage i, our algorithm obtains a matching μ^i for the instance I^i. The single matching of stage 0—the empty matching, is a POM in I^0. The core idea is to modify μ^{i-1} in such way that if μ^{i-1} is a POM with respect to I^{i-1} then μ^i is a POM with respect to I^i. To achieve this, the algorithm relies on the following flow network.

Consider the digraph $D = (V, E)$. Its node set is $V = A \cup T \cup C \cup \{\sigma, \tau\}$ where σ and τ are the source and the sink and vertices in T correspond to the ties in the preference lists of all applicants; i.e., T has a node (a, t) per applicant a' and tie t such that $C^a_t \neq \emptyset$. Its arc set is $E = E_1 \cup E_2 \cup E_3 \cup E_4$ where $E_1 =$

$\{(\sigma, a) : a \in A\}$, $E_2 = \{(a, (a, t)) : a \in A, C_t^a \neq \emptyset\}$, $E_3 = \{((a, t), c) : c \in C_t^a\}$ and $E_4 = \{(c, \tau) : c \in C\}$. The graph D for the instance of Table 1 appears in Fig. 1.

Using digraph $D = (V, E)$, we obtain a flow network N^i at each stage i of the algorithm, i.e., a network corresponding to instance I^i, by appropriately varying the capacities of the arcs. (For an introduction on network flow algorithms see, e.g., [3].) The capacity of each arc in E_3 is always 1 (since each course may be received at most once by each applicant) and the capacity of an arc $e = (c, \tau) \in E_4$ is always $q(c)$. The capacities of all arcs in $E_1 \cup E_2$ are initially 0 and, at stage i, the capacities of only certain arcs associated with applicant a'^i are increased by 1. For this reason, for each applicant a' we use the variable $curr(a')$ that indicates her 'active' tie; initially, $curr(a')$ is set to 1 for all $a' \in A$.

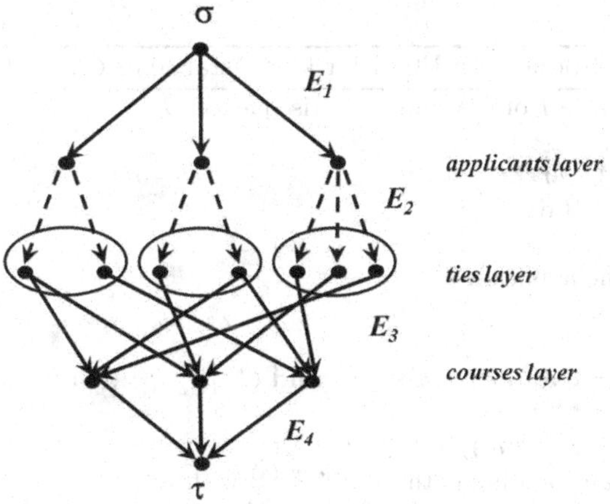

Fig. 1. Digraph D for the instance I from Table 1. An oval encircles all the vertices of T that correspond to the same applicant.

In stage i, the algorithm computes a maximum flow f^i whose saturated arcs in E_3 indicate the corresponding matching μ^i. The algorithm starts with $f^0 = 0$ and $\mu^0 = \emptyset$. Let the applicant $a'^i \in A$ be a copy of applicant a' considered in stage i. The algorithm increases by 1 the capacity of arc $(\sigma, a') \in E_1$ (i.e., the applicant is allowed to receive an additional course). It then examines the tie $curr(a')$ to check whether the additional course can be received from tie $curr(a')$. To do this, the capacity of arc $(a', (a', curr(a'))) \in E_2$ is increased by 1. The network in stage i where tie $curr(a^i)$ is examined is denoted by $N^{i,curr(a^i)}$. If there is an augmenting $\sigma - \tau$ path in this network, the algorithm augments the current flow f^{i-1} to obtain f^i, accordingly augments μ^{i-1} to obtain μ^i (i.e., it sets μ^i to the symmetric difference of μ^{i-1} and all pairs (a', c) for which there is an arc $((a', t), c)$ in the augmenting path) and proceeds to the next stage. Otherwise, it decreases the capacity of $(a', (a', curr(a')))$ by 1 (but not the capacity of arc (σ, a')) and it increases $curr(a')$ by 1 to examine the next tie of a'; if all (non-empty) ties have been examined, the algorithm proceeds to the next stage without augmenting the flow. Note that an augmenting $\sigma-\tau$ path in the network $N^{i,curr(a^i)}$ corresponds to an augmenting path coalition in μ^{i-1} with respect to I^i.

A formal description of GSDT is provided by Algorithm 1, where $w(e)$ denotes the capacity of an arc $e \in E$ and $\oplus$ denotes the operation of augmenting along an augmenting path (either relative to a flow or a matching). Observe that all arcs in E_2 are saturated, except for the arc corresponding to the current applicant and tie, thus any augmenting path has one arc from each of E_1, E_2 and E_4 and all other arcs from E_3; as a consequence, the number of courses each applicant receives at stage i in any tie cannot decrease at any subsequent step. Also, μ^i dominates μ^{i-1} with respect to instance I^i if and only if there is a flow in N^i that saturates all arcs in E_2.

Algorithm 1. Producing a POM for any instance of CA

Input: an instance I of CA and a multisequence Σ
$f^0 := 0$; $\mu^0 := \emptyset$;
for each $a' \in A$, $curr(a') := 1$;
for $i = 1, 2, \ldots, B$ **do**
{
  consider the applicant $a' = a'^i$;
  $w(\sigma, a')$++;
  $\mathfrak{P} := \emptyset$;
  while $\mathfrak{P} = \emptyset$ **and** $curr(a') \leq n_2$ **and** $C^{a'}_{curr(a')} \neq \emptyset$ **do**
  {
    $w(a', (a', curr(a')))$++;
    $\mathfrak{P}$:= augmenting path in $N^{i,curr(a)}$ with respect to f^{i-1};
    if $\mathfrak{P} = \emptyset$ **then** { $w(a', (a', curr(a')))$- -; $curr(a)$++};
  }
  if $\mathfrak{P} \neq \emptyset$ **then** { $f^i := f^{i-1} \oplus \mathfrak{P}$;
    $\mu^i := \mu^{i-1} \oplus \{(a', c) : ((a', t), c) \in \mathfrak{P}$ for some $a' \in$ A and t
$\in$ T}; }
  otherwise { $f^i := f^{i-1}$; $\mu^i := \mu^{i-1}$; }
}
return μ^B;

To prove the correctness of GSDT, we need two intermediate lemmas. Let $e_t \in \mathbb{R}^{n_2}$ be the vector having 1 at entry t and 0 elsewhere.

Lemma 1. *Let $N^{i,t}$ be the network at stage i while tie t of applicant a'^i is examined. Then, there is an augmenting path with respect to f^{i-1} in $N^{i,t}$ if and only if there is a matching μ such that*

$$\chi'_a(\mu(a')) = \chi'_a(\mu^{i-1}(a')) \text{ for each } a' \neq a'^i \text{ and } \chi_{a'^i}(\mu(a'^i)) = \chi_{a'^i}(\mu^{i-1}(a'^i)) + e_t.$$

Lemma 2. *Let $S \succeq_{a'} U$ and $|S| \geq |U|$. If c_S and c_U denote a least preferred course of applicant a in S and U, respectively, then $S \backslash \{c_S\} \succeq_{a'} U \backslash \{c_U\}$.*

Theorem 2. *For each i, the matching μ^i obtained by GSDT is a POM for instance I^i.*

Proof. We apply induction on i. Clearly, $\mu^0 = \emptyset$ is the single matching in I^0 and hence a POM in I^0. We assume that μ^{i-1} is a POM in I^{i-1} and prove that μ^i is a POM in I^i.

Assuming to the contrary that μ^i is not a POM in I^i implies that there is a matching ξ in I^i that dominates μ^i. Then, for all $a' \in A$, $\xi(a') \succeq_{a'} \mu^i(a') \succeq_{a'} \mu^{i-1}(a')$. Since the capacities of all applicants in I^i are as in I^{i-1} except for the capacity of a'^i that has been increased by 1, for all $a' \in A \setminus \{a'^i\}$, $|\xi(a')|$ does not exceed the capacity of a' in instance I^{i-1}, namely $b^{i-1}(a')$, while $|\xi(a'^i)|$ may exceed $b^{i-1}(a'^i)$ by at most 1.

Moreover, it holds that $|\xi(a'^i)| \geq |\mu^i(a'^i)|$. Assuming to the contrary, that $|\xi(a'^i)| < |\mu^i(a'^i)|$ yields that ξ is feasible also in instance I^{i-1}. In addition, $|\xi(a'^i)| < |\mu^i(a'^i)|$ implies that it cannot be $\xi(a'^i) \simeq_{a'^i} \mu^i(a'^i)$ thus, together with $\xi(a'^i) \succeq_{a'^i} \mu^i(a'^i) \succeq_{a'^i} \mu^{i-1}(a'^i)$, it yields $\xi(a'^i) \succ_{a'^i} \mu^i(a'^i) \succeq_{a'^i} \mu^{i-1}(a'^i)$. But then, ξ dominates μ^{i-1} in I^{i-1}, a contradiction to μ^{i-1} being a POM in I^{i-1}.

Let us first examine the case in which GSDT enters the 'while' loop and finds an augmenting path, hence μ^i dominates μ^{i-1} in I^i only with respect to applicant a'^i that receives an additional course. This is one of her worst courses in $\mu^i(a'^i)$ denoted as c_μ. Let c_ξ be a worst course for a'^i in $\xi(a'^i)$. Let also ξ' and μ' denote $\xi \setminus \{(a'^i, c_\xi)\}$ and $\mu^i \setminus \{(a'^i, c_\mu)\}$, respectively. Observe that both ξ' and μ' are feasible in I^{i-1}, while having shown that $|\xi(a'^i)| \geq |\mu^i(a'^i)|$ implies through Lemma 2 that ξ' weakly dominates μ' which in turn weakly dominates μ^{i-1} by Lemma 1. Since μ^{i-1} is a POM in I^{i-1}, $\xi'(a') \simeq_{a'} \mu'(a') \simeq_{a'} \mu^{i-1}(a')$ for all $a' \in A$, therefore ξ dominates μ^i only with respect to a'^i and $c_\xi \succ_{a'^i} c_\mu$. Overall, $\xi(a') \simeq_{a'} \mu^i(a') \simeq_{a'} \mu^{i-1}(a')$ for all $a' \in A \setminus \{a'^i\}$ and $\xi(a'^i) \succ_{a'^i} \mu^i(a'^i) \succ_{a'^i} \mu^{i-1}(a'^i)$.

Let t_ξ and t_μ be the ties of applicant a'^i containing c_ξ and c_μ, respectively, where $t_\xi < t_\mu$ because $c_\xi \succ_{a'^i} c_\mu$. Then, Lemma 1 implies that there is a path augmenting f^{i-1} (i.e., the flow corresponding to μ^{i-1}) in the network N^{i,t_ξ}. Let also t' be the value of $curr(a'^i)$ at the beginning of stage i. Since we examine the case where GSDT enters the 'while' loop and finds an augmenting path, $C_{t'}^{a'^i} \neq \emptyset$. Thus, t' indexes the least preferred tie from which a'^i has a course in μ^{i-1}, the same holding for ξ' since $\xi'(a'^i) \simeq_{a'^i} \mu^{i-1}(a'^i)$. Because ξ' is obtained by removing from a'^i its worst course in $\xi(a'^i)$, that course must belong to a tie of index no smaller that t', i.e., $t' \leq t_\xi$. This together with $t_\xi < t_\mu$ yield $t' \leq t_\xi < t_\mu$, which implies that GSDT should have obtained ξ instead of μ^i at stage i, a contradiction.

It remains to examine the cases where, at stage i, GSDT does not enter the 'while' loop or enters it but finds no augmenting path. For both these cases, $\mu^i = \mu^{i-1}$, thus ξ dominating μ^i means that ξ is not feasible in I^{i-1} (since it would then also dominate μ^{i-1}). Then, it holds that $|\xi(a'^i)|$ exceeds $b^{i-1}(a'^i)$ by 1, thus $|\xi(a'^i)| > |\mu^i(a'^i)|$ yielding $\xi(a'^i) \succ_{a'^i} \mu^i(a'^i)$. Let t_ξ be defined as above and t' now be the most preferred tie from which a'^i has more courses in ξ than in μ^i. Clearly, $t' \leq t_\xi$ since t_ξ indexes the least preferred tie from which

a'^i has a course in ξ. If $t' < t_\xi$ then the matching ξ', defined as above, is feasible in I^{i-1} and dominates μ^{i-1} because $\xi'(a'^i) \succ_{a'^i} \mu^{i-1}(a'^i)$, a contradiction; the same holds if $t' = t_\xi$ and a'^i has in ξ at least two more courses from t_ξ than in μ^i. Otherwise, $t' = t_\xi$ and a'^i has in ξ exactly one more course from t_ξ than in μ^i; that, together with $|\xi(a'^i)| > |\mu^i(a'^i)|$ and the definition of t_ξ, implies that the index of the least preferred tie from which a'^i has a course in μ^{i-1} and, therefore, the value of $curr(a'^i)$ in the beginning of stage i, is at most t'. But then GSDT should have obtained ξ instead of μ^i at stage i, a contradiction. □

The following statement is now direct.

Corollary 1. *GSDT produces a POM for instance I.*

To derive the complexity bound for GSDT, let us denote by L the length of the preference profile in I, i.e. the total number of courses in the preference lists of all applicants. Notice that $|E_3| = L$ and neither the size of any matching in I nor the total number of ties in all preference lists exceeds L.

Within one stage, several searches in the network might be needed to find a tie of the active applicant for which the current flow can be augmented. However, one tie is unsuccessfully explored at most once, hence each search either augments the flow thus adding a pair to the current matching or moves to the next tie. So the total number of searches performed by the algorithm is bounded by the size of the obtained matching plus the number of ties in the preference profile, i.e. it is $O(L)$. A search requires a number of steps that remains linear in the number of arcs in the current network (i.e., $N^{i,curr(a^i)}$), but as at most one arc per E_1, E_2 and E_4 is used, any search needs $O(|E_3|) = O(L)$ steps. This leads to a complexity bound $O(L^2)$ for GSDT.

Next we show that GSDT can produce any POM.

Theorem 3. *Given a CA instance I and a POM μ, there exists a suitable priority ordering over applicants Σ given which GSDT can produce μ.*

5 Truthfulness of Mechanisms for Finding POMs

It is well-known that the SDM for HA is truthful, regardless of the given priority ordering over applicants. We will show shortly that GSDT is not necessarily truthful, but first prove that this property does hold for some priority orderings over applicants.

Theorem 4. *GSDT is truthful given Σ if, for each applicant a', all occurrences of a' in Σ are consecutive.*

Proof. W.l.o.g. let the applicants appear in Σ in the following order

$$\underbrace{a_1, a_1, \ldots, a_1}_{b(a_1)\text{-times}}, \underbrace{a_2, a_2, \ldots, a_2}_{b(a_2)\text{-times}} \ldots, \underbrace{a_{i-1}, a_{i-1}, \ldots, a_{i-1}}_{b(a_{i-1})\text{-times}}, \underbrace{a_i, a_i, \ldots, a_i}_{b(a_i)\text{-times}}, \ldots$$

Assume to the contrary that some applicant benefits from misrepresenting her preferences. Let a_i' be the first such applicant in Σ who reports $P'(a_i')$ instead of $P(a_i')$ in order to benefit and $\mathcal{P}' = (P'(a_i'), \mathcal{P}(-a_i'))$. Let μ denote the matching returned by GSDT using ordering Σ on instance $I = (A, C, \mathcal{P}, b, q)$ (i.e. the instance in which applicant a_i reports truthfully) and ξ the matching returned by GSDT using Σ but on instance $I' = (A, C, \mathcal{P}', b, q)$. Let $s = (\Sigma_{\ell<i} b(a_\ell)) + 1$, i.e., s is the first stage in which our mechanism considers applicant a_i'. Let j be the first stage of GSDT such that a_i' prefers ξ^j to μ^j, where $s \leq j < s + b(a_i')$.

Given that applicants $a_1', \ldots, a_{i-1}'$ report the same in I as in I' and all their occurrences in Σ are before stage j, Lemma 1 yields $\mu^j(a_\ell') \simeq_{a_\ell'} \xi^j(a_\ell')$ for $\ell = 1, 2, \ldots, i-1$. Also $\mu^j(a_\ell') = \xi^j(a_\ell') = \emptyset$ for $\ell = i+1, i+2, \ldots, n_1$, since no such applicant has been considered before stage j. But then, all applicants apart from a_i' are indifferent between μ^j and ξ^j, therefore a_i' preferring ξ^j to μ^j implies that μ^j is not a POM in I^j, a contradiction to Theorem 2. □

The next result then follows directly from Theorem 4.

Corollary 2. *GSDT is truthful if all applicants have quota equal to one.*

There are priority orderings for which an applicant may benefit from misreporting her preferences, even if preferences are strict. This phenomenon has also been observed in a slightly different context [7]. Let us also provide an example.

Example 1. Consider a setting with applicants a_1' and a_2' and courses c_1 and c_2, for which $b(a_1') = 2$, $b(a_2') = 1$, $q(c_1) = 1$, and $q(c_2) = 1$. Let I be an instance in which $c_2 \succ_{a_1'} c_1$ and a_2' finds only c_1 acceptable. This setting admits two POMs, namely $\mu_1 = \{(a_1', c_2), (a_2', c_1)\}$ and $\mu_2 = \{(a_1', c_1), (a_1', c_2)\}$. GSDT returns μ_1 for $\Sigma = (a_1', a_2', a_1')$. If a_1' misreports by stating that she prefers c_1 to c_2, GSDT returns μ_2 instead of μ_1. Since $\mu_2 \succ_{a_1'} \mu_1$, GSDT is not truthful given Σ.

The above observation seems to be a deficiency of GSDT. We conclude by showing that no mechanism capable of producing all POMs is immune to this shortcoming.

Theorem 5. *There is no universally truthful randomized mechanism that produces all POMs in CA, even if applicants' preferences are strict and all courses have quota equal to one.*

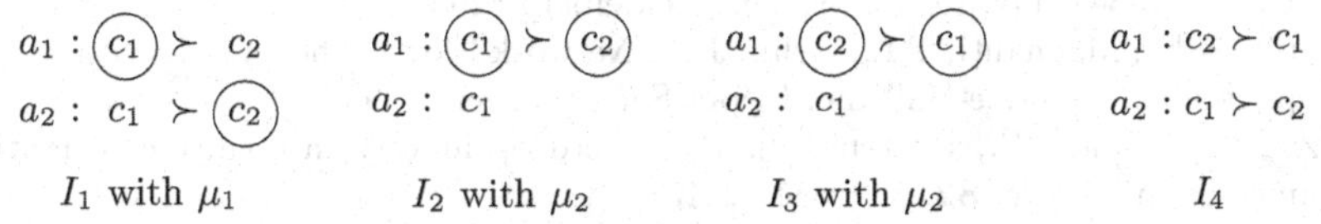

Fig. 2. Four instances of CA used in the proof of Theorem 5. In all four instances $b(a_1') = 2$, $b(a_2') = 1$, $q(c_1) = q(c_2) = 1$. For each of instances I_1 to I_3, a matching is indicated using circles in applicants' preference lists.

Proof. The instance I_1 in Fig. 2 admits three POMs, namely $\mu_1 = \{(a'_1, c_1), (a'_2, c_2)\}$, $\mu_2 = \{(a'_1, c_1), (a'_1, c_2)\}$ and $\mu_3 = \{(a'_1, c_2), (a'_2, c_1)\}$. Assume a randomized mechanism ϕ that produces all these matchings. Therefore, there must be a deterministic realization of it, denoted as ϕ^D, that returns μ_1 given I_1. Let us examine the outcome of ϕ^D under the slightly different applicants' preferences shown in Fig. 2, bearing in mind that ϕ^D is truthful.

- Under I_2, ϕ^D must return μ_2. The only other POM under I_2 is μ_3, but if ϕ^D returns μ_3 then a'_2 under I_1 has an incentive to lie and declare only c_1 acceptable (as in I_2).
- Under I_3, ϕ^D must return μ_2. The only other POM under I_3 is μ_3, but if ϕ^D returns μ_3 then a'_1 under I_3 has an incentive to lie and declare that she prefers c_1 to c_2 (as in I_2).

I_4 admits two POMs, namely μ_2 and μ_3. If ϕ^D returns μ_2, then a'_1 under I_1 has an incentive to lie and declare that she prefers c_2 to c_1 (as in I_4). If ϕ^D returns μ_3, then a'_2 under I_3 has an incentive to lie and declare c_2 acceptable—in addition to c_1—and less preferred than c_1 (as in I_4). Thus overall ϕ^D cannot return a POM under I_4 while maintaining truthfulness. □

6 Future Work

A particularly important problem is to investigate the expected size of the matching produced by the randomized version of GSDT. It is also interesting to characterize priority orderings that induce truthful-telling in GSDT. Should this be possible, it would be interesting to compute the expected size of the matching produced by a randomized GSDT in which the randomization is taken over the priority orderings that give rise to truthfulness.

References

1. Abdulkadiroğlu, A., Sönmez, T.: Random serial dictatorship and the core from random endowments in house allocation problems. Econometrica **66**(3), 689–701 (1998)
2. Abraham, D.J., Cechlárová, K., Manlove, D.F., Mehlhorn, K.: Pareto optimality in house allocation problems. In: Fleischer, R., Trippen, G. (eds.) ISAAC 2004. LNCS, vol. 3341, pp. 3–15. Springer, Heidelberg (2004)
3. Ahuja, R.K., Magnanti, T.L., Orlin, J.B.: Network Flows: Theory, Algorithms, and Applications. Prentice-Hall Inc, Upper Saddle River (1993)
4. Aziz, H., Brandt, F., Harrenstein, P.: Pareto optimality in coalition formation. Games Econ. Behav. **82**, 562–581 (2013)
5. Bogomolnaia, A., Moulin, H.: A new solution to the random assignment problem. J. Econ. Theor. **100**(2), 295–328 (2001)
6. Bogomolnaia, A., Moulin, H.: Random matching under dichotomous preferences. Econometrica **72**(1), 257–279 (2004)
7. Budish, E., Cantillon, E.: The multi-unit assignment problem: theory and evidence from course allocation at harvard. Am. Econ. Rev. **102**(5), 2237–2271 (2012)

8. Cechlárová, K., Eirinakis, P., Fleiner, T., Magos, D., Manlove, D., Mourtos, I., Oceľáková, E., Rastegari, B.: Pareto optimal matchings in many-to-many markets with ties. Technical Report 1507.02866, 2015. Accessed on http://arxiv.org/abs/1507.02866
9. Cechlárová, K., Eirinakis, P., Fleiner, T., Magos, D., Mourtos, I., Potpinková, E.: Pareto optimality in many-to-many matching problems. Discrete Optim. **14**, 160–169 (2014)
10. Fujita, E., Lesca, J., Sonoda, A., Todo, T., Yokoo, M.: A complexity approach for core-selecting exchange with multiple indivisible goods under lexicographic preferences. In: Proceedings of AAAI 2015 (2015)
11. Gärdenfors, P.: Assignment problem based on ordinal preferences. Manage. Sci. **20**(3), 331–340 (1973)
12. Gigerenzer, G., Goldstein, D.G.: Reasoning the fast and frugal way: models of bounded rationality. Psychol. Rev. **103**(4), 650–669 (1996)
13. Hylland, A., Zeckhauser, R.: The efficient allocation of individuals to positions. J. Polit. Econ. **87**(2), 293–314 (1979)
14. Klaus, B., Miyagawa, E.: Strategy-proofness, solidarity, and consistency for multiple assignment problems. Int. J. Game Theor. **30**, 421–435 (2001)
15. Krysta, P., Manlove, D., Rastegari, B., Zhang, J.: Size versus truthfulness in the house allocation problem. Technical report 1404.5245. A shorter version appeared in the Proceedings of EC 2014 (2014)
16. Manlove, D.F.: Algorithmics of Matching Under Preferences. World Scientific, Singapore (2013)
17. Saban, D., Sethuraman, J.: The complexity of computing the random priority allocation matrix. In: Chen, Y., Immorlica, N. (eds.) WINE 2013. LNCS, vol. 8289, pp. 421–421. Springer, Heidelberg (2013)
18. Svensson, L.G.: Queue allocation of indivisible goods. Soc. Choice Welfare **11**(4), 323–330 (1994)

Cost Sharing

Cost-Sharing Models in Participatory Sensing

Georgios Birmpas[1], Costas Courcoubetis[1,2], Ioannis Giotis[1], and Evangelos Markakis[1](✉)

[1] Department of Informatics, Athens University of Economics and Business, Athens, Greece
markakis@gmail.com
[2] Singapore University of Technology and Design, Singapore, Singapore

Abstract. In Smart City and Participatory Sensing initiatives the key concept is for user communities to contribute sensor information and form a body of knowledge that can be exploited by innovative applications and data analytics services. A key aspect in all such platforms is that sensor information is not free but comes at a cost. As a result, these platforms may suffer due to insufficient sensor information made publicly available if applications do not share efficiently the cost of the sensor information they consume.

We explore the design of specialized market mechanisms that match demand to supply while taking into account important positive demand externalities: sensors are digital goods and their cost can be shared by applications. We focus on the buyer side and define different demand models according to the flexibility in choosing sensor data for satisfying application needs. We then investigate the properties of various cost-sharing mechanisms with respect to efficiency and budget balance. In doing so, we also propose and study a new mechanism, which although lacks strategyproofness, it exhibits important efficiency improvement along with certain fairness properties.

1 Introduction

A remarkable incorporation of sensors has occurred in the last few years in a wide range of devices. Starting from the inclusion of GPS receivers, accelerometers and barometers in smartphones, lately we are also seeing a wave of health related sensors being used in the form of fitness bands and smartwatches. Aside from personal devices, home automation and power management devices are distinctively on the rise and include a different variety of sensor data.

Such sensor information can potentially be collected in more precise detail and volume, opening up possibilities for research on unprecedented scales.

Research supported by the European Union (European Social Fund - ESF) and Greek national funds through the Operational Program "Education and Lifelong Learning" of the National Strategic Reference Framework (NSRF) - Research Funding Program: THALES, investing in knowledge society through the European Social Fund.

M. Hoefer (Ed.): SAGT 2015, LNCS 9347, pp. 43–56, 2015.
DOI: 10.1007/978-3-662-48433-3_4

Towards this, participatory sensing initiatives form a natural and promising approach, replacing traditional sensor networks, where user communities can contribute sensor information, that can later be exploited by innovative applications. There are already existing deployments and platforms that support a variety of applications like environmental monitoring (OpenSense), transportation (CrowdPark), fitness (BikeTastic), urban sensing (PulsodelaCiudad), and medical research (Apple's ResearchKit).

Unfortunately, gathering this information from individually owned devices proved to be not a straightforward task. Some of these platforms have suffered from insufficient participation because users that voluntarily submit their sensing data found no interest in remaining active in the system without being rewarded, or at least, have their cost covered. These undesirable facts have already been observed in [4,8,9], which focus on incentive issues arising in the supplier's side. Namely, suppliers may drop out unless there is a positive Return on Investment, which depends on the total cost for collecting data (battery consumption, device resources, privacy, etc.). But potential buyers of data may also be reluctant to participate in the market if for instance the prices are prohibitively high, or if the underlying mechanisms do not aim at economic efficiency. How should applications express their demand for sensor information in such an environment and how should prices be determined? At the same time, one also needs mechanisms for matching (elastic) demand with potential sensor providers, exploiting the fact that once a sensor is turned on it can be used simultaneously by multiple applications.

Contribution. To begin with, we develop a framework for operating a large market of sensor data in participatory sensing environments. On one side of the market, we have buyers interested in obtaining data potentially from multiple sources and for different types of sensors. Their demand can be elastic or inelastic in terms of the number of sensors they require. On the other side of the market, the data suppliers correspond to users or organizations owning sensors, and require a payment to cover their costs. In this work, we are not concerned on how suppliers define their prices. We focus on the buyers' side and the design of mechanisms that match demand with supply. We assume a centralized platform that is able to execute such pricing schemes, as well as collect and distribute the data and payments. An important aspect of such a marketplace is that we can distribute the same sensor data to multiple interested parties, at no extra cost, i.e., we can view sensors as digital goods. This implies positive externalities between buyers, since they can profit from each other by sharing sensor costs where possible.

Within this framework, we introduce 2 simple cost-sharing settings regarding the demand of the users, and study various mechanisms. The first scenario involves single-minded buyers, interested in different subsets of sensor types. The second scenario concerns bidders with multi-unit, elastic demand. In both scenarios, the two important and conflicting objectives we care for are *(i) budget-balance:* the market management platform does not incur economic loss while operating the system, making it self-sustainable, *(ii) economic efficiency, i.e., social welfare maximization:* we would like to satisfy more customer queries if this increases the net surplus in the system. Given the strong impossibility results of [5,12], we cannot achieve both objectives at the same time with strategyproof mechanisms. Instead, we first show how to achieve each one of these separately, and with polynomial time complexity. For economic efficiency we prove that the VCG mechanism can be implemented efficiently, whereas for budget balance, we utilize and adapt ideas from the Moulin-Shenker mechanisms [10,11].

For the second scenario, we also propose a natural hybrid mechanism that improves efficiency under budget balance by relaxing strategyproofness. Despite the loss of strategyproofness, our hybrid mechanism has its own merits. It is simple to implement, and is based on a very natural approach for increasing the social welfare. We prove that this mechanism achieves higher welfare than other established cost-sharing mechanisms. Furthermore, our hybrid mechanism satisfies certain fairness properties, in the sense that wealthier players contribute more to the total cost than poorer ones. Finally we also study welfare properties at the equilibria of this mechanism and exhibit cases where socially optimal equilibria exist.

1.1 Related Work

Regarding mechanism design for participatory sensing, a specialized reverse auction is proposed in [9], to incentivize suppliers to increase their participation. Another reverse auction is also proposed in [8]. The work of [4] on the other hand is limited to using a fixed price approach. All these works focus on the suppliers' side of the market. An issue that is not covered by these works is the modeling of the demand side of the market, which is what we mainly address in this paper.

We have recently performed experimental evaluations for some of the mechanisms we study here, reported briefly in [14]. The main message from these simulations is that certain altruistic versions of budget-balanced mechanisms, where some richer players could contribute a higher payment, may have a practical appeal. The fact that buyers here are able to share the same sensor, implies that in some occasions wealthy buyers may have incentives to help out and contribute a higher cost-share so that the costs are covered and they can still have access to sensors. No theoretical analysis is provided though in [14]. The hybrid mechanism we propose in Sect. 4 is motivated by such observations (though not implemented or suggested in [14]).

The works from the economics literature that are most relevant to ours are the cost-sharing mechanisms of Moulin and Shenker [10,11]. These mechanisms work for a setting where each user is either granted the same identical service with all other users or is declined. We also consider the Marginal Cost Pricing mechanism, see [10], which is the adaptation of the VCG mechanism into the cost-sharing setting.

2 Definitions and Notation

In the models we study, we have a set $N = \{1, ..., n\}$ of potential buyers, who have a demand for some sensor data. We use interchangeably the terms buyer or player, to refer to any $i \in N$. Different types of demand (e.g., elastic vs inelastic, or single tuple vs multiple tuples) are examined in Sects. 3 and 4. We also have a set $M = \{1, ..., m\}$, representing the different sensor *basic types*, e.g., accelerometer, temperature, CO_2, etc. Finally, we have a set of suppliers or providers who own sensor data (via their mobile or any other device). Each supplier may specify a price per sensor type that he needs to be paid for in order to provide access to the value of the sensor. Note also that a value provided by one supplier can be used by many buyers. Finally, suppliers do not all necessarily have the same set of sensor types available.

Our main focus is on the following criteria, and especially on the first two:

- **Budget balance.** A mechanism is *budget-balanced* if for every instance, the payments assigned to the buyers cover exactly the cost of the provider.
- **Social welfare maximization.** Following [10], the *social welfare* or *surplus* in a cost-sharing setting is the sum of the buyers' derived values minus the cost incurred (the payments made by the buyers cancel out with what the providers receive). If x denotes an outcome of a mechanism, the social welfare is $\sum_i v_i(x) - C(x)$, where v_i is the valuation of buyer i and $C(x)$ is the cost incurred.
- **No Positive Transfers (NPT):** The cost shares are always nonnegative.
- **Voluntary Participation (VP):** The welfare level corresponding to not providing service at no cost is guaranteed to each agent if they report truthfully.

In the remainder of the paper, several proofs are omitted due to space constraints.

3 Scenario 1: Single-Minded Buyers

We consider a simple scenario, in which each buyer $i \in N$ is interested in a subset $P_i \subseteq M$ of sensor types. For example P_i could be of the form (speedometer, accelerometer). Furthermore, he requests access to a single tuple with values from these types of sensors, i.e., a tuple (x, y), where x is a value for speed and y is a value for the acceleration. These values do not necessarily need to come from the same provider (but buyers can request that all the data come from providers within a certain geographical region, e.g., the city center, in order to collect information about traffic; we omit such implementation aspects from the description of the mechanisms). Hence, the request specified by each buyer $i \in N$, is in the form (v_i, P_i), where v_i is the value derived by i for receiving this tuple, i.e., his willingness to pay. The demand is inelastic in the sense that buyer i is not deriving any utility if he receives only a strict subset of sensors from P_i. We call such buyers *single-minded*, in analogy to single-minded bidders in combinatorial auctions. Clearly such demands can come and go dynamically in the course of time, but we are interested in a static snapshot, i.e., an instance of our problem may correspond to the demands within a given time window during which the centralized platform needs to make a decision on which users to serve.

The cost function $C(S)$ for serving a set of customers $S \subseteq N$ can be easily computed for any S. For any sensor type $j \in M$, let c_j be the cost for the platform of providing a single value for this type. The values of the sensors can be viewed as digital goods, and since each bidder is interested in receiving a single tuple, we can use just one actual sensor for each type requested, to satisfy all customers. Hence, the cost c_j could be taken to be the cheapest price specified by some supplier of type j (it is not though important for the mechanism how c_j is derived). Therefore, for a set $S \subseteq N$ of buyers, the cost $C(S)$ is the sum of the costs of all sensor types required by S:

$$C(S) = \sum_{j \in P(S)} c_j, \text{ where } P(S) = \bigcup_{i \in S} P_i. \tag{1}$$

3.1 Social Welfare Maximization

We first look at the objective of maximizing the social welfare. Let $\boldsymbol{\theta} = (\theta_1, ..., \theta_n)$ be the vector of the agents' types, i.e., under Scenario 1, $\theta_i = (v_i, P_i)$. If a mechanism chooses $S \subseteq N$, as the set of buyers to be served, then the generated welfare from S is: $SW(S, \boldsymbol{\theta}) = \sum_{i \in S} v_i - C(S)$.

Given the set of all agents, N, and their true type vector $\boldsymbol{\theta}$, let us denote by $SW^*(\boldsymbol{\theta})$ the optimal welfare that can be achieved by N, i.e.:

$$SW^*(\boldsymbol{\theta}) = \max_{S \subseteq N} \{\sum_{i \in S} v_i - C(S)\}$$

Our main result in this section is the following:

Theorem 1. *The problem of social welfare maximization under Scenario 1 can be solved in polynomial time.*

To prove Theorem 1, we need to avoid the exponential search over all subsets of N. Note also that we do not have any monotonicity properties here (larger sets do not necessarily produce higher welfare). To solve our problem, we resort to a linear programming formulation, which turns out to yield a totally unimodular constraint matrix.

Proof of Theorem 1: We begin by writing down an ILP for our problem. For this, we use an integer variable x_i for each buyer $i \in N$ and an integer variable y_j for each sensor type $j \in M$. The rationale is that when $x_i = 1$, agent i receives his requested tuple P_i. When $y_j = 1$, this means that the sensor of type j is allocated. Note that determining the set of players who receive service, also determines the set of sensor types that will be set to 1. We claim that the following is an ILP describing our problem.

$$\begin{aligned}
\text{maximize:} \quad & \sum_{i \in N} v_i x_i - \sum_{j \in M} c_j y_j \\
\text{subject to:} \quad & x_i \leq y_j \ , \qquad \forall i \in N, \forall j \in P_i \\
& x_i \in \{0,1\} \ , \quad \forall i \in N \\
& y_j \in \{0,1\} \ , \quad \forall j \in M
\end{aligned}$$

To see why this suffices, note that if $x_i = 0$, then for $j \in P_i$, the variable y_j could be either 0 or 1, depending on other buyers' demand sets. If $x_i = 1$ however, then we must have that $y_j = 1$. Hence, the only constraint beyond integrality that we need is that $x_i \leq y_j$ for $j \in P_i$. It is easy to see now that every solution to our problem corresponds to a feasible solution of the ILP (there are also some feasible solutions in which we can have $y_j = 1$ without allocating the sensor but these are clearly not optimal solutions).

We relax the ILP to get an LP relaxation, by setting that $x_i, y_j \in [0,1]$. So now we have a linear program, which we can write in the form $\{max\ w^T z \mid Az \leq b, z \geq 0\}$. The rest of the proof is devoted to showing that our constraint matrix A is totally unimodular, which implies that the LP always has an integral optimal solution.

Lemma 1. *The constraint matrix A of the LP relaxation is totally unimodular.*

The proof of Lemma 1 is based on a sufficient condition for total unimodularity as established by [6], see also [13][page 276], and we omit it here. Therefore, we can solve the Social Welfare maximization problem in polynomial time, and the proof of Theorem 1 is complete. □

Theorem 1, implies that we can have strategyproof and efficient mechanisms implemented in polynomial time. For example, we can utilize the VCG mechanism, which we briefly recall for the sake of completeness. The VCG mechanism first computes a set $S^* \subseteq N$, where optimal welfare is attained. Then, if the declared type vector is $\boldsymbol{\theta} = (\theta_1, ..., \theta_n)$, where $\theta_i = (b_i, P_i)$ may not necessarily be equal to the true type of i, the payment for every player $i \in S^*$, can be written in the form:

$$p_i = b_i - (SW^*(\boldsymbol{\theta}) - SW^*(\boldsymbol{\theta}_{-i})) \,. \tag{2}$$

Agents not picked in the optimal set do not pay anything. This is also known as the pivotal mechanism [3], and also referred to in the cost-sharing context, as the Marginal Cost (MC) mechanism in [11]. Hence, we can conclude with the following:

Corollary 1. *Under scenario 1, the VCG mechanism is strategyproof, satisfies NPT and VP, and can be implemented in polynomial time.*

More generally, we can have a family of strategyproof mechanisms by replacing $SW^*(\boldsymbol{\theta}_{-i})$ in (2) with any function of the form $h_i(\boldsymbol{\theta}_{-i})$.

Since VCG is efficient, the impossibility results of [5,12] imply that it cannot be budget-balanced. In fact, we cannot even hope to be "approximately" budget-balanced, since in the cases where no player is pivotal, the VCG payments are all 0.

3.2 Budget-Balanced Mechanisms

We now focus on the design of budget-balanced mechanisms. The family of mechanisms we consider are derived directly from the pioneering work of Moulin and Shenker [10,11]. Their work concerns a setting that differs from ours in 2 respects: first, their model is simpler in terms of the service requested. Namely, they have a binary setup, where there is a single provider, offering the same identical service to everyone, and each agent will be either granted or declined the service. In our case the buyers are interested in different subsets, and hence in a different type of service each. Second, in our model, the cost function is simpler due to the fact that sensors correspond to digital goods and can be shared. This implies that for instances where a set S of buyers requests the same set of sensors, then in our setting $C(S)$ is the same as $C(T)$ for any $T \subseteq S$ with $T \neq \emptyset$. In their work $C(\cdot)$ is an arbitrary submodular set function.

We can easily adapt the approach of Moulin and Shenker for Scenario 1. To do this, we need to define first an underlying *cost-sharing* method. A cost-sharing method is a function $\xi(\cdot, \cdot)$ such that $\xi(i, R)$ determines the cost-share of agent i, when R is the set to be served by the mechanism. We demand that a cost-sharing method satisfies $\sum_{i \in R} \xi(i, R) = C(R)$ for all $R \subseteq N$, i.e., the sum of the payments balance the cost.

We mainly focus on the *egalitarian* cost-sharing method, since this may have more appeal in practice due to its simplicity. To define the share $\xi(i, R)$ for a given set R to be served, we split the cost of each used sensor equally among the people who want it. Let y_j be the number of buyers who have j in their demand set. Egalitarian cost sharing means that each customer i contributes a share c_j/y_j towards the cost of sensor j. Hence for a buyer i, with demand set P_i, his total cost-share is:

$$\xi(i, R) = \sum_{j \in P_i} \frac{c_j}{y_j}. \tag{3}$$

It is obvious that we have: $\sum_{i \in R} \xi(i, R) = C(R)$, for any $R \subseteq N$. Given now any cost-sharing method ξ, one can define parametrically the mechanism below for determining who receives service along with the cost-shares. In the description below, we let $\mathbf{b} = (b_1, ..., b_n)$ be the agents' declared values for their demand sets.

The Mechanism MS(ξ) (Moulin-Shenker mechanism under $\xi(\cdot, \cdot)$):

- Start by trying to serve all agents, with cost-share $\xi(i, N)$. Remove any agent who cannot cover his share, i.e., anyone for which $b_i < \xi(i, N)$. If no one is removed in this step, stop here, otherwise let R^1 be the set of remaining agents.
- Check if we can serve R^1 with a cost-share of $\xi(i, R^1)$ for every $i \in R^1$. Again remove those who cannot afford this price.
- Continue like this and in every round obtain the set $R^{t+1} = \{i \in R^t : b_i \geq \xi(i, R^t)\}$.
- Stop either when we reach the empty set, or when we reach a set in which all agents can afford to pay their cost-share.

This family of mechanisms turns out to have nice properties if the cost function $C(\cdot)$ and the cost-sharing method $\xi(\cdot,\cdot)$ satisfy certain conditions. Regarding $\xi(\cdot,\cdot)$, the following is an important and desirable property, which simply says that the cost-share of an agent should not become higher when more people receive service.

Definition 1. A cost-sharing method is cross-monotonic if for any $T \subseteq N$,

$$\xi(i, R) \geq \xi(i, T) \text{ for any } R \subseteq T \text{ and } i \in R. \tag{4}$$

Claim 1. *The egalitarian cost-sharing method described by* (3) *is cross-monotonic.*

We also need submodularity of our cost function, which is easy to establish. Given Claim 1, the following theorem is a straightforward extension of the results from [10,11] to our setting.

Theorem 2. *Given any cross-monotonic cost-sharing method ξ for single-minded bidders, the Mechanism $MS(\xi)$ is budget-balanced, group-strategyproof and satisfies NPT and VP. In particular, if ξ is the egalitarian cost-sharing according to* (3)*, $MS(\xi)$ satisfies these properties and can also be implemented in polynomial time.*

The obvious question is how do these mechanisms perform with respect to social welfare. Unfortunately, they are far from efficient. We can construct examples where the mechanism generates zero welfare, whereas the optimal welfare is far from zero.

Our discussion in Sects. 3.1 and 3.2 highlights the tradeoff between achieving efficiency and budget-balance. In the next section, we will see a way of achieving better trade-offs in a scenario of multi-unit elastic demand (but not applicable to Scenario 1).

4 Scenario 2: Multiple Units and Elastic Demand

At this orthogonal scenario all players have the same type of demand, i.e., the set P_i is the same for every player. This could involve buyers who are all interested in the same type of information, e.g., traffic in the city center, or environmental sensors within a region. What differentiates the players is that each player i specifies an additional amount d_i, for the maximum number of tuples that he is interested in acquiring. The demand is elastic, so that player i does not mind receiving less than d_i tuples. Each player also specifies his per-tuple willingness to pay v_i. This encodes a submodular[1] valuation, which is additive up to the first

[1] The model of this section can also be adapted for general submodular valuations in the form $v_i = (v_i(1), .., v_i(d_i))$, where $v_i(j)$ is the value for the j-th tuple. We prefer the current exposition, due to its simplicity and more practical appeal for participatory sensing applications.

d_i tuples. We assume that there is a sufficient supply of tuples from the providers, i.e., there are at least d_{max} of them with $d_{max} = \max d_i$. Each tuple has some cost c_k so that we can sort them from the cheapest to the most expensive one, say $c_1 \leq c_2 \cdots \leq c_{d_{max}}$.

We start with showing that maximizing the social welfare can be solved in polynomial time. The important property is that once we decide for allocating a tuple, we do not lose in welfare by giving the tuple to all customers who have demand for it, since we are only adding more value to the current welfare. Hence, if $\boldsymbol{\theta} = (\theta_1, ..., \theta_n)$ is the type vector, with $\theta_i = (v_i, d_i)$, the optimization problem for the social welfare becomes

$$SW^*(\boldsymbol{\theta}) = \max_{1 \leq k \leq d_{max}} \left[\sum_{i \in N} v_i \cdot \min\{k, d_i\} - \sum_{j=1}^{k} c_j\right] \tag{5}$$

We can solve (5) simply by trying all values for k. Hence we have:

Theorem 3. *Under Scenario 2, we can have polynomial time, strategyproof, and efficient cost-sharing mechanisms, that also satisfy NPT and VP.*

4.1 Budget Balance: Sequential Moulin-Shenker Mechanisms

The application of Moulin-Shenker mechanisms is not any more straightforward in the case of multi-unit demand. Each customer i corresponds now to a set of potential service levels, ranging from 0 to d_i tuples. Hence, we cannot just run an analog of $MS(\xi)$ from Sect. 3. One could consider all combinations of service levels to customers, and run $MS(\xi)$ for each such combination (and then choose the one that is more efficient). But this has prohibitively high complexity to be run in practice.

Instead, one can utilize the Moulin-Shenker approach in a sequential manner.

The Mechanism SMS(ξ)(Sequential Moulin-Shenker):

1. Sort the d_{max} cheapest tuples so that $c_1 \leq c_2 \cdots \leq c_{d_{max}}$. Let $A^1 = N$ be the set of active players before the first round (initially all are active).
2. At round r (with r ranging from 1 to d_{max}):
   (a) If A^r is the set of currently active players, run the mechanism MS(ξ) from Sect. 3 on A^r, to determine who receives the r-th cheapest tuple, along with their cost shares for that round.
   (b) Remove from A^r all customers who were not selected to be served. Remove also any customer with $d_i = r$.
   (c) Let A^{r+1} be the set of surviving customers after the previous step. Continue with the next round in the same manner, unless $A^{r+1} = \emptyset$.

For the remainder of the paper, we fix again ξ to be the egalitarian cost sharing method and denote the mechanism as SMS, rather than SMS(ξ). Since everybody is interested in the same tuple, if there are say k active players in a certain run of SMS at a round r, the cost share is defined as c_r/k. The SMS mechanism is (group) strategy proof, which can be shown using the same arguments as in Theorem 2. Hence:

Theorem 4. *Under Scenario 2, the SMS mechanism with egalitarian cost-shares runs in polynomial time, is budget-balanced, group-strategyproof and satisfies NPT and VP.*

However, as in the previous Section, we can easily construct instances where we have a great loss of efficiency, even with 2 players and 1 round.

4.2 Budget-Balance with Better Social Welfare: A Hybrid Mechanism

We propose in this section a different mechanism, as an attempt to maintain budget-balance but achieve higher welfare than the Moulin-Shenker mechanisms. Our mechanism is quite intuitive and uses a very natural approach in order to achieve better welfare. In each round now, we start by running the VCG mechanism for sharing the tuple of that round. To achieve budget-balance, we complement the VCG payments with an egalitarian cost-share for the remaining cost. If this results in high costs for some players, we reject them and repeat for the remaining players.

Assume that the input to the mechanism is $\boldsymbol{\theta} = (\theta_1, \cdots, \theta_n)$ with $\theta_i = (b_i, d_i)$. We define first the per-round VCG mechanism, which is quite simple in this setting. If we run VCG only for the tuple at round r, and A^r is the set of currently active players, then the tuple is allocated if $\sum_{j \in A^r} b_j \geq c_r$. A player i is *pivotal* at round r, if $\sum_{j \in A^r} b_j \geq c_r$ and $\sum_{j \in A^r \setminus \{i\}} b_j < c_r$, i.e., player i has an impact on having the tuple allocated. The only players that pay under VCG are the pivotal players, according to (2). Hence, if $\sum_{j \in A^r} b_j \geq c_r$, the VCG payments are:

$$p_i{}^{VCG} = \begin{cases} c_r - \sum_{j \in A^r \setminus \{i\}} b_j, & \text{if player } i \text{ is pivotal} \\ 0, & \text{if player } i \text{ is not pivotal} \end{cases} \tag{6}$$

Our mechanism runs as follows:

The Hybrid Mechanism

1. Again sort the tuples so that $c_1 \leq c_2 \cdots \leq c_{d_{max}}$. Let $A^1 = N$.
2. At round r (with r ranging from 1 to d_{max}):
   (a) Check if $\sum_{i \in A^r} b_i \geq c_r$, where A^r is the set of currently active players during round r. If not, the mechanism stops.
   (b) Run the VCG mechanism on A^r, for the tuple of round r, and let $p_i{}^{VCG}$ be the VCG payment for each $i \in A^r$, as defined in (6).
   (c) Let $c_r{}'$ be the reduced cost after the VCG payments: $c_r{}' = c_r - \sum_{i \in A^r} p_i{}^{VCG}$.
   (d) Split the cost c_r' equally among A^r, i.e. let $p^E = c_r'/|A^r|$. Define the candidate cost shares as $p_i{}^H = p_i{}^{VCG} + p^E$.
   (e) If there are players with $b_i < p_i^H$, then pick the one with the lowest bid, set $A^r = A^r \setminus \{i\}$, and go to step 2a to repeat the process for round r.
   (f) Otherwise, if $b_i \geq p^H$, for each $i \in A^r$, set $A^{r+1} = A^r \setminus \{i : d_i = r\}$, and continue to round $r+1$, unless $A^{r+1} = \emptyset$.

Remark 1. Note that at step 2e, we remove only one player, even if there can be more players with $b_i < p_i^H$. This turns out to be crucial regarding the total welfare achieved.

Regarding the pros and cons of this new mechanism, on the positive side, we shall prove that it can attain much higher social welfare than SMS. On the negative side, this is not a strategyproof mechanism. We do not view the lack of strategyproofness as a prohibitive disadvantage for such mechanisms. In the recent literature there have been several studies analyzing simple and non-strategyproof mechanisms that have practical appeal. In the context of auctions for example, see e.g., [1,2,7].

Apart from achieving better welfare, the Hybrid mechanism has other merits as well. First, it maintains low complexity like SMS, since the VCG step is very easy to run in this scenario. Second, we consider it a very natural approach towards increasing the welfare of budget-balanced mechanisms and can be applicable to other settings too. Third, it also satisfies certain fairness properties, in the sense that wealthier players contribute more to the total cost than poorer ones, see Claim 2 below. During the VCG step in each round, the set of players who pay are the richer ones, according to (6). The remaining cost is then an egalitarian cost share for all active players. Hence, the mechanism helps the poorer players to satisfy their demand. But in addition to that, the wealthier players are also rewarded. As we will see in Lemma 2, a positive payment at the VCG step for player i, ensures that i is never removed during the execution of the mechanism, and he will thus be able to get the desired tuples (as long as the cost of a tuple is covered by the sum of bids).

Claim 2. *In the Hybrid mechanism, wealthier players have higher payments, i.e., if $b_i \geq b_j$ then ${p_i}^H \geq {p_j}^H$, at every round of the mechanism.*

The main positive result for the Hybrid mechanism is that it dominates the SMS mechanism as follows:

Theorem 5. *For any type vector $\boldsymbol{\theta}$, if we run both the Hybrid and the SMS mechanism on input $\boldsymbol{\theta}$, then the Hybrid mechanism always achieves at least as good social welfare as the SMS mechanism, w.r.t. $\boldsymbol{\theta}$.*

proof. The proof is based on two auxiliary lemmas stated below. The following lemma shows that players who are asked to pay something at the VCG run of a certain round cannot be rejected at that step of the mechanism (in fact this implies that they will not be rejected from any future round where the sum of active bids covers the cost).

Lemma 2. *Consider a round r in the Hybrid mechanism and let A^r be the set of active players just before an execution of step 2b within round r. If $\sum_{j \in A^r} b_j \geq c_r$, then for every player $i \in A^r$ for which ${p_i}^{VCG} > 0$, the mechanism cannot remove i from A^r during that step, i.e., $b_i \geq {p_i}^H$ in the execution of that iteration.*

Using Lemma 2, we can then prove the following fact.

Lemma 3. *Consider a run of the SMS and the Hybrid mechanism on the same instance. At every round r of each mechanism, let N_r^S and N_r^H be the set of players who receive the r-th tuple by the SMS mechanism and by the Hybrid mechanism respectively. Then $N_r^S \subseteq N_r^H$, for every r.*

Lemma 3 implies that the Hybrid mechanism produces at least as good social welfare as the SMS mechanism in each round. Hence, this completes our proof. □

We would like to stress here that for Scenario 1, we can construct examples showing that an analog of the Hybrid mechanism does not necessarily produce better social welfare than $MS(\xi)$. More details on this will be provided in the full version.

Equilibria Under the Hybrid Mechanism: An obvious question is whether we can have a price of anarchy analysis for this cost-sharing setting. Do all the Nash equilibria of the Hybrid mechanism achieve good social welfare? The answer is generally negative as there may exist many "unreasonable" equilibria. E.g., in instances where everybody has a value lower than the total cost of a round, it is an equilibrium if everybody declares 0, and this is inherent in most cost-sharing mechanisms, since no player would be willing to cover the cost of a service on his own. Nevertheless, these are equilibria that are not expected to be attained in practice.

The next step is towards a price of stability analysis, and the existence of equilibria with better guarantees. The Hybrid mechanism is promising in that direction. We briefly summarize some results here for the existence of socially optimal pure equilibria.

Theorem 6. *Consider a set of players with the same demand $d_i = d$, for $i \in N$. Then, there is a Nash equilibrium producing optimal social welfare when $d = 1$ or when all the tuples have the same cost, $c_1 = ... = c_d$. In both cases, if the optimal welfare is positive ($\sum_i v_i > c_1$), then every vector $\mathbf{b}$ with $\sum_{j=1}^n b_j = c_1$ and $b_i \leq v_i$, is a Nash equilibrium which produces optimal social welfare (w.r.t. the true valuation vector).*

As we see, there can be a plethora of optimal equilibria in the above cases. Next, we identify some more conditions that enable the existence of socially optimal equilibria. For simplicity, we stick to the case where all players have the same demand d and the optimal welfare is achieved by allocating all d tuples. Note then, that at an equilibirum, we need to have $\sum b_j = c_d$, i.e., if the bids exceed the cost of the last round, then there are incentives for people to deviate. Second, to enforce an efficient equilibrium, we also need some relation between the values v_i, the parameter d, and possibly the marginal cost increase between rounds. The following conditions that we have identified say that as long as we do not have very poor players (otherwise some people will have incentives to shade their bids), socially optimal equilibria do exist.

Theorem 7. *Consider an instance with players having the same demand d as before and let $\delta = max_i\{c_i - c_{i-1}\}$. If the following 2 conditions hold, there exists a socially optimal equilibrium.*

1. $v_i > 2(d-1)\delta$, *for every* $i \in N$,
2. $c_d \in \left(n(d-1)\delta,\ \sum_i v_i - n(d-1)\delta\right]$.

We defer a further discussion on equilibria to the full version of our work.

5 Concluding Remarks

We have proposed a general framework for market operation in participatory sensing environments, and studied various mechanisms under this framework. To our knowledge, a marketplace tailored to the specificities of participatory sensing applications, is missing today. We conjecture that with the wider adoption of devices containing sensors and new types of micro-payments, such marketplaces for data originating from individually owned devices will be eventually developed.

There are still many directions that one can explore in the context of sensor-data markets, depending on the criteria that one wants to optimize. For example, are there simple budget-balanced mechanisms that achieve a constant factor approximation to the social welfare in either of the scenarios presented here? Even without theoretical guarantees on the social welfare, are there budget-balanced mechanisms that perform better on average than our hybrid mechanism. Note also that variations of the hybrid mechanism can be defined for other scenarios as well, not just what we studied here. It would be interesting to further explore its properties in more general settings.

References

1. Bhawalkar, K., Roughgarden, T.: Welfare guarantees for combinatorial auctions with item bidding. In: Proceedings of the 22nd ACM-SIAM Symposium on Discrete Algorithms (SODA 2011), pp. 700–709 (2011)
2. Christodoulou, G., Kovács, A., Schapira, M.: Bayesian combinatorial auctions. In: Aceto, L., Damgård, I., Goldberg, L.A., Halldórsson, M.M., Ingólfsdóttir, A., Walukiewicz, I. (eds.) ICALP 2008, Part I. LNCS, vol. 5125, pp. 820–832. Springer, Heidelberg (2008)
3. Clarke, E.H.: Multipart pricing of public goods. Public Choice **11**, 17–33 (1971)
4. Danezis, G., Lewis, S., Anderson, R.J.: How much is location privacy worth? In: 4th Workshop on the Economics of Information Security (WEIS 2005) (2005)
5. Green, J., Kohlberg, E., Laffont, J.J.: Partial equilibrium approach to the free rider problem. J. Public Econ. **6**, 375–394 (1976)
6. Heller, I., Tompkins, C.B.: An extension of a theorem of Dantzig's. In: Kuhn, H.W., Tucker, A.W. (eds.) Linear Inequalities and Related Systems, pp. 247–254. Princeton University Press, Princeton (1956)
7. de Keijzer, B., Markakis, E., Schäfer, G., Telelis, O.: Inefficiency of standard multi-unit auctions. In: Bodlaender, H.L., Italiano, G.F. (eds.) ESA 2013. LNCS, vol. 8125, pp. 385–396. Springer, Heidelberg (2013)
8. Koutsopoulos, I.: Optimal incentive-driven design of participatory sensing systems. In: Proceedings of the 32nd IEEE International Conference on Computer Communications (INFOCOM 2013), pp. 1402–1410 (2013)

9. Lee, J.S., Hoh, B.: Dynamic pricing incentive for participatory sensing. Pervasive Mob. Comput. **6**(6), 693–708 (2010)
10. Moulin, H.: Incremental cost sharing: characterization by coalition strategy-proofness. Soc. Choice Welfare **16**, 279–320 (1999)
11. Moulin, H., Shenker, S.: Strategyproof sharing of submodular costs: budget balance vs efficiency. Econ. Theory **18**, 511–533 (2001)
12. Roberts, K.: The characterization of implementable choice rules. In: Laffont, J.J. (ed.) Aggregation and Revelation of Preferences. Elsevier, Amsterdam (1979)
13. Schrijver, A.: Theory of Linear and Integer Programming. Wiley, New York (1986)
14. Thanos, G.A., Courcoubetis, C., Markakis, E., Stamoulis, G.D.: Design and experimental evaluation of market mechanisms for participatory sensing environments. In: Proceedings of the 13th International Conference on Autonomous Agents and Multi-agent Systems (AAMAS 2014), pp. 1515–1516 (2014)

Further Results on Capacitated Network Design Games

Thomas Erlebach and Matthew Radoja(✉)

Department of Computer Science, University of Leicester, Leicester, UK
{te17,mr193}@leicester.ac.uk

Abstract. In a capacitated network design game, each of n players selects a path from her source to her sink. The cost of each edge is shared equally among the players using the edge. Every edge has a finite capacity that limits the number of players using the edge. We study the price of stability for such games with respect to the max-cost objective, i.e., the maximum cost paid by any player. We show that the price of stability is $O(n)$ for symmetric games, and this bound is tight. Furthermore, we show that the price of stability for asymmetric games can be $\Omega(n \log n)$, matching the previously known upper bound. We also prove that the convergence time of best response dynamics cannot be bounded by any function of n.

1 Introduction

The quantification of the inefficiency of Nash equilibria has received considerable attention in recent years. The concept of the *price of anarchy*, measuring the inefficiency of the worst Nash equilibrium (NE) of a given game compared to a social optimum, was introduced by Koutsoupias and Papadimitriou [5], who called it the *coordination ratio*. The *price of stability*, measuring the inefficiency of the best NE of a given game, was first studied by Schulz et al. [9], under the name *optimistic price of anarchy*. Games for which these measures have been studied include scheduling games [5], routing games [8], network design games [2], and capacitated network design games [4]. Apart from the study of the inefficiency of NE, one is also interested in the convergence time of best response dynamics (BRD), i.e., the process that starts with an arbitrary strategy profile and iteratively allows one of the players to update her strategy to one that optimises her cost given the current strategies of all the other players.

In a capacitated network design game, we are given an undirected graph with edge costs and edge capacities, and each of the n players selects a path from her source to her destination. The cost of an edge is shared equally among the players using the edge. Each player aims to minimise her own cost. A capacitated network design game is symmetric if all players share the same source and the same destination, and asymmetric otherwise. As the social optimum, one usually considers the best strategy profile with respect to sum-cost (total cost of all players) or max-cost (maximum cost of any player).

M. Hoefer (Ed.): SAGT 2015, LNCS 9347, pp. 57–68, 2015.
DOI: 10.1007/978-3-662-48433-3_5

Feldman and Ron [4] studied symmetric capacitated network design games and considered instances where the underlying graph is a set of parallel links, a series-parallel graph, or an arbitrary graph. They gave tight bounds on the maximum price of stability and the maximum price of anarchy for all cases for both the max-cost objective and the sum-cost objective, except for the price of stability with respect to max-cost for arbitrary graphs. For the latter case, they showed an upper bound of $O(n \log n)$ and a lower bound of $\Omega(n)$, and they posed closing this gap as an open problem. They also analysed BRD and showed that there are symmetric capacitated network design games where convergence requires $\Omega(n^{3/2})$ steps, contrary to the uncapacitated version of symmetric network design games where BRD always converge in at most n steps.

Our Contribution. For symmetric games with n players, we show that the price of stability with respect to max-cost is $O(n)$. This bound is tight, as implied by the matching lower bound from [4], and hence resolves the open problem posed by Feldman and Ron. A standard proof technique for bounding the price of stability is to bound the increase in social cost during best response dynamics starting from the optimal strategy profile. We show that this technique does not work in our case, as best response dynamics starting from the optimal strategy profile can actually increase the max-cost by a factor of $\Theta(n \log n)$. Therefore, we use a different approach to bound the price of stability, which may be of independent interest. For asymmetric games with n players, we show that the price of stability can be $\Omega(n \log n)$, matching the previously known upper bound. We also analyse BRD and show that the number of update steps required to converge to a NE cannot be bounded by any function of n, even for symmetric games. Our construction does not depend on the order in which players are allowed to update their strategies. Furthermore, we observe that the cost of a player can grow by an arbitrary factor (not bounded by any function of n) during BRD.

Outline. The remainder of the paper is structured as follows. Section 2 discusses related work. Section 3 gives formal definitions and other preliminaries. Our results on BRD and on the price of stability with respect to max-cost are presented in Sects. 4 and 5, respectively. Section 6 suggests possible directions for future research.

2 Related Work

We discuss only related work on network design games and refer to [7] for general background on algorithmic game theory and the inefficiency of equilibria for different types of games. Network design games with fair cost sharing, where the cost of an edge is distributed to all players using the edge in equal shares, were first studied by Anshelevich et al. [2]. They observe that these games are potential games [6] and therefore always have a NE in pure strategies, and BRD converge to such a NE. For asymmetric, uncapacitated network design games on directed graphs, they show that the price of stability with respect to sum-cost is at most $\mathrm{H}(n) = \Theta(\log n)$, where $\mathrm{H}(n) = \sum_{i=1}^{n} 1/i$ denotes the n-th harmonic number.

They prove this result by considering a potential function that decreases with every improving move of a player and using it to show that BRD from an optimal strategy profile must lead to a NE whose sum-cost is at most $\mathrm{H}(n)$ times the sum-cost of the starting profile. We will use the same potential function several times in this paper. They also show that the upper bound of $\mathrm{H}(n)$ on the price of stability for sum-cost holds for several generalisations, including capacitated network design games. Regarding BRD, they construct a network design game with n players where the convergence to a NE may take a number of steps that is exponential in n (if players make their improving moves in a certain order). It is also known that in symmetric uncapacitated network design games, BRD converge to a NE in at most n steps, as the best response for the first update will also be the best response for all other players [4].

The price of stability of uncapacitated network design games with respect to sum-cost for *undirected* networks is still open. The best known lower bounds are constant and the best known upper bound is $(1 - \Theta(1/n^4))\mathrm{H}(n)$, showing that the maximum price of stability for undirected networks is smaller than it is for directed networks, see [3] and the references given there.

As already noted in [4], it is easy to see that the price of stability is 1 for both sum-cost and max-cost for symmetric network design games without capacities, since the strategy profile where all players choose the same minimum-cost path from the common source to the common destination is a NE and also the social optimum.

Feldman and Ron [4] present a comprehensive study of symmetric capacitated network design games in undirected networks. They show that the price of anarchy is unbounded for both sum-cost and max-cost in general networks, but is bounded by $O(n)$ for parallel links and series-parallel networks. For the price of stability with respect to sum-cost, they show a bound of $O(\log n)$ that is tight even for parallel links. For the price of stability with respect to max-cost, they give tight bounds of $O(n)$ for parallel links and series-parallel networks, but for arbitrary networks their upper bound of $O(n \log n)$ leaves a gap to the lower bound of $\Omega(n)$.

3 Model and Definitions

Capacitated Network Design Games. We consider capacitated network design games, also known as capacitated cost sharing (CCS) games and referred to as CCS games in the following. These games are discrete. All players (or agents) have perfect knowledge of their strategy space and the cost, ceterus paribus, associated with each strategy. For some directed or undirected graph $G = (V, E)$, each player in a set of n must establish a connection between their *source* and *sink* nodes. Every edge $e \in E$ has cost $p(e) \in \mathbb{R}^{\geq 0}$ and capacity $c(e) \in \mathbb{N}$. We also write p_e for $p(e)$ and c_e for $c(e)$. Let $[n]$ denote the set $\{1, 2, \ldots, n\}$. The game can be represented as the tuple

$$\Delta = \langle n, G = (V, E), \{s_i\}_{i \in [n]}, \{t_i\}_{i \in [n]}, \{p_e\}_{e \in E}, \{c_e\}_{e \in E} \rangle .$$

The set of agent i's strategies is the set of s_i-t_i paths in G. We usually denote the strategy of agent i by S_i and the strategy profile of all n players as S. By S_{-i} we denote the joint action of all agents except i in some profile S. As we are considering capacitated networks, a feasibility issue arises. A strategy profile $S = (S_1, \ldots, S_n)$ is *feasible* if $x_e(S) \leq c_e$ for all $e \in E$, where $x_e(S) = |\{i : e \in S_i\}|$ denotes the number of agents that use e in their path. Throughout this paper we will only consider feasible games, i.e., games for which there is at least one feasible strategy profile. We do not impose any restrictions on the network topology, i.e., we allow arbitrary graphs. If we require that all n players have the same source and the same destination, we call the game *symmetric*, and *asymmetric* otherwise.

The price of an edge to an individual player using the edge is an equal slice of its cost which is shared among all the players using the edge. This fair cost division scheme is derived from the Shapley value, and is one of the most widely studied protocols [6]. The price of an individual's strategy S_i, with respect to the strategy profile S, is defined as $p_i(S) = \sum_{e \in S_i} \frac{p_e}{x_e(S)}$.

A profile S is said to be a *Nash equilibrium* (NE) if no agent can improve their cost by a unilateral deviation from the profile, that is, for every player i we have that for all s_i-t_i paths S_i', it holds that $p_i(S) \leq p_i(S_i', S_{-i})$. We consider two social cost functions: the *sum-cost* of a profile S, denoted by $sc_\Delta(S) = \sum_{i \in [n]} p_i(S)$, is the total cost to all agents in S, while the *max-cost* of a profile S, denoted by $mc_\Delta(S) = \max_{i \in [n]} p_i(S)$, is the maximum cost of any agent in S. We omit the subscript Δ if the game is clear from the context.

Note that a game Δ with undirected graph G can be transformed into an equivalent game in directed graph G' using the following construction: Every undirected edge $\{u, v\}$ of G is replaced by the directed edges (u, x_1), (v, x_1), (x_1, x_2), (x_2, u), and (x_2, v), where x_1 and x_2 are two new nodes created for the transformation of $\{u, v\}$. The capacity and cost of (x_1, x_2) are set equal to those of $\{u, v\}$, the remaining edges have infinite capacity and cost 0. As a consequence of this transformation, any construction of undirected CCS games establishing a lower bound on the price of stability (or on the convergence time of BRD) automatically yields an equivalent construction of directed CCS games. Similarly, any upper bound on the price of stability proved for directed CCS games automatically yields the same upper bound for undirected CCS games. When it is clear from the context that we are considering undirected graphs, we also write undirected edges in the form (u, v) instead of $\{u, v\}$.

Best Response Dynamics. If a strategy profile S is not a NE, there will be a cheaper alternative to some player's path. We assume agents have full knowledge of the paths available to them, as well as their opponents' strategies, so they know the cost of all alternatives with respect to S_{-i}. Being self-motivated, players will update their strategies to the cheapest path available at any given point in what is known as *best response dynamics* (BRD). We do not specify the order in which updates are made, only that they are sequential and that the choice of strategy of the player making the update must be the best response to her opponents' current strategies.

Existence of Nash Equilibria. The CCS games we consider fall into the class of *congestion games* studied by Monderer and Shapley [6], who show that all such games have pure Nash equilibria. They do this by defining a potential function Φ, which in the context of our model is

$$\Phi(S) = \sum_{e \in E} \sum_{i=1}^{x_e(S)} \frac{p_e}{i}. \tag{1}$$

Note that $\Phi(S)$ is bounded by $\mathrm{H}(n)$ times the sum-cost of S. As players only make improving moves, best response dynamics will strictly reduce the potential of the solution with each step, meaning a profile cannot be revisited. As the strategy space of a game is finite, any sequence of updates will terminate at a profile where no player can make a unilateral improvement, which must be a NE.

Quality of Nash Equilibria. When measuring the quality of a NE we will compare its cost, by either the sum-cost or max-cost objective, to that of the optimal solution. The ratio between the objective value of the worst NE and the optimal objective value is called the *price of anarchy*, while the ratio between the objective value of the cheapest NE and the optimal objective value is called the *price of stability*, abbreviated to PoA and PoS, respectively. We refer to the optimal objective value with respect to max-cost as OPT_{mc}, and that with respect to sum-cost as OPT_{sc}. Furthermore, we write $\mathrm{PoS}_{mc}(\Delta)$ for the price of stability with respect to max-cost, and similarly for the other cases. For a particular CCS game Δ whose set of Nash equilibria is denoted by $NE(\Delta)$, the prices of anarchy and stability with respect to max-cost are defined as

$$\mathrm{PoA}_{mc}(\Delta) = \frac{\max_{S \in NE(\Delta)} mc_\Delta(S)}{OPT_{mc}(\Delta)} \quad \mathrm{PoS}_{mc}(\Delta) = \frac{\min_{S \in NE(\Delta)} mc_\Delta(S)}{OPT_{mc}(\Delta)}$$

with analogous calculations for sum-cost.

4 Cost Increase and Convergence Time of BRD

Best response dynamics are of interest both as a method to discover equilibria and for the effect they can have on an individual's cost. In potential games, the number of updates required to reach a stable solution is bounded by the cardinality of the strategy set, which is the set of all possible strategy combinations for all players. The size of the strategy set for a game depends on the size and topology of the underlying graph, the number of players, and the distribution of their source and sink nodes.

Examining the effect of BRD on a single player's cost, we now show that, with an arbitrary number of updates, an arbitrary increase in cost for that player is possible, within the limits of a factor $\mathrm{H}(n)$ increase in sum-cost. This result is of particular interest as it illustrates that within a game, a start profile which is cheap for a particular player is no guarantee of a good NE for that individual.

Theorem 1. *There exists a symmetric CCS game Δ with 2 players and a start profile for Δ such that BRD increase the cost of a player to an arbitrary factor times the player's cost in the start profile.*

Proof. For $m \in \mathbb{N}$ with $m \geq 3$, consider the CCS game Δ with $n = 2$ players and underlying undirected graph $G = (V, E)$, defined as follows (see Fig. 1 for an illustration for $m = 3$):

$$V = \{x_i, z_i \mid 0 \leq i \leq m\} \cup \{y_i \mid 1 \leq i \leq m\}$$
$$E = \{(x_i, x_{i-1}), (z_i, z_{i-1}), (x_i, y_i), (y_i, z_{i-1}), (y_i, z_i) \mid 1 \leq i \leq m\}$$

We denote the two players by a and b. Their source and sink nodes are $s_a = z_0$, $t_a = x_m$, $s_b = x_0$, and $t_b = z_m$. We will discuss in the end how to make the game symmetric. A horizontal path from x_i to x_j for some $j \geq i$ is denoted by $x_i \to x_j$, and similarly for $z_i \to z_j$. (We will use this convention for denoting horizontal paths throughout the remainder of the paper.)

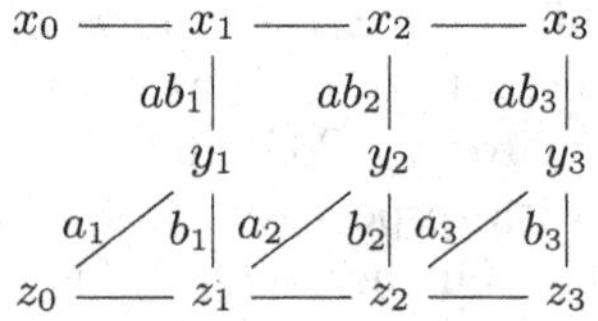

Fig. 1. Underlying graph of game in proof of Theorem 1

All edges have capacity 1, except those connecting an x node and a y node, which have capacity 2. Only edges incident with y nodes have non-zero cost. For any node y_i, the costs of the connections to z_{i-1}, z_i, x_i are denoted by a_i, b_i, ab_i, respectively. These costs are defined as follows (where $\varepsilon > 0$ is a positive constant satisfying $\varepsilon < 1/m^2$):

$$\begin{aligned} a_1 &= 2^m & a_i &= 2^m - 2^i + 2^{i-2} + 1 - i\varepsilon \quad \text{for } i > 1 \\ b_1 &= 1 + \varepsilon & b_i &= 0 \quad \text{for } i > 1 \\ ab_1 &= 0 & ab_i &= 2^{i-1} + \varepsilon \quad \text{for } i > 1 \end{aligned}$$

Let the start profile be $S = ((z_0, y_1, x_1 \to x_m), (x_0, x_1, y_1, z_1 \to z_m))$. Our aim is to enable a sequence of $2m - 2$ best response moves such that the cost of player b increases by an arbitrary factor (depending on m). In the start profile S, players a and b share the edge (x_1, y_1) and their costs are 2^m and $1 + \varepsilon$, respectively. Player a's best response to player b's strategy is now the path $z_0, z_1, y_2, x_2 \to x_m$ with cost $a_2 + ab_2 = 2^m - 2 - 2\varepsilon + 2 + \varepsilon = 2^m - \varepsilon$, so player a will update to that path. Player b's best response to a's new path is now the path $x_0 \to x_2, y_2, z_2 \to z_m$ with cost $ab_2/2 + b_2 = 1 + \frac{\varepsilon}{2}$, so player b will update to that path. As the edge (x_2, y_2) is now shared, this reduces the cost of player a to $2^m - 1 - 2\varepsilon + \varepsilon/2$.

Claim. In the profile reached after $2(i-1)$ best response moves, for $2 \leq i \leq m$, player a uses path $z_0 \to z_{i-1}, y_i, x_i \to x_m$ and player b uses path $x_0 \to x_i, y_i, z_i \to z_m$. Player a's cost is $a_i + ab_i/2 = 2^m - 2^i + 2^{i-2} + 1 - i\varepsilon + 2^{i-2} + \varepsilon/2 = 2^m + 2^{i-1} + 1 - 2^i - i\varepsilon + \varepsilon/2$, and player b's cost is $2^{i-2} + \varepsilon/2$.

After $2(m-1)$ best response moves, player a's path is $(z_0 \to z_{m-1}, y_m, x_m)$ with cost $2^{m-1} + 1 - m\varepsilon + \varepsilon/2$ and player b's path is $(x_0 \to x_m, y_m, z_m)$ with cost $2^{m-2} + \varepsilon/2$. Denote this strategy profile by S^*. We claim that S^* is a NE. First, note that player a does not have an improving move: As the edges (x_{m-1}, x_m) and (z_m, y_m) have capacity 1 and are used by player b, player a can reach x_m only via the edges (z_{m-1}, y_m) and (y_m, x_m), and the path that a uses in S^* contains only zero-cost edges in addition to these two edges. Player b's only alternative paths that are potential improving moves are of the form $(x_0 \to x_i, y_i, z_i, y_{i+1}, z_{i+1}, \ldots, y_{m-1}, z_{m-1}, z_m)$ for some $i < m-1$. Any such path would contain the edge (z_{m-2}, y_{m-1}) with cost $a_{m-1} = 2^m - 2^{m-1} + 2^{m-3} + 1 - (m-1)\varepsilon = 2^{m-1} + 2^{m-3} + 1 - (m-1)\varepsilon > 2^{m-2} + \varepsilon/2$, so it would not be an improving move for b. Therefore, S^* is a NE.

The cost of player b is $1 + \varepsilon$ in the start profile S and $2^{m-2} + \varepsilon/2$ in the NE S^* that is reached by BRD from S. Hence, the cost of player b has increased by a factor arbitrarily close to 2^{m-2}. As m can be chosen arbitrarily large, we have shown that the cost of a player can increase by an arbitrary factor during BRD.

Finally, we observe that the game can be made symmetric by adding edges (s, z_0), (s, x_0), (x_m, t) and (z_m, t) with cost 0 and capacity 1, where s and t are two new nodes that represent the common source and destination, respectively. If player a uses edges (s, z_0) and (x_m, t) and player b uses edges (s, x_0) and (z_m, t) in the initial profile, this property must be maintained in every improving move, and BRD in this symmetric game behave in the same way as in the asymmetric game discussed above. □

The symmetric CCS game defined in the proof of Theorem 1 has $n = 2$ players and the convergence time of BRD is $2(m-1)$, where m can be chosen arbitrarily large. This gives the following corollary, which is in contrast to the uncapacitated symmetric case where BRD converge in at most n steps.

Corollary 1. *There exists a symmetric CCS game and a strategy profile S where BRD converge to NE in an arbitrarily high number of steps, with respect to n.*

5 Price of Stability for Max-Cost

In Sect. 5.1 we show that the PoS is $\Theta(n \log n)$ in the worst case for asymmetric CCS games. In Sect. 5.2 we show that the PoS is bounded by n for symmetric CCS games.

5.1 Asymmetric Games

Theorem 2. *There exists an asymmetric CCS game Δ with n players and*

$$\mathrm{PoS}_{mc}(\Delta) = \Theta(n \log n) .$$

Using the potential function Φ defined in (1) in Sect. 3, it can be shown that the price of stability for max-cost is upper bounded by $O(n \log n)$ for any CCS game: Consider a strategy profile S' that minimises the max-cost. Let C' denote the sum-cost of S', and let M' denote the max-cost of S'. It follows that the optimal max-cost M' satisfies $M' \geq C'/n$. Furthermore, the potential of S' is at most $\Phi(S') \leq C' \cdot \mathrm{H}(n) = C'/n \cdot n\mathrm{H}(n) \leq M'n\mathrm{H}(n)$. BRD starting with S' converge to a NE S^* without increasing the potential. Hence, the sum-cost of S^*, and therefore also the max-cost of S^*, is at most $M'n\mathrm{H}(n)$ [4].

In the following, we will construct a game Δ with an odd number $n \geq 3$ of players where

$$\min_{S^* \in NE(\Delta)} \max_{i \in [n]} p_i(S^*) \approx \frac{n}{2} \mathrm{H}(\lfloor n/2 \rfloor) \cdot OPT_{mc}(\Delta)\,.$$

The construction uses parameters $m \in \mathbb{N}$ and $\varepsilon > 0$, where ε is sufficiently small, e.g., $\varepsilon < 0.1$, and m is sufficiently large. It is useful to think of m as approaching infinity. Furthermore, for $2 \leq i \leq n$, let $k(i)$ denote the value $\lfloor (i+2)/2 \rfloor$. Let Δ be the CCS game with underlying graph $G = (V, E)$ defined as follows (see Fig. 2 for an illustration of the structure of G):

$$V = \begin{array}{l} \{s_1, t_1, x_{[1,m]}, z_{[1,m]}\} \quad \cup \\ \{s_i, t_i, z_{[i,0]} \mid 1 < i \leq n\} \quad \cup \\ \{x_{[i,j]}, y_{[i,j]}, z_{[i,j]} \mid 1 < i \leq n, 1 \leq j \leq m\} \end{array}$$

$$E = \begin{array}{l} \{(s_1, x_{[1,m]}), (t_1, z_{[n,m]}), (x_{[1,m]}, z_{[1,m]})\} \quad \cup \\ \left\{ \begin{array}{l} (x_{[i,j]}, x_{[i,j-1]}), (x_{[i,j]}, y_{[i,j]}), \\ (y_{[i,j]}, z_{[i,j]}), (y_{[i,j]}, z_{[i,j-1]}), (z_{[i,j]}, z_{[i,j-1]}) \end{array} \mid 1 < i \leq n, 1 < j \leq m \right\} \quad \cup \\ \left\{ \begin{array}{l} (z_{[i,0]}, z_{[i-1,m]}), (z_{[i,0]}, z_{[i,1]}), (z_{[i,0]}, y_{[i,1]}), (x_{[i,1]}, y_{[i,1]}), \\ (z_{[i,1]}, y_{[i,1]}), (x_{[1,m]}, x_{[i,1]}), (z_{[i,0]}, s_i), (s_i, z_{[1,m]}), (t_i, x_{[i,m]}) \end{array} \mid 1 < i \leq n \right\} \end{array}$$

$$c(e) = \begin{cases} n & \text{if } e = (x_{[1,m]}, z_{[1,m]}) \\ 2 & \text{if } e = (x_{[i,j]}, y_{[i,j]}) : 1 < i \leq n, 1 \leq j \leq m \\ 1 & \text{otherwise} \end{cases}$$

$$p(e) = \begin{cases} 2 + 2\varepsilon & \text{if } e = (x_{[1,m]}, z_{[1,m]}) \\ \frac{1}{k(i)2^{j-1}} & \text{if } e = (x_{[i,j]}, y_{[i,j]}) : 1 < i \leq n, 1 \leq j \leq m \\ \mathrm{H}(k(i)) - \frac{3}{k(i)2^j} + \frac{\varepsilon}{k(i)m+j} & \text{if } e = (y_{[i,j]}, z_{[i,j]}) : 1 < i \leq n, 1 \leq j \leq m \\ \frac{\varepsilon}{k(i)+j} & \text{if } e = (y_{[i,j]}, z_{[i,j-1]}) : 1 < i \leq n, 1 \leq j \leq m \\ 0 & \text{otherwise} \end{cases}$$

The source and sink of player i, for $1 \leq i \leq n$, are s_i and t_i, respectively. We refer to the nodes of the form $x_{[i,j]}$ as the x-row, to the nodes of the form $y_{[i,j]}$ as the y-row, and to the nodes of the form $z_{[i,j]}$ as the z-row. We divide the main part of the graph into grids as follows: For any $i \geq 2$, the *i-th grid* is the induced subgraph of all x, y, z nodes with subscript $[i, j]$ for any j. Note that the edge costs in pairs of consecutive grids, namely the $(2k-2)$-th and $(2k-1)$-th grid, are the same for $2 \leq k \leq \lceil n/2 \rceil$. For fixed i and j, we refer to the subgraph induced by $x_{[i,j]}$, $y_{[i,j]}$, $z_{[i,j]}$ and $z_{[i,j-1]}$ as *column* j of the i-th grid. For simplicity we will refer to the costs of the connections from $y_{[i,j]}$ to $x_{[i,j]}, z_{[i,j]}, z_{[i,j-1]}$ as $ab_{[k(i),j]}, a_{[k(i),j]}, b_{[k(i),j]}$, respectively.

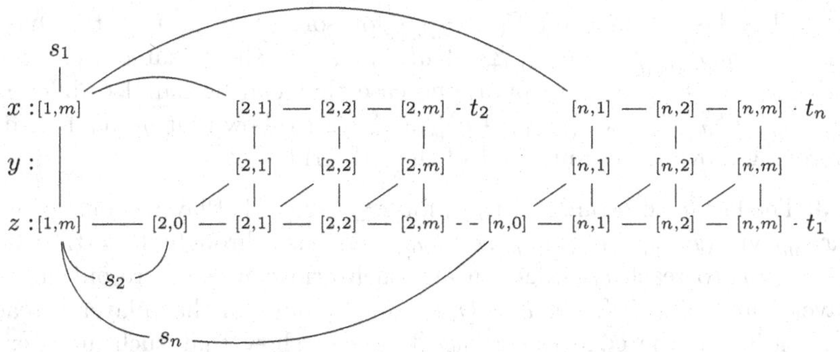

Fig. 2. Game Δ with $\mathrm{PoS}_{mc}(\Delta) = \Theta(n \log n)$ for $m = 3$

Lemma 1. $OPT_{mc}(\Delta) \leq \frac{2+2\varepsilon}{n}$

Proof. Consider the profile S' where player 1 uses the path $(s_1, x_{[1,m]}, z_{[1,m]} \rightarrow z_{[n,m]}, t_1)$ and player i, for $2 \leq i \leq n$, uses the path $(s_i, z_{[1,m]}, x_{[1,m]}, x_{[i,1]} \rightarrow x_{[i,m]}, t_i)$. All n players share the edge $(x_{[1,m]}, z_{[1,m]})$ and use no other edge with non-zero cost. Each player has cost $(2+2\varepsilon)/n$. The optimal max-cost is therefore at most $(2+2\varepsilon)/n$. □

Lemma 2. $\min_{S \in NE(\Delta)} mc_\Delta(S) \geq a_{[k(n),m]} \geq \mathrm{H}(\lfloor n/2 \rfloor) - \frac{3}{k(n)2^m} + \frac{\varepsilon}{(k(n)+1)m}$

Proof (Sketch). We claim that player 1 must pass through $x_{[n,m]}$ or $x_{[n-1,m]}$ and hence use edges $(x_{[i,m]}, y_{[i,m]})$ and $(y_{[i,m]}, z_{[i,m]})$ for $i = n-1$ or for $i = n$ in any NE (note that using edges $(y_{[i,m]}, z_{[i,m-1]})$ and $(z_{[i,m-1]}, z_{[i,m]})$ would block player i from reaching t_i), thus paying at least $a_{[k(n-1),m]} = a_{[k(n),m]}$. To establish that player 1 must pass through $x_{[n,m]}$ or $x_{[n-1,m]}$ in any NE, we show that in all other cases some player has an improving move.

Let S be a NE. Assume that the path S_1 of player 1 does not pass through $x_{[n,m]}$ or $x_{[n-1,m]}$. Consider the last x node (i.e., node in the x-row) that the path S_1 of player 1 visits. Let $x_{[i,j]}$ be that node. Note that $i < n-1$ or $j < m$. If $i \geq 2$, note that $x_{[i,j]}$ must be followed directly by $y_{[i,j]}$ and $z_{[i,j]}$ on S_1 because using the subpath $(y_{[i,j]}, z_{[i,j-1]}, z_{[i,j]})$ would block player i from reaching her destination t_i. If $i = 1$, the path S_1 must use the edge $(x_{[1,m]}, z_{[1,m]})$. There are three possible cases for the location of the last x node on S_1, and it can be shown that some player has an improving move in each case.

Case 1: The last x node on S_1 is $x_{[1,m]}$. S_1 must travel from $x_{[1,m]}$ to $z_{[1,m]}$ and then reach t_1 by visiting all nodes $z_{[i,j]}$ from left to right, possibly visiting some y nodes in between adjacent z nodes. One can show that no other player i can share the edge $e_1 = (x_{[1,m]}, z_{[1,m]})$ with player 1, as player i would have an improving move via $z_{[i,0]}$ in that case due to $\frac{2+2\varepsilon}{i} > ab_{[k(i),1]} + b_{[k(i),1]}$. Hence, player 1 pays the full price $2 + 2\epsilon$ for e_1 and one can show that she has an improving move via $x_{[2,1]}$.

Case 2: The last x node on S_1 is $x_{[i,j]}$ for some $i \geq 2$, $j < m$. One can use $b_{[k(i),j]} + ab_{[k(i),j]}/2 > b_{[k(i),j+1]} + ab_{[k(i),j+1]}$ to show that player i passes through $(x_{[i,j+1]}, y_{[i,j+1]})$ (except in one case that can be handled differently) and $a_{[k(i),j]} + ab_{[k(i),j]} > a_{[k(i),j+1]} + ab_{[k(i),j+1]}/2$ to show that player 1 then has an improving move to column $j+1$ of the i-th grid.

Case 3: The last x node on S_1 is $x_{[i,m]}$ for $2 \leq i < n-1$. Player 1 must continue from $x_{[i,m]}$ via $(x_{[i,m]}, y_{[i,m]}, z_{[i,m]}, z_{[i+1,0]})$ and pass through the i'-th grid for all $i \leq i' \leq n$ to reach t_1. Edges in any such grid can therefore only be used by player 1 and player i'. Let $i' \in \{i+1, i+2\}$ be such that player i' reaches the i'-th grid via the edge $(s_{i'}, z_{[i',0]})$. One can show that such an i' exists. Player 1 must visit $z_{[i',0]}$. Using $a_{[k(i),m]} + ab_{[k(i),m]}/2 > a_{[k(i'),1]} + ab_{[k(i'),1]}/2$ for $i' \in \{i+1, i+2\}$, player 1 has an improving move by choosing a path starting with $(s_1, x_{[1,m]}, x_{[i',1]}, y_{[i',1]})$. □

Proof (of Theorem 2). We have constructed a CCS game Δ with an optimal strategy profile with a max-cost of at most $\frac{2+2\varepsilon}{n}$ by Lemma 1, while the max-cost of the best NE is arbitrarily close to $\mathrm{H}(\lfloor n/2 \rfloor)$ by Lemma 2. Hence, $PoS_{mc}(\Delta)$ approaches $n\mathrm{H}(\lfloor n/2 \rfloor)/(2+\varepsilon) = \Theta(n \log n)$ arbitrarily closely. □

5.2 Symmetric Games

We now consider the case where all players have the same source and sink, and find that the upper bound of $O(n \log n)$ on the price of stability is not tight. First, we note that BRD starting with a strategy profile with optimal max-cost can increase the max-cost by a factor of $\Omega(n \log n)$ even in the symmetric case. The game constructed for the proof of Theorem 2 in Sect. 5.1 can be made symmetric by attaching a common source s that is made adjacent to $\{s_1, \ldots, s_n\}$ and a common destination t that is made adjacent to $\{t_1, \ldots, t_n\}$. Starting with the strategy profile where player i visits s_i, the edge $(x_{[1,m]}, z_{[1,m]})$, and t_i, BRD will converge to a NE that corresponds to a NE for the asymmetric game and has max-cost $\Omega(n \log n)$ times the optimal max-cost. Hence, in order to bound the price of stability in the symmetric case, we cannot use the standard proof technique of analysing BRD starting with the optimal strategy profile. Instead, we use a different approach that may be of independent interest. We iteratively discard a single expensive path from the NE reached by BRD and recombine the remaining $n-1$ paths with the optimal strategy profile, until a NE with small max-cost is obtained. In this way we are able to show that for every symmetric CCS game Δ there is always a NE where no player pays more than n times $OPT_{mc}(\Delta)$.

Theorem 3 *For any symmetric CCS game Δ in directed or undirected networks,* $\mathrm{PoS}_{mc}(\Delta) \leq n$.

Proof We present the proof for directed networks. The result for undirected networks follows using the standard transformation of an undirected network into an equivalent directed network discussed in Sect. 3.

Let Δ be a symmetric CCS game with directed graph $G = (V, E)$, n players, common source s and common destination t. Let S be the optimal strategy profile with respect to max-cost. Without loss of generality, we can scale the edge costs so that the sum-cost of S is n. This implies $mc(S) \geq 1$.

Consider the NE S^* that is obtained from S using BRD. If $mc(S^*) \leq n$, then $\text{PoS}(\Delta) \leq n$ and we are done. Otherwise, we have $n < mc(S^*) \leq sc(S^*) \leq \Phi(S^*) < \Phi(S)$. Let $\Phi(S) = n + \alpha$ and $mc(S^*) = n + \beta$ for some $\alpha, \beta > 0$, and let $\Phi(S^*) = \Phi(S) - \delta$ for some $\delta > 0$. Note that $0 < \beta \leq \alpha - \delta$. The following table illustrates these quantities:

	mc	Φ
S	≥ 1	$n + \alpha$
S^*	$n + \beta$	$n + \alpha - \delta$

Now consider the $(n-1)$-player profile S^*_{-1} that consists of the $n-1$ cheapest (in terms of cost to the respective player) strategies in S^*. As the change in potential function equals the cost to an individual player when making some change in strategy, we have that $\Phi(S^*_{-1}) = \Phi(S^*) - mc(S^*) = n + \alpha - \delta - (n + \beta) = \alpha - \beta - \delta$.

We construct a new n-player strategy profile S' by combining S and S^*_{-1} using an augmentation step in a suitably defined flow network. (We refer the reader to [1] for background on network flow, residual networks, and augmenting paths.) First, define the capacitated network $\bar{G} = (V, \bar{E})$ from $G = (V, E)$ by letting $\bar{E} = \{e \in E \mid x_e(S) > 0 \text{ or } x_e(S^*_{-1}) > 0\}$ and setting the capacity $\bar{c}(e)$ for each $e \in \bar{E}$ to $\bar{c}(e) = \max\{x_e(S), x_e(S^*_{-1})\}$. The strategy profile S^*_{-1} induces a flow f of value $n-1$ from s to t in $\bar{G}$. The network $\bar{G}$ admits a flow of value n from s to t as the profile S induces such a flow. Hence, the residual network $\bar{G}_f$ of $\bar{G}$ with respect to flow f admits an augmenting path P from s to t. Let f' be the flow of value n obtained by augmenting f with P. Decompose the flow f' into n paths from s to t, and let S' be the strategy profile corresponding to these n paths.

In going from f to f', the flow on any edge increases by at most 1, and every edge on which the flow increases satisfies $x_e(S) > 0$. Let X be the set of edges on which the flow increases. Observe that $\Phi(S') \leq \Phi(S^*_{-1}) + p(X)$, because increasing the number of players on an edge e by 1 adds at most $p(e)$ to the potential. As $X \subseteq \{e \in E \mid x_e(S) > 0\}$, we have $p(X) \leq sc(S) = n$ and hence $\Phi(S') \leq \alpha - \beta - \delta + n < \Phi(S^*)$.

Let S^{**} be the NE obtained from S' via BRD. Note that $\Phi(S^{**}) \leq \Phi(S') < \Phi(S^*)$. If $mc(S^{**}) \leq n$, we have found a NE with max-cost at most n times the optimal max-cost and we are done. Otherwise, we can repeat the construction that we used to create S^{**} from S^*, but starting with S^{**} in place of S^*. Each time we repeat the construction and obtain a NE with max-cost greater than n, that NE has strictly smaller potential than the previous NE. As the number of strategy profiles is finite, we must eventually obtain a NE whose max-cost is at most n. This shows $\text{PoS}_{mc}(\Delta) \leq n$. □

6 Further Research

As we have shown in Corollary 1 that the convergence time of BRD cannot be bounded by any function of n for symmetric CCS games, an interesting question is in what other settings it is possible for BRD convergence to be unbounded in n. In our setting the effect of capacitated edges is to allow the reachable strategy space of an individual player to be limited by their opponent's choice of path, thus increasing the number of states which can be passed before a stable solution is reached. It may be interesting to identify other games where this behaviour could be observed.

To prove that the price of stability with respect to max-cost is bounded by n for symmetric CCS games, we iteratively combined a NE with large max-cost with the optimal strategy profile. It would be interesting to explore whether this method could be turned into an efficient procedure for constructively finding a good NE. As the approach mainly relies on arguments about the reduction in potential of the strategy profiles constructed, it may be possible to apply it to other potential games.

References

1. Ahuja, R.K., Magnanti, T.L., Orlin, J.B.: Network Flows: Theory, Algorithms, and Applications. Prentice Hall, New Jersey (1993)
2. Anshelevich, E., Dasgupta, A., Kleinberg, J.M., Tardos, É., Wexler, T., Roughgarden, T.: The price of stability for network design with fair cost allocation. SIAM J. Comput. **38**(4), 1602–1623 (2008)
3. Disser, Y., Feldmann, A.E., Klimm, M., Mihalák, M.: Improving the H_k-bound on the price of stability in undirected shapley network design games. Theoret. Comput. Sci. **562**, 557–564 (2015)
4. Feldman, M., Ron, T.: Capacitated network design games. In: Serna, M. (ed.) SAGT 2012. LNCS, vol. 7615, pp. 132–143. Springer, Heidelberg (2012)
5. Koutsoupias, E., Papadimitriou, C.: Worst-case equilibria. In: Meinel, C., Tison, S. (eds.) STACS 1999. LNCS, vol. 1563, p. 404. Springer, Heidelberg (1999)
6. Monderer, D., Shapley, L.S.: Potential games. Games Econ. Behav. **14**, 124–143 (1996)
7. Nisan, N., Roughgarden, T., Tardos, E., Vazirani, V.V.: Algorithmic Game Theory. Cambridge University Press, New York (2007)
8. Roughgarden, T., Tardos, É.: How bad is selfish routing? J. ACM **49**(2), 236–259 (2002)
9. Schulz, A.S., Moses, N.E.S.: On the performance of user equilibria in traffic networks. In: Proceedings of the Fourteenth Annual ACM-SIAM Symposium on Discrete Algorithms (SODA 2003), pp. 86–87 (2003)

Cost-Sharing Scheduling Games on Restricted Unrelated Machines

Guy Avni[1] and Tami Tamir[2](✉)

[1] School of Computer Science and Engineering, The Hebrew University, Jerusalem, Israel
[2] School of Computer Science, The Interdisciplinary Center Herzliya, Herzliya, Israel
tami@idc.ac.il

Abstract. We study a very general cost-sharing scheduling game. An instance consists of k jobs and m machines and an arbitrary weighed bipartite graph denoting the job strategies. An edge connecting a job and a machine specifies that the job may choose the machine; edge weights correspond to processing times. Each machine has an activation cost that needs to be covered by the job assigned to it. Jobs assigned to a particular machine share its cost proportionally to the load they generate.

Our game generalizes singleton cost-sharing games with weighted players. We provide a complete analysis of the game with respect to equilibrium existence, computation, convergence and quality – with respect to the total cost. We study both unilateral and coordinated deviations.

We show that the main factor in determining the stability of an instance and the quality of a stable assignment is the machines' activation-cost. Games with unit-cost machines are potential games, and every instance has an optimal solution which is also a pure Nash equilibrium (PNE). On the other hand, with arbitrary-cost machines, a PNE is guaranteed to exist only for very limited instances, and the price of stability is linear in the number of players. Also, the problem of deciding whether a given game instance has a PNE is NP-complete.

In our analysis of coordinated deviations, we characterize instances for which a strong equilibrium exists and can be calculated efficiently, and show tight bounds for the SPoS and the SPoA.

1 Introduction

In job-scheduling applications, jobs are assigned to machines to be processed. Many interesting combinatorial optimization problems arise in this setting, which is a major discipline in operation research. A centralized scheduler should assign the jobs in a way that achieves load balancing, an effective use of the system's resources, or a target quality of service [12]. Many modern systems provide service to multiple strategic users, whose individual payoff is affected by the decisions made by others. As a result, non-cooperative game theory has become an essential tool in the analysis of job-scheduling applications. We assume that each job is controlled by a player which has strategic considerations and act to minimize

M. Hoefer (Ed.): SAGT 2015, LNCS 9347, pp. 69–81, 2015.
DOI: 10.1007/978-3-662-48433-3_6

his own cost, rather than to optimize any global objective. Practically, this means that the jobs *choose* a machine instead of being assigned to one by a centralized scheduler. In this paper we study the corresponding cost-sharing scheduling game (CSSGs, for short) on restricted unrelated parallel machines.

An instance of CSSG is given by an arbitrary weighted bipartite graph whose vertex set consists of job-vertices and machine-vertices. The scheduling is restricted in a sense that not all machines are feasible to all jobs: each job is connected by edges to the machines that are capable to process it. Edge weights specify the processing times, reflecting the load generated by the job on the machine. Scheduling on restricted unrelated machines is the most general model of scheduling on parallel machines.

In the corresponding game, the strategy space of a job is the set of machines that can process it. Each machine has an activation cost that needs to be covered by the jobs assigned to it. Cost-sharing games, in which players' strategies are subsets of resources and the resource's activation cost is covered by its users, arise in many applications, and are well-studied. Our game is different from previously studied games in several ways, each arising new challenges. Previous work on cost-sharing scheduling games assume that either the activation-cost of a resource is shared uniformly by its users, or that players are weighted. To the best of our knowledge, this is the first time that this most-general scheduling model is analyzed as a non-cooperative cost-sharing game.

1.1 Preliminaries

An instance of CSSG is given by an arbitrary weighted bipartite graph G whose vertex set is $\mathcal{J} \cup \mathcal{M}$, where $\mathcal{J}$ is a set of k jobs, and $\mathcal{M}$ is a set of m machines. Not all machines are feasible to all jobs: each $i \in \mathcal{J}$, has a set $M_i \subseteq \mathcal{M}$ of machines that may process it. For every job i and machine $j \in M_i$, it is known what the processing time $p_{j,i}$ of i on machine j is. The feasible sets and the processing times are given by the edges of the bipartite graph. Specifically, there is an edge (i, j) whose weight is $p_{j,i}$ for every $j \in M_i$.

Job i is controlled by Player i whose strategy space is the set of machines in M_i. Each machine $j \in \mathcal{M}$ has an activation cost, $c(j)$, which is shared by the jobs assigned to it, where the share is proportional to the load generated by the job.

A profile of a CSSG game is a vector $P = \langle s_1, s_2, \ldots, s_k \rangle \in (M_1 \times M_2 \times \ldots \times M_k)$ describing the machines selected by the players. For a machine $j \in \mathcal{M}$, we define the *load* on j in P, denoted $L_j(P)$, as the total processing times of the jobs assigned to machine j in P, that is, $L_j(P) = \sum_{\{i|s_i=j\}} p_{j,i}$. When P is clear from the context we omit it. The cost of Player i in the profile P is $cost_i(P) = \frac{p_{s_i,i}}{L_{s_i}(P)} \cdot c(s_i)$ and the cost of the profile P is $cost(P) = \sum_{1 \leq i \leq k} cost_i(P)$. Note that $cost(P)$ also equals the total activation-cost of non-idle machines, that is, $cost(P) = \sum_{j \in \cup_i s_i} c(j)$.

Consider a game G. For a profile P, a job $i \in \mathcal{J}$, and a strategy $s'_i \in M_i$, let $P[i \leftarrow s'_i]$ denote the profile obtained from P by replacing the strategy of Player i by s'_i. That is, the profile resulting from a migration of job i from machine s_i to machine s'_i. A profile P is a *pure Nash equilibrium* (NE) if no job i can benefit from unilaterally deviating from his strategy in P to another strategy; i.e., for every player i and every strategy $s'_i \in M_i$ it holds that $cost_i(P[i \leftarrow s'_i]) \geq cost_i(P)$.

Best-Response Dynamics (BRD) is a local-search method where in each step some player is chosen and plays its best improving deviation (if one exists), given the strategies of the other players. Since BRD corresponds to actual dynamics in real life applications, the question of BRD convergence and the quality of possible BRD outcomes are major issues in our study.

It is well known that decentralized decision-making may lead to sub-optimal solutions from the point of view of society as a whole. We denote by OPT the cost of a social-optimal (SO) solution; i.e., $OPT = \min_P cost(P)$. We quantify the inefficiency incurred due to self-interested behavior according to the *price of anarchy* (PoA) [9,11] and *price of stability* (PoS) [2,14] measures. The PoA is the worst-case inefficiency of a Nash equilibrium, while the PoS measures the best-case inefficiency of a Nash equilibrium. Formally,

Definition 1. *Let $\mathcal{G}$ be a family of games, and let G be a game in $\mathcal{G}$. Let $\Upsilon(G)$ be the set of Nash equilibria of the game G. Assume that $\Upsilon(G) \neq \emptyset$.*

- *The* price of anarchy *of G is the ratio between the* maximal *cost of a PNE and the social optimum of G. That is, $PoA(G) = \max_{P \in \Upsilon(G)} cost(P)/OPT(G)$. The* price of anarchy *of the family of games $\mathcal{G}$ is $PoA(\mathcal{G}) = sup_{G \in \mathcal{G}} PoA(G)$.*
- *The* price of stability *of G is the ratio between the* minimal *cost of a PNE and the social optimum of G. That is, $PoS(G) = \min_{P \in \Upsilon(G)} cost(P)/OPT(G)$. The* price of stability *of the family of games $\mathcal{G}$ is $PoS(\mathcal{G}) = sup_{G \in \mathcal{G}} PoS(G)$.*

A firmer notion of stability requires that a profile is stable against *coordinated deviations.* A set of players $\Gamma \subseteq \mathcal{J}$ forms a *coalition* if there exists a move where each job $i \in \Gamma$ strictly reduces its cost. A profile P is a *Strong Equilibrium* (SE) if there is no coalition $\Gamma \subseteq \mathcal{J}$ that has a beneficial move from P [3]. The *strong price of anarchy* (SPoA) and the *strong price of stability* (SPoS) introduced in [1] are defined similarly, where $\Upsilon(G)$ refers to the set of strong equilibria.

In our study of CSSGs, we distinguish between unit-cost instances, in which all machines have the same activation cost, say $c(j) = 1$ for all $j \in \mathcal{M}$, and the general case, where $c(j)$ is arbitrary. We say that an instance has *machine-independent* processing-times if for every job i there is $p_i > 0$ such that $p_{j,i} = p_i$ for all $j \in M_i$.

1.2 Related Work and Our Results

Game-theoretic analysis became an important tool for analyzing huge systems that are controlled by users with strategic consideration. In particular, systems in which a set of resources is shared by selfish users.

Congestion games [13] consist of a set of resources and a set of players who need to use these resources. Players' strategies are subsets of resources. Each resource has a latency function which, given the load generated by the players on the resource, returns the cost of the resource. We refer to the setting in which the latency functions are increasing as *congestion games* (the more congested the resource, the higher the waiting time), and we focus on *cost-sharing games* in which each resource has an activation cost that is shared by the players using it according to some sharing mechanism. For example, in network formation games, players have reachability objectives and strategies are subsets of edges, each inducing a simple path from the source to the target [2]. Players that use an edge uniformly share its cost. Such games always have a PNE and the PoS is logarithmic in the number of players.

Weighted cost-sharing games are cost-sharing games in which each player i has a *weight* $w_i \in \mathbb{N}$, and his contribution to the load of the resources he uses as well as his payments are multiplied by w_i. In [2] the authors study the counterpart of network formation games in the weighted cost-sharing setting. They show that every two-player game admits a PNE and that the PoS is an order of the number of players. Later, [5] closed the problem of PNE existence in these games by showing an example of a three-player game with no PNE.

In a more general setting, players' strategies are multisets of resources. Thus, a player may need multiple uses of the same resource and his cost for using the resource depends on the number of times he uses the resource [4]. Such *multiset cost-sharing games* are less stable than classical cost-sharing games. Even very simple instances may not have a PNE and an equilibrium may be extremely inefficient (the PoS may equal the number of players) [4].

A lot of attention has been given to scheduling congestion games (for a survey, see [16]), which can be thought of as a special case of congestion games in which the players' strategies are singletons. Most previous work assumes that the cost of a player is simply the load on the machine, and is thus independent of the job's length. Scheduling congestion games that do take the length into an account, were defined and studied in [10] (there, defined and studied as weighted congestion games with separable preferences) and [17].

The SPoA and SPoS measures where introduced by [1], which study a similar game to ours only with congestion effects rather than cost-sharing, and with a different definition of the social optimum; namely the cost of the highest paying player (which is the *makespan* in their setting). The SPoA and SPoS where studied in [6] for network formation games in the cost-sharing setting.

In this work we complete the picture and study scheduling cost-sharing games; i.e., when jobs have an incentive to be assigned to a heavily loaded machine. CSSGs can be viewed as a generalization of classical cost-sharing games with weighted players [2]. The latter corresponds to the special case in which all the machines are identical; i.e., all machines are feasible to all jobs and the processing time of a job on a machine is independent of the machine.

The paper [15] studies the complexity of equilibria in a wide range of cost sharing games. Their results on singleton cost sharing games correspond to our model with unit-length jobs (and therefore also fair cost-sharing).

In this paper we provide a complete analysis of the game with respect to equilibrium existence, computation, convergence and quality. We study both unilateral and coordinated deviations, distinguishing between instance having unit or arbitrary machine-activation costs. Our results are detailed in Table 1.

Due to space constraints, some proofs are omitted.

Table 1. Summary of our results. (†) Deciding whether a PNE exists is NP-complete. (‡) Adopted to our model from [2]. (§) Extension of [15].

Activation costs	Processing times	Pure Nash equilibrium			Strong equilibrium		
		∃	PoA	PoS	∃	SPoA	SPoS
Unit	arbitrary	yes	$\min\{m,k\}$	1	no	$\min\{m,\frac{k}{2}+\frac{1}{2}\}$	$\min\{\frac{m}{2},\frac{k}{4}+\frac{1}{2}\}$
	machine-indp.	yes	$\min\{m,k\}$	1	yes	$\min\{\frac{m}{2},\frac{k}{4}+\frac{1}{2}\}$	$\min\{\frac{m}{2},\frac{k}{4}+\frac{1}{2}\}$
Arbitrary	arbitrary	no†	k	k	no	k	k
	machine-indp.	yes	k‡	k	yes§	k	k

2 Instances with Unit-Cost Machines

In this section we study game instances in which all machines have the same activation cost, say $c(j) = 1$ for all $j \in \mathcal{M}$. We suggest a non-standard potential function to show that an CSSG with unit costs is a potential game. Hence, a PNE exists. We also provide tight bounds for the PoA and PoS. Let P be a profile of an CSSG with unit costs. Recall that with unit-cost machines, $cost(P)$ gives the number of active machines in P, that is, $cost(P) = |\{j \in \mathcal{M} | L_j > 0\}|$.

Theorem 1. *A CSSG with unit-cost machines is a potential game.*

Proof. Let G be an CSSG with unit-cost machines. Let P be a profile of G. Consider the function

$$\Phi(P) = (cost(P), \Pi_{\{j \in \mathcal{M} | L_j > 0\}} L_j),$$

that maps a profile to a 2-dim vector. The first entry in the vector specifies the number of active machines in P; The second entry is the product of these machines' loads.

We show that Φ is a potential function for the game. Specifically, we show that every migration of a job in best response dynamics reduces the lexicographic order of the potential. Consider a profile P and assume, w.l.o.g, that Player 1 migrates from machine u to machine w, and the resulting profile is P'. Denote by L_u, L'_u, L_w, and L'_w the loads on machines u and w before and after the deviation of Player 1, respectively, that is, $L_u = \sum_{\{i|s_i=u\}} p_{u,i}$ and $L'_u = L_u - p_{u,1}$, $L_w = \sum_{\{i|s_i=w\}} p_{w,i}$ and $L'_w = L_w + p_{w,1}$.

Clearly, the migration is be beneficial only if $L_w > 0$. Thus, $cost(P') \leq cost(P)$. If $L'_u = 0$, then $cost(P') = cost(P) - 1$ and $\Phi(P) \succ \Phi(P')$. Otherwise, $cost(P') = cost(P)$. We show that the second entry in the potential vector strictly decreases by showing that $\Phi(P)_2 / \Phi(P')_2 > 1$.

Since the loads on machines other than u, w do not change, we have

$$\frac{\Phi(P)_2}{\Phi(P')_2} = \frac{L_u \cdot L_w}{L'_u \cdot L'_w}.$$

Note that the above fraction is well-defined as $L'_w > p_{w,1} > 0$ and $L'_u > 0$ since we analyze the case $cost(P') = cost(P)$.

Multiply both numerator and denominator by $p_{u,1}$ and $p_{w,1}$ and rearrange to get

$$\frac{\Phi(P)_2}{\Phi(P')_2} = \frac{L_u}{p_{u,1}} \cdot \frac{p_{w,1}}{L'_w} \cdot \frac{L_w}{p_{w,1}} \cdot \frac{p_{u,1}}{L'_u}.$$

Note that

$$\frac{L_u}{p_{u,1}} = \frac{1}{cost_1(P)} \quad \text{and} \quad \frac{p_{w,1}}{L'_w} = cost_1(P'). \tag{1}$$

Also,

$$\frac{1}{cost_1(P)} = \frac{L_u}{p_{u,1}} = \frac{L'_u}{p_{u,1}} + 1 \quad \text{and} \quad \frac{1}{cost_1(P')} = \frac{L'_w}{p_{w,1}} = \frac{L_w}{p_{w,1}} + 1.$$

Thus,

$$\frac{p_{u,1}}{L'_u} = \frac{cost_1(P)}{1 - cost_1(P)} \quad \text{and} \quad \frac{L_w}{p_{w,1}} = \frac{1 - cost_1(P')}{cost_1(P')}. \tag{2}$$

Combining (1) and (2), we have

$$\frac{\Phi(P)_2}{\Phi(P')_2} = \frac{cost_1(P')}{cost_1(P)} \cdot \frac{cost_1(P)}{1 - cost_1(P)} \cdot \frac{1 - cost_1(P')}{cost_1(P')} = \frac{1 - cost_1(P')}{1 - cost_1(P)}.$$

Since the migration is beneficial, $cost_1(P) > cost_1(P')$. Since both costs are positive and strictly lower than 1, we conclude that $\Phi(P)_2/\Phi(P')_2 > 1$. Thus, $\Phi(P) \succ \Phi(P')$, as required. □

We turn to study the equilibrium inefficiency. Recall that our measurement for a profile P is the total players' cost, which is equal to the number of active machines.

Theorem 2. *Every CSSG instance with unit-cost machines has $PoS = 1$. If $k < m$, then $PoA = k$. If $m \leq k < 2m - 1$, then $PoA = m - 1$. If $k \geq 2m - 1$, then $PoA = m$.*

Proof. Consider a BRD sequence that starts from the social optimum profile (SO). By Theorem 1, the sequence reaches a PNE. Note that the maximal cost of a player in the SO is 1. Therefore, during the BRD process, when a player deviates, he will never activate a new machine (at cost 1). It follows that the number of active machines in the resulting PNE is at most the social optimum. Thus, PoS = 1.

We turn to analyze the PoA. Assume first that $k < m$. We describe a family of game instances for which PoA = k. Let $\mathcal{M} = \{0, 1, \ldots, k, k+1, \ldots, m-1\}$. For $1 \leq i \leq k$, the capable machines for Player i are $\{0, i\}$. Thus, machines $k+1, \ldots, m-1$ are dummy machines and not capable for any player. The social optimum is 1 and it is attained when all players are assigned to machine 0.

The worst PNE is when for all $1 \leq i \leq k$, Player i is assigned to machine i. This is indeed a PNE since no player can reduce his payment by deviating to machine 0 - as this machine is not used by any player in this profile and the machine costs are equal. Thus, PoA $= \frac{k}{1} = k$. Clearly, this bound is tight as PoA $\leq k$ trivially holds.

The analysis for $m \leq k < 2m - 1$ is omitted.

Assume $k \geq 2m - 1$. We show a family of instances in which the SO is 1 and the worst PNE uses m machines, and thus PoA $= m$. This is clearly a tight bound as the SO is at least 1 and any schedule uses at most m machines. We continue to describe the family. The only capable machines for Player 1 is machine 0. For $i = 2, 4, \ldots, 2m-2$, Players i and $i+1$ have $M_i = M_{i+1} = \{0, \frac{i}{2}\}$. For $2m - 1 < i \leq k$, we have $M_i = \{0, m-1\}$. The processing times of the players on all the machines is equal. The SO is clearly 1 and it is achieved when all players choose machine 0. We claim that the profile in which all players (except for Player 1) choose their "second" machine is a PNE. Indeed, note that in this profile there are at least two players using machines $1, \ldots, m-1$ and the share of the machines' cost is divided equally. Since only Player 1 uses machine 0, a player cannot reduce his payment by deviating to that machine. □

3 Instances with Arbitrary Cost Machines

In this section we extend the model and consider instances with arbitrary cost machines. As we show, a PNE may not exist even in very small instances. Moreover, it is NP-hard to decide whether a given instance has a PNE. On the other hand, a PNE is guaranteed to exist and can be calculated efficiently for instances with machine-independent processing times.

Theorem 3. *A PNE is guaranteed to exist in every CSSG in which $m \leq 2$ or $k \leq 3$. There is an CSSG with $m = 3$ and $k = 4$ with no PNE.*

Proof. The PNE-existence proof for $m \leq 2$ or $k \leq 3$ is omitted. We show that there exists an instance with $m = 3$ machines and $k = 4$ players that has no PNE. Consider an instance, I_{noNE}, with three machines having activation costs 30, 12 and 14, and four jobs having processing times as given in the table. Note that Job d must be assigned to m_1 and each of the other jobs has two feasible machines. Figure 1 presents a loop of beneficial moves that covers six out of the eight possible configurations. The payment vector is given below each configuration. The job that has a beneficial move is darker and it deviates to the next configuration (the leftmost configuration follows the rightmost one). It is easy to see that the two other configurations (in which no machine accommodates two jobs from a, b, c) are not stable either. □

The next natural question is whether it is possible to decide efficiently whether a given instance has a PNE. We show that this is an NP-complete problem.

Theorem 4. *The question whether a game instance with arbitrary-cost machines has a PNE is NP-complete.*

	m_1	m_2	m_3
a	2	--	13
b	1	1	–
c	--	2	1
d	2	--	--

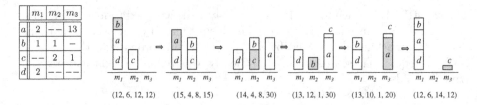

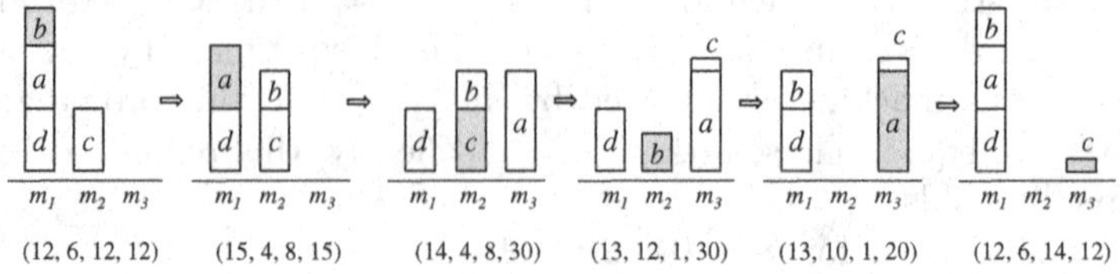

Fig. 1. For $c(m_1) = 30, c(m_2) = 12$, and $c(m_3) = 14$, the instance has no PNE.

Proof (Sketch). Checking stability of a given profile can be done efficiently, therefore the problem is clearly in NP. We prove hardness by showing a reduction from the 3-dimensional matching problem (3DM), which is known to be NP-hard [7]. The input to the 3DM problem is a set of triplets $T \subseteq X_1 \times X_2 \times X_3$, where $|X_1| = |X_2| = |X_3| = n$. The number of triplets is $|T| \geq n$. The desired output is a 3-dim matching $T' \subseteq T$ such that $|T'| = n$ and any element in $X_1 \cup X_2 \cup X_3$ appears exactly once in T'.

Given an instance of 3DM, we construct a game G with $|T| + 9n$ machines and $12n$ jobs. The first $|T|$ machines, denoted triplet-machines, correspond to the 3DM triplets. The additional $9n$ machines form $3n$ copies of machines m_1, m_2, m_3 introduced in the instance I_{noNE} in the proof of Theorem 3. Each copy is associated with one element of $X_1 \cup X_2 \cup X_3$.

For each element in $X_1 \cup X_2 \cup X_3$, there are four jobs. The first job corresponds to the element itself, and three additional jobs are copies of jobs a, b and c from I_{noNE}.

The main idea is that a 3DM corresponds to a schedule in which the element-jobs are assigned in triplets to triplet-machines – each paying one third of a triplet-machine cost. On the other hand, if a 3DM does not exist, then every unmatched element pays at least half of a triplet-machine cost, and prefers migrating to a corresponding copy of m_1, generating an instance of I_{noNE} as a sub-instance that has no stable-assignment. Thus, a 3DM matching exists if and only if G has a PNE schedule. □

Remark 1. An interesting open question along the lines of [8,17] is to suggest a different cost-sharing mechanism, whose application induces a potential game. It is not possible to adopt the approach suggested in [8] for weighted cost-sharing games, since in our game the Shapley values of the players are not well defined.

3.1 Machine-Independent Processing Times

We show that when the processing times are machine independent, a PNE is guaranteed to exist, can be found efficiently, and BRD converges from any initial configuration. The BRD convergence proof builds on the proof for weighted symmetric cost-sharing games [2]. Our game is different since in the setting of [2], all machines are feasible to all jobs, that is, for all i, we have $M_i = \mathcal{M}$. An efficient algorithm for calculating a Strong NE for machine-independent processing times can be derived by generalizing the algorithm in [15] for fair cost sharing (unweighted jobs).

Theorem 5. *If the processing times are independent of the machines, then a PNE can be found efficiently and BRD converges to a PNE.*

Remark 2. A different restricted class of instances assumes job-independent processing times. That is, for every machine j there exists a $p_j > 0$ be such that for all jobs for which $j \in M_i$, we have $p_{j,i} = p_j$. Since the cost of a machine is shared evenly by the jobs assigned to it, a PNE can be computed in polynomial time by the general algorithm for finding a PNE in fair cost-sharing games with singleton strategies [15]. Moreover, $\Phi(P) = \sum_{j \in \mathcal{M}} c(j) \cdot H(L_j(P)/p_j)$, where $H(0) = 0$, and $H(k) = 1 + 1/2 + \ldots + 1/k$, is a potential function whose value reduces with every improving step of a player.

3.2 Equilibrium Inefficiency

We show that stability might lead to an extremely inefficient outcome with respect to the total players' cost. Similar to classic congestion games, the PoA equals the number of players. On the other hand, the PoS might also be linear in the number of players (compared to O(log k) in classical cost-sharing games). Specifically,

Theorem 6. *The PoA of CSSGs equals the number of players.*

Theorem 7. *CSSGs with $m > 3$ machines and $k < m$ players have PoS $= k$.*

Proof. Since PoS $\leq$ PoA, Theorem 6 implies that PoS $\leq k$. For the lower bound, consider the following game in which the unique PNE has cost $k - \varepsilon'$ while the social optimum has cost 1. The jobs have lengths $1, \varepsilon, \varepsilon^2, \ldots, \varepsilon^{k-1}$, independent of the machine they are assigned to. Assume that a single machine, having cost 1 is feasible to all jobs. There are $k-1$ additional machines each having cost $\frac{1-\varepsilon}{1+\varepsilon}$. Each of these machines is feasible to a single job among the $k-1$ longer jobs. The unique PNE is when each machine accommodates a single job. If two or more jobs are assigned together to the first machine, then the longer will escape to its dedicated machine. The PNE's cost is $k - \varepsilon'$, thus PoA $=$ PoS $= k$, and we are done. □

For some special cases of CSSGs the PoS can be bounded as follows.

Theorem 8. *CSSGs with $m \in \{2, 3\}$ machines have PoS $= m$. CSSGs with k players and $m < k$ machines have PoS $= \Theta(k)$.*

4 Coordinated Deviations

Recall that a strong equilibrium (SE) is a configuration in which no *coalition* of players can deviate in a way that benefits all its members. We show that for machine-independent processing times, a SE is guaranteed to exist, and we present a poly-time algorithm to find one. We also prove that SPoS $=$ SPoA $= \frac{m}{2}$. On the other hand, we show that a SE may not exist when jobs have arbitrary processing times. In fact, even with unit-cost machines and if just a single job is allowed to have two variable processing times, there exists an instance, with $m = 3$ machines and $k = 5$ jobs that has no SE. The inefficiency of the general case decreases; we show that SPoS $= \frac{m}{2}$ and SPoA $= m$.

We start with the simpler class of machine-independent processing times. Recall that for every job i there is $p_i > 0$ such that $p_{j,i} = p_i$ for all $j \in M_i$. We show that any sequence of beneficial coordinated deviations converges to a SE. Moreover, a simple greedy algorithm for finding a SE exist (omitted from this extended abstract) – even for instances with arbitrary cost machines.

Theorem 9. *For any instance with unit-cost machines and machine-independent processing times, any sequence of beneficial coordinated deviations converges to a SE.*

We turn to study the inefficiency of a strong equilibrium. We show that even with machine-independent processing times, an optimal solution may be significantly better than any stable one. Thus, in systems where coordinated deviations are allowed, we may end-up with an extremely poor outcome.

Theorem 10. *CSSGs with unit-costs machines and machine-independent processing times have SPoS = SPoA* $= \min\{\frac{m}{2}, \frac{k}{4} + \frac{1}{2}\}$.

Proof. We first show the bounds w.r.t. the number of machines. We show that SPoS $\geq m/2$ and SPoA $\leq m/2$. The statement will follow since SPoS $\leq$ SPoA. We start with the upper bound and show SPoA $\leq m/2$. For any profile P it clearly holds that $cost(P) \leq m$. If the SO assigns the jobs on two or more machines, then SPoA $\leq m/2$ as required. Assume that the SO assigns all the jobs on a single machine m_0. We show that only one machine is active in any SE, implying that in this case, SPoA $= 1$. Consider any profile P with more than a single active machine. We claim that all the jobs assigned on $\mathcal{M} \setminus \{m_0\}$ form a coalition whose beneficial move is to join m_0. By the assumption, this is a valid migration. Let S_i be the set of jobs assigned in P to an active machine M_i. In P, their total cost is 1. After the deviation, their total cost is strictly less than 1 and since the relative cost of every job in S_i remains the same, all the coalition members benefit from the deviation. We conclude that if $SO = 1$ then any SE has cost 1, and if $SO \geq 2$ then the $m/2$-ratio clearly holds, thus, SPoA $\leq m/2$.

For the lower bound, we describe an instance with unit-cost machines achieving SPoS $= m/2$. An example for $m = 5$ is given in Fig. 2. Given m, there are $n = 2(m-1)$ jobs consisting of $m-1$ pairs, $a_1, b_1, \ldots, a_{m-1}, b_{m-1}$. Let $\mathcal{M} = \{m_0, \ldots, m_{m-1}\}$. For $1 \leq k \leq m-1$, the processing time of jobs a_k and b_k is 2^k. Job a_1 is restricted to machine m_0, Job b_1 is restricted to machine m_1. For $2 \leq k \leq m-1$, Job a_k is restricted to m_0 or m_k, and Job b_k is restricted to m_1 or m_k.

The SO assigns all the jobs $\{a_k\}$ on machine m_0, and all the jobs $\{b_k\}$ on machine m_1. This optimal profile is not an SE. Note that since $2^k > \sum_{i=1}^{k-1} 2^i$, each of a_{m-1} and b_{m-1} has cost more than 1/2. This pair would benefit from migrating to machine m_{m-1}, where each will have cost exactly 1/2. After this deviation, by the same argument, each of a_{m-2} and b_{m-2} has cost more than 1/2.

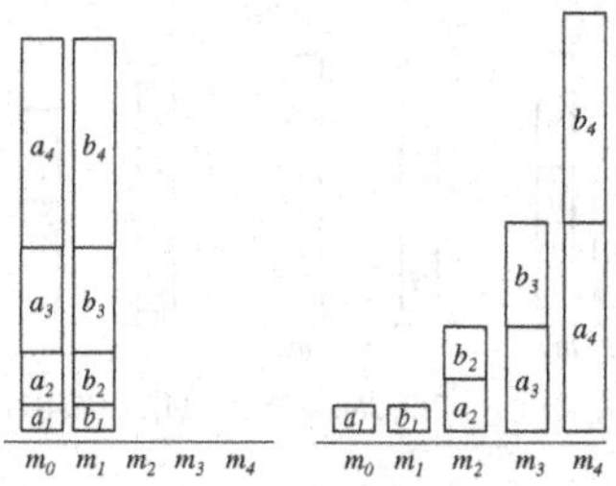

Fig. 2. The social optimum (left) and the only SE (right) of a unit-cost instance achieving SPoS $= m/2$.

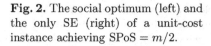

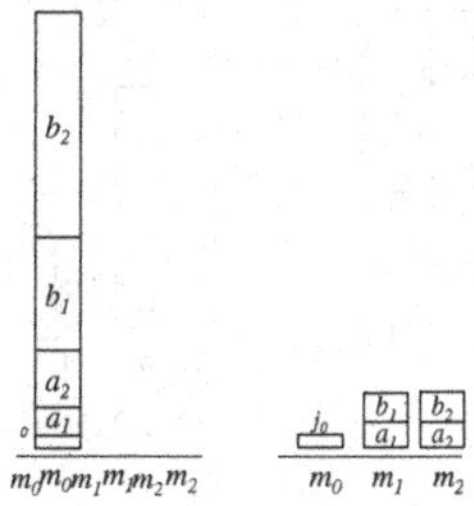

Fig. 3. The social optimum (left) and the worst SE (right) of a unit-cost instance achieving SPoA $= m$.

This pair would benefit from migrating to machine m_{m-2}. Next, in turn, every pair will deviate to a new machine, resulting in the only SE of this instance in which Job a_1 is alone on m_0, Job b_1 is alone on m_1, and for every $2 \leq k \leq m-1$, the pair of jobs a_k and b_k is on m_k. There are m active machines in this SE, while only two machines are active in the SO.

We proceed to prove the bounds w.r.t. the number of players. First, we show that SPoA $\leq \frac{k}{4} + \frac{1}{2}$. We start with the following claim. Assume the SO uses x machines, where $x > 1$ as otherwise the analysis above shows SPoA$= 1$. We claim that in any SE, the number of machines that accommodate a single job is at most x. Otherwise, there is a SE P in which at least $x + 1$ machines accommodate a single job each. Then, there are (at least) two jobs who share the same machine in the SO and use a machine by themselves in P. These two jobs can deviate to their machine in the SO and decrease their cost from 1 in P to less than 1, contradicting the fact that P is a SE. A corollary of the claim is that any SE costs at most $x + \frac{k-x}{2}$. Thus, SPoA $= \frac{x+(k-x)/2}{x} = 1 + \frac{k}{2x} - \frac{1}{2}$, which gets the maximal value of $\frac{k}{4} + \frac{1}{2}$ when $x = 2$. The lower bound is identical to the one described above: the SO costs 2 and the only SE costs $\frac{k-2}{2} + 2 = \frac{k}{2} + \frac{1}{2}$, thus SPoS $\geq \frac{k}{4} + \frac{1}{2}$. □

While a SE is guaranteed to exist for any instance with machine-independent processing times, we show that even the slightest relaxation in this condition may result in an instance with no SE. Specifically, in Fig. 4, we present an instance with unit-cost machines that has no SE. Note that all jobs except for a single one have machine-independent processing times.

Theorem 11. *There is an instance with $m = 3$ unit-cost machines and $k = 5$ jobs that has no SE.*

For instances with arbitrary processing times, a SE is not guaranteed to exist. For instances having a SE, the bounds on the equilibrium inefficiency depend on the processing environment: For instances with arbitrary activation costs, the analysis in Theorem 7 is valid also for coordinated deviations. Thus, SPoA $=$ SPoS $= k$. For instances with unit-cost machines, we prove the following.

	m_1	m_2	m_3
a	2	--	2
b	6	4	--
c	--	4	4
d	2	--	--
e	--	--	3

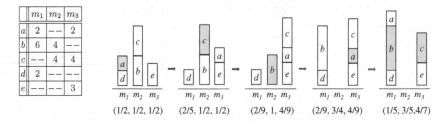

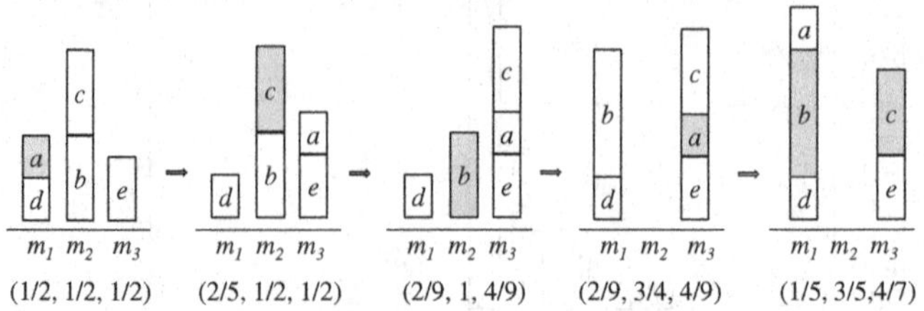

Fig. 4. An instance that has no SE. The table on the left gives the processing times. The payment vector of jobs a, b and c is given below each configuration. The coalition that has a beneficial move is darker and its members deviate to the next configuration. The leftmost configuration follows the rightmost one, creating a loop. It is easy to see that the three other configurations, in which either b or c are alone on m_2, are not stable either.

Theorem 12. *CSSGs with unit-cost machines and arbitrary processing times have SPoS* $= \min\{\frac{m}{2}, \frac{k}{4} + \frac{1}{2}\}$ *and SPoA* $= \min\{m, \frac{k}{2} + \frac{1}{2}\}$.

Proof. We show the SPoA lower bound w.r.t the number of machines. The rest of the proof is omitted. We describe an instance with unit-cost machines achieving SPoA $= m$. An example for $m = 3$ is given in Fig. 3. Given m, let $\mathcal{M} = \{m_0, ...m_{m-1}\}$. There are $n = 2m - 1$ jobs consisting on $m - 1$ pairs, $a_1, b_1, \ldots, a_{m-1}, b_{m-1}$, and a job j_0, which is restricted to go to machine m_0. The processing time of j_0 on m_0 is $\varepsilon < 1$. For $1 \leq k \leq m-1$, jobs a_k and b_k are restricted to go to m_0 or m_k. The processing time of either a_k or b_k on m_k is 1. The processing times of the $2(m-1)$ jobs $\{a_k, b_k\}$ on m_0 are arbitrary *distinct* powers of 2.

The SO assigns all the jobs on m_0. It is easy to verify that the SO is a SE. Only one job (whose processing time is the highest power of 2) pays more than half. This job will not benefit from migrating by itself, and no job would join it and pay at least half after the deviation. However, the SO is not the only SE. Consider the profile P' in which j_0 is on m_0 and for $1 \leq k \leq m-1$, jobs a_k and b_k are on m_k. The cost of P' is m, where j_0 has cost 1 and each of the other jobs has cost $1/2$. We claim that P' is a SE. The only possible deviation is into m_0. However, since the processing times on m_0 are distinct powers of 2, some job will cause more than half of the load on m_0, resulting in cost more than $1/2$. Since all the coalition members have cost $1/2$ in P', the deviation is not beneficial for this job. We conclude that a SE whose cost is m exists and SPoA $\geq m$. □

References

1. Andelman, N., Feldman, M., Mansour, Y.: Strong price of anarchy. Games Econ. Behav. **65**(2), 289–317 (2009)
2. Anshelevich, E., Dasgupta, A., Kleinberg, J., Tardos, E., Wexler, T., Roughgarden, T.: The price of stability for network design with fair cost allocation. SIAM J. Comput. **38**(4), 1602–1623 (2008)

3. Aumann, R.: Acceptable points in games of perfect information. Contrib. Theor. Games **4**, 287–324 (1959)
4. Avni, G., Kupferman, O., Tamir, T.: Network-formation games with regular objectives. In: Muscholl, A. (ed.) FOSSACS 2014 (ETAPS). LNCS, vol. 8412, pp. 119–133. Springer, Heidelberg (2014)
5. Chen, H., Roughgarden, T.: Network design with weighted players. Theor. Comput. Syst. **45**(2), 302–324 (2009)
6. Epstein, A., Feldman, M., Mansour, Y.: Strong equilibrium in cost sharing connection games. Games Econ. Behav. **67**(1), 51–68 (2009)
7. Karp, R.M.: Reducibility among combinatorial problems. In: Miller, R.E., Thatcher, J.W., Bohlinger, J.D. (eds.) Complexity of Computer Computations, pp. 85–103. Springer, USA (1972)
8. Kollias, K., Roughgarden, T.: Restoring pure equilibria to weighted congestion games. In: Aceto, L., Henzinger, M., Sgall, J. (eds.) ICALP 2011, Part II. LNCS, vol. 6756, pp. 539–551. Springer, Heidelberg (2011)
9. Koutsoupias, E., Papadimitriou, C.: Worst-case equilibria. Comput. Sci. Rev. **3**(2), 65–69 (2009)
10. Milchtaich, I.: Weighted congestion games with separable preferences. Games Econ. Behav. **67**(2), 750–757 (2009)
11. Papadimitriou, C.H.: Algorithms, games, and the internet. In: Proceedings of the 33rd STOC, pp. 749–753 (2001)
12. Pinedo, M.L.: Scheduling: Theory, Algorithms, and Systems. Springer, New York (2008)
13. Rosenthal, R.W.: A class of games possessing pure-strategy Nash equilibria. Int. J. Game Theor. **2**, 65–67 (1973)
14. Schulz, A.S., Stier, N.E.: Moses On the performance of user equilibria in traffic networks. In: Proceedings of the 14th SODA, pp. 86–87 (2003)
15. Syrgkanis, V.: The complexity of equilibria in cost sharing games. In: Saberi, A. (ed.) WINE 2010. LNCS, vol. 6484, pp. 366–377. Springer, Heidelberg (2010)
16. Vöcking, B.: Algorithmic Game Theory, Chapter 20: Selfish Load Balancing. Cambridge University Press, New York (2007)
17. von Falkenhausen, P., Harks, T.: Optimal cost sharing protocols for scheduling games. In: Proceedings of the 12th EC, pp. 285–294 (2011)

Mechanism Design and Social Choice

Efficient Money Burning in General Domains

Dimitris Fotakis[1], Dimitris Tsipras[1(✉)], Christos Tzamos[2], and Emmanouil Zampetakis[2]

[1] National Technical University of Athens, 157 80 Athens, Greece
fotakis@cs.ntua.gr, dtsip@corelab.ntua.gr
[2] Massachusetts Institute of Technology, Cambridge, MA 02139, USA
{tzamos,mzampet}@mit.edu

Abstract. We study mechanism design where the payments charged to the agents are not in the form of monetary transfers, but are effectively burned. In this setting, the objective is to maximize *social utility*, i.e., the social welfare minus the payments charged. We consider a general setting with m discrete outcomes and n multidimensional agents. We present two essentially orthogonal randomized truthful mechanisms that extract an $O(\log m)$ fraction of the maximum welfare as social utility. Moreover, the first mechanism achieves a $O(\log m)$-approximation for the social welfare, which is improved to an $O(1)$-approximation by the second mechanism. An interesting feature of the second mechanism is that it optimizes over an appropriately "smoothed" space, thus achieving a nice and smooth tradeoff between welfare approximation and the payments charged.

1 Introduction

The extensive use of monetary transfers in the Algorithmic Game Theory is due to the fact that so little can be implemented truthfully in their absence (see e.g., [14]). On the other hand, if monetary transfers are available (and acceptable for the particular application), the famous Vickrey-Clarke-Groves (VCG) mechanism (see e.g., [14]) succeeds in truthfully maximizing the *social welfare*, i.e., the total value generated for the agents, albeit with possible very large monetary transfers from the agents to the center. This is acceptable as long as the payments generate revenue for the center (e.g., the government for public good allocation or the auctioneer for allocation of private goods), since the funds are not lost, but are transferred to the center. Then, the funds could be redistributed among the agents (see e.g., [9,10]) or invested in favor of the society.

However, there are settings where the payments required for truthful implementation take the form of wasted resources, a.k.a. *money burning*, instead of

This research was supported by the project AlgoNow, co-financed by the European Union (European Social Fund - ESF) and Greek national funds, through the Operational Program "Education and Lifelong Learning" of the National Strategic Reference Framework (NSRF) - Research Funding Program: THALES, investing in knowledge society through the European Social Fund.

M. Hoefer (Ed.): SAGT 2015, LNCS 9347, pp. 85–97, 2015.
DOI: 10.1007/978-3-662-48433-3_7

actual monetary transfers. One could think of "computational" challenges (e.g., captcha), waiting times (e.g., waiting lists in hospitals [2] or in popular events or places), or reduction in service quality (see also [4,11] for more examples). In these settings, the natural objective is to maximize the net gain of the society, measured by the social welfare minus the payments charged and usually referred to as the *social utility* (or the *social surplus*).

In the AGT community, the general idea of money burning and social utility maximization by truthful mechanisms was first considered by Hartline and Roughgarden [11]. They considered single-unit and k-unit (unit demand) auctions and presented a family of truthful prior-free mechanisms that guarantee at least a constant fraction of the optimal (wrt. the social utility) Bayesian mechanism. Their mechanisms randomize among a VCG auction and a randomized posted price mechanism. To show that these mechanisms achieve an $O(1)$-approximation to the social utility extracted by an optimal Bayesian mechanism with knowledge of the agents' distribution (under the i.i.d. assumption), Hartline and Roughgarden used Myerson's theorem and characterized the optimal Bayesian mechanism for single-parameter agents. They also proved that if we compare the social utility of a truthful mechanism to the maximum social welfare, then the best possible approximation guarantee for k-unit auctions is $\Theta(1 + \log \frac{n}{k})$, where n is the number of agents.

Contribution. In this work, we consider social utility maximization by truthful mechanisms in a general mechanism design setting with m discrete possible outcomes and multidimensional agents with positive valuations. Due to the fact that social utility maximization is closely related to revenue maximization, coming up with a characterization of the optimal (wrt. the social utility) truthful Bayesian mechanism, as in [11], is a daunting task and far beyond the scope of this work. Instead, we evaluate the performance of our mechanisms by comparing their social utility to the maximum social welfare (achievable by an optimal algorithm that does not need to be truthful). In fact, we seek for mechanisms that achieve nontrivial approximation guarantees wrt. both social utility and social welfare. Our main contribution is two randomized truthful mechanisms, based on essentially orthogonal approaches, that approximate social utility within a best possible factor of $O(\log m)$, thus extending the last result of [11] to our general mechanism design setting.

Probably the simplest candidate mechanisms for utility maximization are the random allocation, where each outcome is implemented with probability $1/m$, and the VCG mechanism. Clearly, the approximation ratio of random allocation for both the social utility and the social welfare is m, while VCG cannot approximate within a ratio of m even for the natural case of uniform i.i.d. bidders. A natural way to approximate social utility is through a careful tradeoff between VCG, which optimizes welfare but may result in poor utility due to high payments, and random allocation on appropriate sets of outcomes, which is truthful without payments and thus, translates all welfare into utility.

Exploiting this intuition and building on the mechanism of [11, Theorem 5.2], we present a randomized truthful mechanism that approximates both the social utility and the social welfare within a factor of $O(\log m)$. The idea is to select a random integer j from 0 to $\log m$, and then, select a random outcome i among the best (in total value) 2^j outcomes, and apply VCG payments. The key step in establishing the approximation guarantee is to show that in terms of utility maximization, the worst-case instances correspond to single item auctions. Then, the upper bound of [11, Theorem 5.2] carries over to our more general setting. Moreover, since the single item auction is a special case of our setting, the lower bound of [11, Proposition 5.1] implies that our approximation ratio is aymptotically tight.

Our second mechanism optimizes the social welfare (using VCG) over a carefully defined subspace of the unit simplex with all probability distributions over the outcomes. Intuitively, if we optimized over the unit simplex, we would have optimal welfare but probably poor utility, due to high payments when the two best outcomes are close in total value. So, we define a subspace that is slightly curved close to the vertices of the unit simplex, thus achieving a significant reduction in the payments if the best outcomes are close in total value. Due to this fact, this mechanism is *partial*, in the sense that with probability $1-\epsilon$ it may not implement any outcome (see [5] for another use of partial allocation to induce truthfulness). For any $\epsilon > 0$, the approximation ratio is $1+\epsilon$ for the social welfare and $O(\epsilon^{-1} \log m)$ for the social utility. Hence, this mechanism achieves a best possible approximation ratio for the social utility and a constant approximation for the social welfare, thus significantly improving on our first mechanism. The main idea behind this mechanism is to "smoothen" the solution space so that we achieve a smooth tradeoff between welfare approximation and the payments charged, where for mechanisms close to the optimal, payments are reduced significantly faster that social welfare. On the technical side, this mechanism bears a resemblance to proper scoring rules in [8]. We believe that such mechanisms, which are based on carefully chosen "smoothed" subspaces and provide smooth tradeoffs between approximation and payments, are of independent interest and may find other applications in mechanism design settings with restricted payments.

Our mechanisms run in time polynomial in the total number of outcomes m and in the number of agents n. In domains that allow for succinct input representation (e.g., Combinatorial Auctions, Combinatorial Public Projects), m is usually exponential in the size of the input. This is not surprising, since our approximation guarantees are significantly better than known lower bounds on the polynomial time approximability of several **NP**-hard problems. In certain domains, we can combine our mechanisms with existing Maximal-in-Range mechanisms so that everything runs in polynomial time (e.g., for subadditive Combinatorial Public Projects, we can use the Maximal-in-Range mechanism of [15, Sect. 3.2] and obtain a randomized polynomial-time truthful mechanism that with $O(\min\{k, \sqrt{u}\})$-approximation for the social welfare and $O(\min\{k, \sqrt{u}\} \log u)$-approximation for the social utility, where u is the number of items and k is the size of the project).

Related Work. There is much work on (mostly polynomial-time) truthful mechanisms with monetary transfers that seek to maximize (exactly or approximately) the social welfare. In this general agenda, our work is closest in spirit to mechanisms with frugal payments (see e.g., [1,6]). In addition to [11], Chakravarty and Kaplan [4] characterized the Bayesian mechanism of maximum social utility in multi-unit (unit demand) auctions. More recently, Braverman et al. [2] considered utility optimization in health care service allocation, but they focused on the complexity of computing efficient equilibrium allocations, instead of approximate truthful mechanisms.

An orthogonal direction is that of revenue redistribution (see e.g., [3,9,10] and the references therein). Although most of the literature focuses on maximizing the amount of redistributed VCG payments, some positive results in this direction concern social utility optimization relaxing the requirement for social welfare maximization (see e.g., [10]). Our viewpoint and results are incomparable, both technically and conceptually, to those in the area of redistribution mechanisms. A crucial difference is that in any efficient redistribution mechanism, certain agents should receive payments (this is unavoidable if one insists on efficiency and individual rationality, see e.g., [12]). This is infeasible our setting, where money-burning payment schemes (e.g., computational challenges or waiting time) make redistribution infeasible.

2 Preliminaries and Notation

For any integer m, $[m] \equiv \{1, \dots, m\}$. We denote the j-th coordinate of a vector $\boldsymbol{x}$ by x_j. For a vector $\boldsymbol{x} = (x_1, \dots, x_m)$ and $i \in [m]$, $\boldsymbol{x}_{-i}$ is $\boldsymbol{x}$ without coordinate i. For a vector $\boldsymbol{x} \in \mathbb{R}^m$ and some $\ell \geq 0$, $\boldsymbol{x}^\ell = (x_1^\ell, \dots, x_m^\ell)$ is the coordinate-wise power of $\boldsymbol{x}$ and $\|\boldsymbol{x}\|_\ell = (\sum_{j=1}^m x_j^\ell)^{1/\ell}$ is the ℓ-norm of $\boldsymbol{x}$. For convenience, we let $\|\boldsymbol{x}\|_1 = |\boldsymbol{x}|$. Moreover, $\|\boldsymbol{x}\|_\infty = \max_{j \in [m]}\{x_j\}$ is the infinity norm of $\boldsymbol{x}$.

The Setting. There is a finite set of possible outcomes O and we denote $|O| = m$. We consider a set of n strategic agents, each with a *private, non-negative value* for each outcome. For agent i, we denote his *valuation* as a vector $\boldsymbol{x}_i \in \mathbb{R}_+^m$, that is, agent i receives value x_{ij} for outcome j. We call the vector of all valuations $\boldsymbol{x} = (\boldsymbol{x}_1, \dots, \boldsymbol{x}_n)$ a *valuation profile.* For a valuation profile $\boldsymbol{x}$, $\boldsymbol{w}(\boldsymbol{x}) = \boldsymbol{x}_1 + \ldots + \boldsymbol{x}_n$ is the vector of *weights* for the outcomes. We will write $\boldsymbol{w}$ instead of $\boldsymbol{w}(\boldsymbol{x})$ and $\boldsymbol{w}_{-i}$ instead of $\boldsymbol{w}(\boldsymbol{x}_{-i})$ when $\boldsymbol{x}$ is clear from the context.

Allocation Rules and Mechanisms. For a finite set S, $\Delta(S)$ denotes the unit simplex over S. A (randomized) allocation rule is a function $f : (\mathbb{R}_+^m)^n \to \Delta(O)$, mapping valuation profiles to probability distributions over outcomes. Then $f_j(\boldsymbol{x})$ is the probability of outcome j on valuation profile $\boldsymbol{x}$. It follows that the expected value of agent i is $\boldsymbol{x}_i \cdot f(\boldsymbol{x})$. We consider allocation rules that are *strongly anonymous,* in the sense that $f(\boldsymbol{x})$ depends only on $\boldsymbol{w}(\boldsymbol{x})$, and we therefore write the allocation rule only in terms of the weight vector.

A *payment rule* is a function $p : (\mathbb{R}_+^m)^n \to \mathbb{R}^n$ mapping valuation profiles to *payment vectors.* A mechanism is a pair $\mathcal{M} = (f, p)$ that given some valuation profile $\boldsymbol{x}$ outputs the probability distribution $f(\boldsymbol{x})$ and charges agent i the amount $p_i(\boldsymbol{x})$. We focus on symmetric payment rules, and we therefore represent the amount charged to agent i as $p(\boldsymbol{x}_{-i}, \boldsymbol{x}_i)$. The expected *utility* of agent i on valuation profile $\boldsymbol{x}$ under mechanism $\mathcal{M} = (f, p)$ is

$$\boldsymbol{x}_i \cdot f(\boldsymbol{x}) - p(\boldsymbol{x}_{-i}, \boldsymbol{x}_i)$$

and is the amount he aims to maximize.

We require that our mechanisms are truthful and individually rational in expectation. A mechanism (f, p) is *truthful* (in expectation) if for any agent i, valuation profile $\boldsymbol{x}$ and valuation $\boldsymbol{x}'_i$,

$$\boldsymbol{x}_i \cdot f(\boldsymbol{x}) - p(\boldsymbol{x}_{-i}, \boldsymbol{x}_i) \geq \boldsymbol{x}_i \cdot f(\boldsymbol{x}_{-i}, \boldsymbol{x}'_i) - p(\boldsymbol{x}_{-i}, \boldsymbol{x}'_i)$$

and *individually rational (IR)* if for any agent i and valuation profile $\boldsymbol{x}$,

$$\boldsymbol{x}_i \cdot f(\boldsymbol{x}) - p(\boldsymbol{x}_{-i}, \boldsymbol{x}_i) \geq 0$$

Objectives and Approximation. Let some mechanism $\mathcal{M} = (f, p)$ and valuation profile $\boldsymbol{x}$. We denote the total payments of $\mathcal{M}$ on input $\boldsymbol{x}$ by $P[\boldsymbol{x}] = \sum_i p(\boldsymbol{x}_{-i}, \boldsymbol{x}_i)$. The quantities we are interested in maximizing are the social welfare and the social utility. The *social welfare* of $\mathcal{M}$ on $\boldsymbol{x}$ is $SW[\boldsymbol{x}] = \sum_i \boldsymbol{x}_i \cdot f(\boldsymbol{x}) = \boldsymbol{w} \cdot f(\boldsymbol{x})$ and the *social utility* of $\mathcal{M}$ on $\boldsymbol{x}$ is $U[\boldsymbol{x}] = SW[\boldsymbol{x}] - P[\boldsymbol{x}]$. The maximum possible social utility and social welfare of the mechanism (ignoring truthfulness constraints) on input $\boldsymbol{x}$ is $\|\boldsymbol{w}\|_\infty$. We say that mechanism $\mathcal{M}$, ρ-approximates social welfare (resp. social utility) if for any input $\boldsymbol{x}$, $SW[\boldsymbol{x}] \geq \frac{1}{\rho}\|\boldsymbol{w}(\boldsymbol{x})\|_\infty$ (resp. $U[\boldsymbol{x}] \geq \frac{1}{\rho}\|\boldsymbol{w}(\boldsymbol{x})\|_\infty$). For a mechanism $\mathcal{M}$ that ρ_1-approximates social welfare and ρ_2-approximates social utility, we say that it approximates *social efficiency* within (ρ_1, ρ_2).

Implementable Rules. For every set $S \subseteq R^m_+$, the mechanism $\mathcal{M} = (f, p)$ such that $f(\boldsymbol{x}) = \arg\max_{\boldsymbol{s} \in S} \boldsymbol{s} \cdot \boldsymbol{w}$ and $p(\boldsymbol{x}_{-i}, \boldsymbol{x}_i) = \boldsymbol{w}_{-i} \cdot f(\boldsymbol{x}_{-i}) - \boldsymbol{w}_{-i} \cdot f(\boldsymbol{x})$ is truthful and individually rational. This follows directly from the analysis of the VCG mechanism [14]. We refer to such mechanisms as *Maximal in Distributional Range* (MIDR) and to the corresponding payment rule as the *VCG payment scheme.*

3 Best-Possible Guarantees for Social Utility

In contrast to social welfare maximization, where monetary transfers can be used freely to truthfully elicit the agents' preferences, in the case of social utility maximization, the transfers needed for the implementation of some mechanisms may be a significant part of the social welfare, thus prohibiting any non-trivial approximation guarantees.

Since the model we consider is so rich, the single item auction is a special case of it, when we restrict the domain to m outcomes and m agents, where agent i has a value $v_i \geq 0$ for outcome i and zero for the rest. By proving lower bounds to the approximation of social utility maximization in this special case, we get the same lower bounds for the general model. Our main tool here is Myerson's characterization of the revenue of any truthful auction in the single parameter environment.

Theorem 1 (Myerson [13]). *For any truthful mechanism* $\mathcal{M} = (f, p)$ *and valuation profile* $\boldsymbol{x}$*, where agent* i *has some value* $v_i \geq 0$ *only for outcome* i *and* v_i *is drawn independently from distribution* $\mathcal{F}$ *with cumulative distribution function* $F_{\mathcal{F}}(v)$ *and probability density function* $f_{\mathcal{F}}(v)$*,* $\mathbb{E}[P(\boldsymbol{x})] = \mathbb{E}[\boldsymbol{\phi} \cdot f(\boldsymbol{x})]$*, where* $\phi_i = v_i - \frac{1-F_{\mathcal{F}}(v_i)}{f_{\mathcal{F}}(v_i)}$*.*

Theorem (1) completely deftermines the expected amount of payments for any truthful allocation rule. This in turn determines the expected utility in terms of the allocation rule. By plugging in an appropriate distribution we can come up with lower bounds to the social utility of truthful mechanisms.

Corollary 1. *The Vickrey Auction when bidders are drawn i.i.d. from the uniform distribution* $\mathcal{U}(0, 1)$*, cannot approximate social utility within a factor better than* m*.*

This shows that the VCG mechanism for the natural case of uniform i.i.d. bidders performs no better that a random allocation. By aiming to maximize the social welfare, it has to charge every bidder his critical price which results to a high amount of payments, negating the welfare it produces. We therefore need to come up with mechanisms that instead of maximizing social welfare, employ suboptimal allocations to reduce payments, while preserving some amount of welfare. Our goal is to achieve the best possible worst-case guarantee for social utility maximization. A lower bound on the best approximation ratio in our setting can be obtained from [11, Proposition 5.1], which we prove here for completeness.

Corollary 2 ([11]). *No truthful mechanism can approximate social utility within a factor of* $o(\log m)$*.*

Proof. If agents are drawn from the exponential distribution, that is $f_{\mathcal{E}}(x) = e^{-x}$, $F_{\mathcal{E}}(x) = 1 - e^{-x}$, then $\phi_i = v_i - 1$ and by applying Theorem (1) we get that

$$\mathbb{E}[P(\boldsymbol{x})] = \mathbb{E}[\boldsymbol{w} \cdot f(\boldsymbol{x}) - |f(\boldsymbol{x})|] = \mathbb{E}[SW[\boldsymbol{x}] - |f(\boldsymbol{x})|]$$

and by linearity of expectation $\mathbb{E}[U[\boldsymbol{x}]] \leq 1$ It is straightforward to show that the expectation of the maximum of m i.i.d. exponential random variables equals H_m where H_m the m-th harmonic number. Then

$$\mathbb{E}[U(\boldsymbol{x})] \leq \mathbb{E}\left[\frac{\|\boldsymbol{w}\|_\infty}{H_m}\right]$$

and by the probabilistic method we get that there is some profile $\boldsymbol{x}$ for which the approximation ratio is logarithmic. □

We will now describe a mechanism that matches this lower bound in the general domain.

Definition 1. *For some* $k \in [m]$, *the* Top_k *allocation rule on input* $\boldsymbol{x}$, *orders outcomes in decreasing weight order,* $w_1 \geq \ldots \geq w_m$ *(breaking ties arbitrarily) and assigns probability* $\frac{1}{k}$ *to the first* k. *Formally,* $\text{Top}_k(\boldsymbol{x}) = \arg\max_{\boldsymbol{s} \in S_k} \boldsymbol{s} \cdot \boldsymbol{w}$, *where* S_k *is the set of vectors in* $\mathbb{R}_+^m$ *with exactly* k *coordinates equal to* $\frac{1}{k}$ *and* $m - k$ *equal to* 0.

Since Top_k are welfare maximizers, they can be turned into truthful and IR mechanisms with the VCG payment scheme. We denote mechanisms of this family by $\mathcal{M}_k = (\text{Top}_k, p_k)$. Each of these mechanism achieves different approximation guarantees with respect to social welfare and social utility in different settings, with respect to k. Thus by randomizing over them we can provide worst-case guarantees. In Mechanism 1 we achieve such an optimal social utility approximation guarantee by randomizing over exponentially increasing values of k. For simplicity we assume that m is a power of 2.

Mechanism 1. A $\log m$-approximate mechanism for Social Utility

Choose j uniformly at random from $\{0, 1, 2, \ldots, \log m\}$
Let $k \leftarrow 2^j$
Output the probability distribution $\text{Top}_k(\boldsymbol{x})$ over outcomes
Charge agent i the amount $\boldsymbol{w}_{-i} \cdot \text{Top}_k(\boldsymbol{x}_{-i}) - \boldsymbol{w}_{-i} \cdot \text{Top}_k(\boldsymbol{x})$

The complete mechanism is randomization over $\mathcal{M}_k$ for some k independent of the input. As a result, Mechanism 1 is truthful and IR as a whole. In order to quantify the efficiency of the mechanism in terms of utility maximization, we first show that the worst-case instances are those of the single item auction, that is for each outcome i there is exactly one single-minded agent with valuation v_i for it (a bidder i is called single-minded if his utility is $v_i \geq 0$ for some outcome $j \in [m]$ and zero for the rest).

Lemma 1. *For any valuation profile* $\boldsymbol{x} = (\boldsymbol{x}_1, \ldots, \boldsymbol{x}_n)$, *the utility of Mechanism 1 on* $\boldsymbol{x}$ *is higher than the utility on the valuation profile* $\boldsymbol{y} = (\boldsymbol{y}_1, \ldots, \boldsymbol{y}_m)$, *where* $\boldsymbol{y}_i$ *is a single-minded agent for outcome* i ($y_{ij} = 0$ *for any* $i \neq j$).

Proof (sketch). Since the complete mechanism is a randomization over mechanisms $\mathcal{M}_k$ is suffices to show this property for each $\mathcal{M}_k$ separately. We prove the claim in two steps:

- First we show that if an agent has positive value for multiple outcomes, splitting this agent into single-minded agents (one for each outcome) can only decrease the total utility. This holds since the "competition" between agents is increased, and as a result, so do the payments, thus decreasing the total utility (the social welfare remains unaltered since the mechanism depends only on the weight of each outcome). By induction we transform any input to one with single-minded agents without increasing the utility.

- Then we show that if there are multiple single-minded bidders for the same outcome, joining their values into a single agent can only decrease the total utility. The reason for this is that the value the agents must "prove" (in the form of payments) to the mechanism is initially split amongst them, and can only increase as they aggregate their values. A single agent with high value is more critical for the auction than many agents with small values. Again by induction we can transform any input with single-minded agents to an input with one single-minded agent per outcome.

The technical details can be found in the full version of the paper. □

Theorem 2. *Mechanism 1 is a* $(O(\log m), O(\log m))$ *approximation to the social efficiency.*

Proof. By Proposition (1) and the analysis of [11], Mechanism 1, is a $O(\log m)$-approximation to social utility (and therefore social welfare). For the instance $\boldsymbol{x} = (\boldsymbol{x}_1, 0, \ldots, 0)$, where i is a single minded agent with value v_1 for outcome 1, the approximation ratio of $\log m$ is tight for both the welfare and the utility. □

4 Optimizing Social Utility Without Sacrificing Social Welfare

The mechanism of Sect. 3, approximates utility within an optimal logarithmic factor. However, it also approximates Social Welfare within the same logarithmic factor. The impossibility of Corollary 2 implies that no mechanism can do better than $(O(1), O(\log m))$-approximate social efficiency. So the question of simultaneously optimizing social welfare stands. We answer this question affirmatively by presenting a mechanism that optimizes welfare on a smooth probability space.

Theorem 3. *For any* $\epsilon > 0$, *there is a mechanism* $\mathcal{M}$ *that* $\left((1+\epsilon), \frac{(1+\epsilon)^2}{\epsilon} \ln m\right)$-*approximates social efficiency.*

Remark 1. We can $(O(1), O(\log m))$-approximate social efficiency simply by randomizing, with constant probability, between the VCG mechanism and Mechanism 1. However, the mechanism of Theorem 3 follows from a more general approach that yields a smooth mechanism and may be of independent interest.

4.1 The Mechanism

Similarly to the previous mechanism, we need a careful tradeoff between the VCG mechanism and suboptimal allocations close to the uniform mechanism. We note that the VCG mechanism optimizes the expected welfare by selecting the best outcome in the unit simplex $\Delta(O)$. Here, we optimize on a surface that is close to the unit simplex, but slightly curved towards the corners, in order to reduce the payments when the best outcomes are close in weight. To this end, we define a mechanism by optimizing on the following family of surfaces:

$$S_k = \left\{ \boldsymbol{s} \in \mathbb{R}_+^m \mid \|\boldsymbol{s}\|_k \leq \frac{1}{m^{1-1/k}} \right\} \tag{1}$$

For any $k \geq 1$ or for $k = \infty$, we define the mechanism $f_k(\boldsymbol{x}) = \arg\max_{\boldsymbol{s} \in S_k} \boldsymbol{s} \cdot \boldsymbol{w}(\boldsymbol{x})$.

The reason VCG is not working for utility maximization is that if the weight vector for e.g. 2 outcomes is $(1, 1+\epsilon)$, the mechanism will output the second outcome instead of a mixture of both. Such a mechanism requires a high amount of payments in order to truthfully distinguish between the outcomes, leading to minimal utility. In contrast, the mechanism with allocation f_k outputs a "smooth max" over outcomes leading to a reduced amount of payments (Fig. 1).

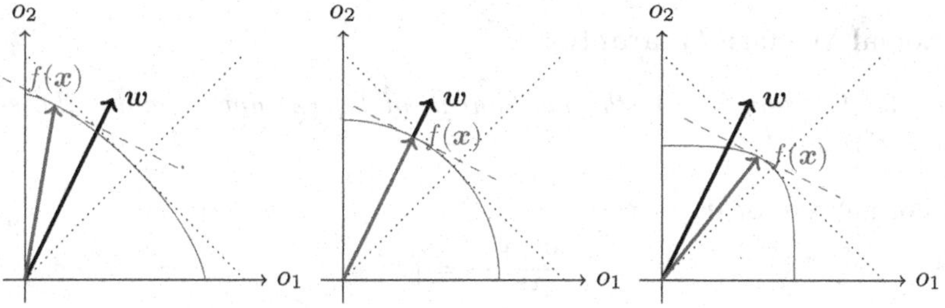

Fig. 1. Optimizing on the curved surfaces for $m = 2$

Lemma 2. *The closed form of the mechanism* f_k *is*

$$f_k(\boldsymbol{x}) = \frac{1}{m^{1-1/k}} \frac{\boldsymbol{w}^{\frac{1}{k-1}}}{\|\boldsymbol{w}^{\frac{1}{k-1}}\|_k}$$

Proof. The outcome of the mechanism is the vector $\boldsymbol{s}$ the optimizes $\boldsymbol{w} \cdot \boldsymbol{s}$ subject to $\|\boldsymbol{s}\|_k \leq m^{-\frac{k-1}{k}}$. By the Minkowski inequality, Eq. (1) defines a strictly convex space. Therefore the optimal point will lie on the boundary of the space, at the extremal point in the direction of $\boldsymbol{w}$. The boundary is defined by

$$\|\boldsymbol{s}\|_k = \frac{1}{m^{1-\frac{1}{k}}} \iff \|\boldsymbol{s}\|_k^k = \frac{1}{m^{k-1}}$$

and since we seek the extremal point in the direction of $\boldsymbol{w}$, $\boldsymbol{w}$ must be perpendicular to the boundary at the optimal point. Therefore at the optimal point $\boldsymbol{s}_*$ the gradient of the surface is in the direction of $\boldsymbol{w}$, that is there is some c such the

$$\nabla(\|\boldsymbol{s}_*\|_k^k) = c\boldsymbol{w} \iff \boldsymbol{s}_* = \left(\frac{c}{k}\right)^{\frac{1}{k-1}} \boldsymbol{w}^{\frac{1}{k-1}}$$

Moreover $\boldsymbol{s}_*$ needs to be to be on the surface, and therefore

$$\|\boldsymbol{s}_*\|_k^k = \frac{1}{m^{k-1}} \iff \left(\frac{c}{k}\right)^{\frac{1}{k-1}} = \frac{1}{m^{\frac{k-1}{k}} \|\boldsymbol{w}\|_{\frac{k}{k-1}}^{\frac{1}{k-1}}}$$

Substituting in the equation for $\boldsymbol{s}_*$ concludes the proof. □

We are interested in mechanisms with S_k close to S_1, so we set $k = \ell/(\ell-1)$ for some integer $\ell \geq 1$. The resulting mechanism is

$$f_\ell(\boldsymbol{x}) = \frac{1}{m^{1/\ell}} \frac{\boldsymbol{w}^{\ell-1}}{||\boldsymbol{w}^{\ell-1}||_{\frac{\ell}{\ell-1}}} \tag{2}$$

The reader is invited to verify that the mechanism exhibits a smooth transition between the VCG mechanism (for $\ell \to \infty$) and the uniform mechanism (for $\ell = 1$). Moreover, the mechanism is partial in the sense that for $\ell \in (1, \infty)$, $|f_\ell(\boldsymbol{x})| < 1$ and there is a positive probability that f_ℓ does not implement any outcome.

4.2 Social Welfare Guarantees

Lemma 3. *For any $\ell \geq 1$, the mechanism of Eq. (2) approximates the social welfare within $m^{1/\ell}$.*

Proof. For any vector $\boldsymbol{a}$

$$\frac{\|\boldsymbol{a}^\ell\|_1}{\|\boldsymbol{a}^{\ell-1}\|_{\frac{\ell}{\ell-1}}} = \|\boldsymbol{a}\|_\ell \tag{3}$$

The approximation ratio follows from

$$\frac{\boldsymbol{w} \cdot f(\boldsymbol{x})}{\|\boldsymbol{w}\|_\infty} = \frac{1}{m^{1/\ell}} \frac{\boldsymbol{w} \cdot \boldsymbol{w}^{\ell-1}}{\|\boldsymbol{w}^{\ell-1}\|_{\frac{\ell}{\ell-1}} \|\boldsymbol{w}\|_\infty} \overset{\text{Equation (3)}}{=} m^{1/\ell} \frac{\|\boldsymbol{w}\|_\ell}{\|\boldsymbol{w}\|_\infty} \geq \frac{1}{m^{1/\ell}}$$

The analysis is tight since when $\boldsymbol{x}$ consists of a single-minded agent with unit value, $\boldsymbol{w} \cdot f(\boldsymbol{x}) = \frac{1}{m^{1/\ell}}$ and $\|\boldsymbol{w}\|_\infty = 1$. □

4.3 Bounds to the Revenue of the Mechanism

We will now study the amount of payments charged by the mechanism. The payments of player i are computed as follows

$$\begin{aligned} p(\boldsymbol{x}_{-i}, \boldsymbol{x}_i) &= \boldsymbol{w}_{-i} \cdot f(\boldsymbol{x}_{-i}) - \boldsymbol{w}_{-i} \cdot f(\boldsymbol{x}) \\ &= \frac{1}{m^{1/\ell}} \left(\|\boldsymbol{w}_{-i}\|_\ell - \|\boldsymbol{w}\|_\ell + \frac{\boldsymbol{x}_i \cdot \boldsymbol{w}^{\ell-1}}{\|\boldsymbol{w}\|_\ell^{\ell-1}} \right) \end{aligned}$$

Therefore we can now bound the total amount of payments

Lemma 4. *For any integer $\ell \geq 1$, the mechanism of Eq. (2) charges the set of agents at most*

$$P[\boldsymbol{x}] \leq \frac{1}{m^{1/\ell}} \left(1 - \frac{1}{\ell}\right) \|\boldsymbol{w}(\boldsymbol{x})\|_\ell \tag{4}$$

Proof. By summing up the individual payments.

$$\sum_{i=1}^{n} p(\boldsymbol{x}_{-i}, \boldsymbol{x}_i) = \frac{1}{m^{1/\ell}} \left(\frac{(\sum_i \boldsymbol{x}_i) \cdot \boldsymbol{w}}{\|\boldsymbol{w}\|_\ell^{l-1}} - \sum_i (\|\boldsymbol{w}\|_\ell - \|\boldsymbol{w}_{-i}\|_\ell) \right)$$
$$= \frac{1}{m^{1/\ell}} \left(\|\boldsymbol{w}\|_\ell - \sum_i (\|\boldsymbol{w}\|_\ell - \|\boldsymbol{w}_{-i}\|_\ell) \right)$$

Therefore it suffices to show that

$$\sum_i (\|\boldsymbol{w}\|_\ell - \|\boldsymbol{w}_{-i}\|_\ell) \geq \frac{\|\boldsymbol{w}\|_\ell}{\ell}$$

The ℓ-th power difference is bound as follows

$$\|\boldsymbol{w}\|_\ell^\ell - \|\boldsymbol{w}_{-i}\|_\ell^\ell = (\|\boldsymbol{w}\|_\ell - \|\boldsymbol{w}_{-i}\|_\ell) \cdot \sum_{k=0}^{l-1} (\|\boldsymbol{w}\|_\ell^{l-1-k} \|\boldsymbol{w}_{-i}\|_\ell^k)$$
$$\leq (\|\boldsymbol{w}\|_\ell - \|\boldsymbol{w}_{-i}\|_\ell) \cdot \ell \|\boldsymbol{w}\|_\ell^{\ell-1}$$

For the rest of the proof, for a vector $\boldsymbol{a}$ we denote its j-th coordinate by $\boldsymbol{a}[j]$. Then

$$\sum_i \frac{\|\boldsymbol{w}\|_\ell^\ell - \|\boldsymbol{w}_{-i}\|_\ell^\ell}{\ell \|\boldsymbol{w}\|_\ell^{\ell-1}} \geq \frac{\|\boldsymbol{w}\|_\ell}{\ell} \iff \sum_i (\|\boldsymbol{w}\|_\ell^\ell - \|\boldsymbol{w}_{-i}\|_\ell^\ell) \geq \|\boldsymbol{w}\|_\ell^\ell$$
$$\iff \sum_{i=1}^{n} \left(\sum_{j=1}^{m} \boldsymbol{w}^\ell[j] - \sum_{j=1}^{n} \boldsymbol{w}_{-i}^\ell[j] \right) \geq \sum_{j=1}^{m} \boldsymbol{w}^\ell[j]$$
$$\iff \sum_{j=1}^{m} \sum_{i=1}^{n} (\boldsymbol{w}^\ell[j] - \boldsymbol{w}_{-i}^\ell[j]) \geq \sum_{j=1}^{m} \boldsymbol{w}^\ell[j]$$

We will prove that the inequality holds for each term separately. It holds that

$$\boldsymbol{w}^\ell[j] - \boldsymbol{w}_{-i}^\ell[j] \geq (\boldsymbol{w}[j] - \boldsymbol{w}_{-i}[j]) \boldsymbol{w}^{\ell-1}[j] = \boldsymbol{x}_i[j] \boldsymbol{w}^{\ell-1}[j]$$

and summing over i gives us

$$\sum_i (\boldsymbol{w}^\ell[j] - \boldsymbol{w}_{-i}^\ell[j]) \geq \sum_i \boldsymbol{x}_i[j] \boldsymbol{w}^{\ell-1}[j] = \boldsymbol{w}[j] \boldsymbol{w}^{\ell-1}[j] = \boldsymbol{w}^\ell[j]$$

concluding the proof. □

4.4 Maximizing Utility

The utility of the mechanism is therefore

$$U[\boldsymbol{x}] = \boldsymbol{w} \cdot f(\boldsymbol{x}) - P[\boldsymbol{x}] \geq \frac{\|\boldsymbol{w}\|_\ell}{\ell m^{1/\ell}} \geq \frac{\|\boldsymbol{w}\|_\infty}{\ell m^{1/\ell}}$$

We summarize our results in the following theorem.

Theorem 4. *For every integer $\ell \geq 1$, there is a truthful mechanism that $(m^{1/\ell}, \ell m^{1/\ell})$-approximates social efficiency.*

The optimal point of this tradeoff in terms of utility maximization is when $\ell = \ln m$ (for simplicity, we assume in this section that if ℓ is not an integer, it is rounded to the smallest integer exceeding the given value).

Corollary 3. *There is a truthful mechanism that $(e, e \ln m)$-approximates social efficiency.*

Alternatively by setting $\ell = \frac{\ln m}{\ln(1+\epsilon)}$ we get the following.

Corollary 4. *There is a truthful mechanism, that for any $\epsilon > 0$, $\left(1+\epsilon, \frac{(1+\epsilon)^2}{\epsilon} \ln m\right)$- approximates social efficiency.*

An interesting property of our mechanism, is that the set of outcomes can be a priori restricted to some subset of the original outcome space. These mechanism are known as *Maximal in Range (MIR)*, and are tailored to obtain suboptimal welfare guarantees in polynomial time for NP-hard problems. Our mechanisms can be run on these modified outcome spaces with no modification preserving welfare guarantees and providing social utility logarithmic to the number of outcomes.

Corollary 5. *Let some MIR mechanism with outcome space S, that a-approximates social welfare. Then, it can be modified to $((1+\epsilon)a)$-approximate social welfare and $\left(\frac{(1+\epsilon)^2}{\epsilon} a \ln |S|\right)$-approximate social utility.*

Remark 2. We have shown that the mechanism is IR in expectation, however there are examples where players net negative utility for certain random outcomes. Nonetheless the mechanism can be modified to be universally IR. Consider some agent i. Let $P_i = \boldsymbol{w} \cdot f(\boldsymbol{x}_{-i}) - \boldsymbol{w} \cdot f(\boldsymbol{x})$ denote the payments that induce truthfulness. If outcome j is realized we charge this agent $p_{ij} = \frac{P_i}{\boldsymbol{x}_i \cdot f(\boldsymbol{x})} x_{ij}$. It is easy to verify that the expected payments are unaltered so truthfulness is preserved. Moreover, the mechanism is now universally IR. A similar technique can be found in [7].

References

1. Archer, A., Tardos, É.: Frugal path mechanisms. ACM Transactions on Algorithms **3**(1) (2007)
2. Braverman, M., Chen, J., Kannan, S.: Optimal provision-after-wait in healthcare. In: Proceedings of the 5th Conference on Innovations in Theoretical Computer Science (ITCS 2014), pp. 541–542 (2014)
3. Cavallo, R.: Efficiency and redistribution in dynamic mechanism design. In: Proceedings of the 9th ACM Conference on Electronic Commerce (EC 2008), pp. 220–229 (2008)
4. Chakravarty, S., Kaplan, T.R.: Manna from heaven or forty years in the desert: optimal allocation without transfer payments. Social Science Research Network (2006). http://dx.doi.org/10.2139/ssrn.939389
5. Cole, R., Gkatzelis, V., Goel, G.: Mechanism design for fair division: allocating divisible items without payments. In: Proceedings of the 14th ACM Conference on Electronic Commerce (EC 2013), pp. 251–268 (2013)
6. Elkind, E., Sahai, A., Steiglitz, K.: Frugality in path auctions. In: Proceedings of the 15th ACM-SIAM Symposium on Discrete Algorithms (SODA 2004), pp. 701–709 (2004)
7. Eso, P., Futo, G.: Auction design with a risk averse seller. Economics Letters **65**(1), 71–74 (1999)
8. Gneiting, T., Rafterys, A.E.: Strictly proper scoring rules, prediction, and estimation. J. Am. Stat. Assoc. **102**(477), 359–378 (2007)
9. Guo, M., Conitzer, V.: Worst-case optimal redistribution of VCG payments in multi-unit auctions. Games Econ. Behav. **67**(1), 69–98 (2009)
10. Guo, M., Conitzer, V.: Better redistribution with inefficient allocation in multi-unit auctions. Artif. Intell. **216**, 287–308 (2014)
11. Hartline, J.D., Roughgarden, T.: Optimal mechanism design and money burning. In: Proceedings of the 40th ACM Symposium on Theory of Computing (STOC 2008), pp. 75–84 (2008)
12. Lesca, J., Todo, T., Yokoo, M.: Coexistence of utilitarian efficiency and false-name-proofness in social choice. In: International Conference on Autonomous Agents and Multi-Agent Systems (AAMAS 2014), pp. 1201–1208 (2014)
13. Myerson, R.: Optimal auction design. Math. Oper. Res. **6**(1), 58–73 (1981)
14. Nisan, N.: Introduction to mechanism design (for computer scientists). Algorithmic Game Theor. **9**, 209–242 (2007)
15. Schapira, M., Singer, Y.: Inapproximability of combinatorial public projects. In: Papadimitriou, C., Zhang, S. (eds.) WINE 2008. LNCS, vol. 5385, pp. 351–361. Springer, Heidelberg (2008)

Towards More Practical Linear Programming-Based Techniques for Algorithmic Mechanism Design

Khaled Elbassioni[1], Kurt Mehlhorn[2], and Fahimeh Ramezani[2](✉)

[1] Masdar Institute of Science and Technology, Abu Dhabi, UAE
kelbassioni@masdar.ac.ae
[2] Max Planck Institute for Informatics, Campus E1 4, 66123 Saarbrucken, Germany
{mehlhorn,ramezani}@mpi-inf.mpg.de

Abstract. R. Lavy and C. Swamy (FOCS 2005, J. ACM 2011) introduced a general method for obtaining truthful-in-expectation mechanisms from linear programming based approximation algorithms. Due to the use of the Ellipsoid method, a direct implementation of the method is unlikely to be efficient in practice. We propose to use the much simpler and usually faster multiplicative weights update method instead. The simplification comes at the cost of slightly weaker approximation and truthfulness guarantees.

1 Introduction

Algorithmic mechanism design studies optimization problems in which part of the input is not directly available to the algorithm; instead, this data is collected from self-interested players. It quests for *polynomial-time* algorithms that *(approximately) optimize* a global objective function (usually called *social welfare*), subject to the *strategic requirement* that the best strategy of the players is to *truthfully* report their part of the input. Such algorithms are called *truthful mechanisms.*

If the underlying optimization problem can be efficiently solved to optimality, the celebrated VCG mechanism (see, e.g., [17]) achieves truthfulness, social welfare optimization, and polynomial running time.

In general, the underlying optimization problem can only be solved approximately. Lavi and Swamy ([15,16]) showed that certain linear programming based approximation algorithms for the social welfare problem can be turned into randomized mechanisms that are truthful-in-expectation, i.e., reporting the truth maximizes the expected utility of an player. The LS-mechanism is powerful (see [4,11,15,16] for applications), but unlikely to be efficient in practice because of its use of the Ellipsoid method. *We show how to use the multiplicative weights update method instead. This results in simpler algorithms at the cost of somewhat weaker approximation and truthfulness guarantees.*

M. Hoefer (Ed.): SAGT 2015, LNCS 9347, pp. 98–109, 2015.
DOI: 10.1007/978-3-662-48433-3_8

We next review the LS-mechanism. It applies to integer linear programming problems of the packing type[1] for which the linear programming relaxation can be solved exactly and for which an α-integrality gap verifier is available (definition below). Let $\mathcal{Q} \subseteq \mathbb{R}^d_{\geq 0}$ be a *packing polytope*, i.e., $\mathcal{Q}$ is the intersection of finitely many halfspaces, and if $y \in \mathcal{Q}$ and $x \leq y$ then $x \in \mathcal{Q}$. We use $\mathcal{Q}_\mathcal{I} := \mathcal{Q} \cap \mathbb{Z}^d$ for the set of integral points in $\mathcal{Q}$, x^j for a typical element of $\mathcal{Q}_\mathcal{I}$, and $\mathcal{N}$ for the index set of all elements in $\mathcal{Q}_\mathcal{I}$. The mechanism consists of three main steps:

1. Let $v_i \in \mathbb{R}^d_{\geq 0}, 1 \leq i \leq n$, be the valuation of the i-th player and let $v = \sum_i v_i$ be the accumulated valuation. Solve the LP-relaxation, i.e., find a maximizer $x^* = \text{argmax}_{x \in Q} v^T x$ for the social welfare of the fractional problem, and determine the VCG prices[2] $p_1, \ldots, p_n$. The allocation x^* and the VCG-prices are a truthful mechanism for the fractional problem.
2. Write $\alpha \cdot x^*$ as a *convex combination of integral solutions* in $\mathcal{Q}$, i.e., $\alpha \cdot x^* = \sum_{j \in \mathcal{N}} \lambda_j x^j$, $\lambda_j \geq 0$, $\sum_{j \in \mathcal{N}} \lambda_j = 1$, and $x^j \in \mathcal{Q}_\mathcal{I}$. This step requires an *α-integrality-gap-verifier* for $\mathcal{Q}_\mathcal{I}$ for some $\alpha \in [0, 1]$. On input $\bar{v} \in \mathbb{R}^d_{\geq 0}$ and $x^* \in \mathcal{Q}$, an α-integrality-gap-verifier returns an $x \in \mathcal{Q}_\mathcal{I}$ such that
$$\bar{v}^T x \geq \alpha \bar{v}^T x^*.$$
3. Pick the integral solution x^j with probability λ_j, and charge the i-th player the price $p_i \cdot (v_i^T x^j / v_i^T x^*)$ and If $v_i^T x^* = 0$, charge zero.

The LS-mechanism approximates social welfare with factor α and guarantees truthfulness-in-expectation, i.e., it converts a truthful fractional mechanism into an α-approximate truthful-in-expectation integral mechanism. With respect to practical applicability, steps 1 and 2 are the two major bottlenecks. Step 1 requires solving a linear program; an exact solution requires the use of the Ellipsoid method (see e.g. [10]), if the dimension is exponential. Furthermore, up to recently, the only method known to perform the decomposition in Step 2 is through the Ellipsoid method. An alternative method avoiding the use of the Ellipsoid method was recently given by Kraft, Fadaei, and Bichler [14]. We comment on their result in the next section.

1.1 Our Results

Our result concerns the design and analysis of a practical algorithm for the LS-scheme. We first consider the case where the LP-relaxation of SWM (social

[1] An example is the combinatorial auction problem. There is a set of m items to be sold to a set of n players. The (reported) value of a set S of items to the i-th player is $v_i(S)$ with $v_i(\emptyset) = 0$ and $v_i(S) \leq v_i(T)$ whenever $S \subseteq T$. Let $x_{i,S}$ be a 0–1 variable indicating that set S is given to player i. Then $\sum_S x_{i,S} \leq 1$ for every player i as at most one set can be given to i, and $\sum_i \sum_{S; j \in S} x_{i,S} \leq 1$ for every item j as any item can be given away only once. The social welfare is $\sum_{i,S} v_i(S) x_{i,S}$. The polytope $\mathcal{Q}$ is obtained by replacing the integrality constraints for $x_{i,s}$ by $0 \leq x_{i,S} \leq 1$. Note that the number d of variables is $n2^m$.

[2] $p_i = \sum_{j \neq i} v_j^T (\hat{x} - x^*)$, where $\hat{x} = \text{argmax}_{x \in Q} \sum_{j \neq i} v_j^T x$.

welfare maximization) in Step 1 of the LS-scheme can be solved exactly and efficiently and then our problem reduces to the design of practical algorithm for the Step 2. In what follows we present an algorithm that is fast and practical for the convex decomposition (i.e., Step 2). Afterwards, we consider a more general problem where the LP-relaxation in Step 1 of the LS-scheme cannot be solved exactly and we only have an approximate solution that maximizes the LP-relaxation within a factor of $1 - \varepsilon$.

Convex Decomposition. Over the past 15 years, simple and fast methods [2,8,9,12,13,18,19] have been developed for solving packing and covering linear programs within an arbitrarily small error guarantee ε. These methods are based on the multiplicative weights update (MWU) method [1], in which a very simple update rule is repeatedly performed until a near-optimal solution is obtained. We show how to replace the use of the Ellipsoid method in Step 2 by an approximation algorithm for covering linear programs. This result is the topic of Sect. 2.

Theorem 1. *Let $\varepsilon > 0$ be arbitrary. Given a fractional point $x^* \in \mathcal{Q}$, and an α-integrality-gap verifier for $\mathcal{Q}_\mathcal{I}$, we can find a convex decomposition*

$$\frac{\alpha}{1+4\varepsilon} \cdot x^* = \sum_{j \in \mathcal{N}} \lambda_j x^j .$$

The convex decomposition has size (= number of nonzero λ_j) at most $s(1 + \lceil \varepsilon^{-2} \ln s \rceil)$, where s is the size of the support of x^* (= number of nonzero components). The algorithm makes at most $s\lceil \varepsilon^{-2} \ln s\rceil$ calls to the integrality-gap-verifier.

Kraft, Fadaei, and Bichler [14] obtained a related result independently. However, their construction is less efficient in two aspects. First, it requires $O(s^2\varepsilon^{-2})$ calls of the oracle. Second, the size of their convex decomposition might be as large as $O(s^3\varepsilon^{-2})$. In the combinatorial auction problem, $s = n + m$. Theorem 1 together with Steps 1 and 3 of the LS scheme implies a mechanism that is truthful-in-expectation and has $(\alpha/(1 + 4\varepsilon))$-social efficiency. A mechanism has *γ-social efficiency*, where $\gamma \in [0, 1]$, if the expected social welfare of the allocation returned by the mechanism is at least γ times the maximum possible social value.

Approximately Truthful-in-Expectation Mechanism. In contrast to Lavi-Swamy mechanism, let us assume that we do not want to solve the LP-relaxation exactly but instead, we want to use an ε-approximation algorithm $\mathcal{A}$ for it. Garg and Könemann [8] showed that there is an FPTAS for the packing problem and hence $\mathcal{A}$ exists, for every $\varepsilon > 0$. Using this, we show how to construct a fractional randomized mechanism for given $\varepsilon_0 \in (0, 1/2]$ and $\varepsilon = \Theta(\frac{\varepsilon_0^5}{n^4})$ that satisfies:

1. *No positive transfer* (i.e., prices are non-negative).
2. *Individually rational with probability* $1-\varepsilon_0$ (i.e., the utility of any truth-telling player is non-negative with probability at least $1 - \varepsilon_0$).

3. $(1-\varepsilon_0)$-*truthful-in-expectation* (i.e., reporting the truth maximizes the expected utility of an player up to a factor $1-\varepsilon_0$)
4. $(1-\varepsilon)(1-\varepsilon_0)$-social efficiency.

Now, let us assume that x is a fractional allocation obtained from the above mechanism. We apply our convex decomposition technique and Step 3 of the Lavi-Swamy mechanism to obtain an integral randomized mechanism that satisfies the aforementioned conditions 1 to 3 and has $\alpha(1-\varepsilon)(1-\varepsilon_0)/(1+4\varepsilon)$-social efficiency. We show this result in Sect. 4.

Note that our fractional mechanism refines the one given in [5], where the dependency of ε on n and ε_0 is as $\varepsilon = \Theta(\varepsilon_0/n^9)$. A recent experimental study of our mechanism on Display Ad Auctions [6] shows the applicability of these methods in practice.

2 A Fast Algorithm for Convex Decompositions

Let $x^* \in \mathcal{Q}$ be arbitrary. Carr and Vempala [3] showed how to construct a convex combination of points in $\mathcal{Q}_\mathcal{I}$ dominating αx^* using a polynomial number of calls to an α-integrality-gap-verifier for $\mathcal{Q}_\mathcal{I}$. Lavi and Swamy [16] modified the construction to get an exact convex decomposition $\alpha x^* = \sum_{i\in\mathcal{N}} \lambda_i x^i$ for the case of packing linear programs. The construction uses the Ellipsoid method. We show an approximate version that replaces the use of the Ellipsoid method by the multiplicative weights update (MWU) method. For any $\varepsilon > 0$, we show how to obtain a convex decomposition of $\alpha x^*/(1+\varepsilon)$. Let s be the number of non-zero components of x^*. The size of the decomposition and the number of calls to the α-integrality gap verifier are $O(s\varepsilon^{-2}\ln s)$.

This section is structured as follows. We first review Kkandekar's FPTAS for covering linear programs (Subsect. 2.1). We then use it and the α-integrality gap verifier to construct a dominating convex combination for $\alpha x^*/(1+4\varepsilon)$, where $x^* \in \mathcal{Q}$ is arbitrary (Subsect. 2.2). In Subsect. 2.3, we show how to convert a dominating convex combination into an exact convex decomposition. Finally, in Subsect. 2.4, we put the pieces together.

2.1 An FPTAS for Covering Linear Programs

Consider a covering linear program:

$$\min c^T x \;\; s.t. \;\; \{Ax \ge b, \;\; x \ge 0\} \tag{1}$$

where $A \in \mathbb{R}^{m\times n}_{\ge 0}$ is an $m \times d$ matrix with non-negative entries and $c \in \mathbb{R}^n_{\ge 0}$ and $b \in \mathbb{R}^m_{\ge 0}$ are non-negative vectors. We assume the availability of a *κ-approximation oracle* for some $\kappa \in (0,1]$.

$\mathcal{O}_\kappa(z)$: Given $z \in \mathbb{R}^m_{\ge 0}$, the oracle finds a column j of A that maximizes $\frac{1}{c_j}\sum_{i=1}^m \frac{z_i a_{ij}}{b_i}$ within a factor of κ:

$$\frac{1}{c_j}\sum_{i=1}^m \frac{z_i a_{ij}}{b_i} \ge \kappa \cdot \max_{j'\in[n]} \frac{1}{c_{j'}}\sum_{i=1}^m \frac{z_i a_{ij'}}{b_i} \tag{2}$$

Algorithm 1. Covering($\mathcal{O}_\kappa$)

Require: a covering system (A, b, c) given by a κ-approximation oracle $\mathcal{O}_\kappa$, where $A \in \mathbb{R}^{m\times n}_{\geq 0}$, $b \in \mathbb{R}^m_{>0}$, $c \in \mathbb{R}^n_{>0}$, and an accuracy parameter $\varepsilon \in (0, 1/2]$

Ensure: A feasible solution $\hat{x} \in \mathbb{R}^n_{\geq 0}$ to (1) s.t. $c^T\hat{x} \leq \frac{(1+4\varepsilon)}{\kappa} z^*$

```
x(0) := 0; t := 0; and T := ln m / ε²
while M(t) < T do
    t := t + 1
    Let j(t) := O_κ(p(t)/||p(t)||_1)
    x_{j(t)}(t) := x_{j(t)}(t − 1) + δ(t) and x_j(t) = x_j(t − 1) for j ≠ j(t)
end while
return x̂ = x(t) / M(t)
```

For an exact oracle $\kappa = 1$, Khandekar [12] gave an algorithm which computes a feasible solution $\hat{x}$ to covering LP (1) such that $c^T\hat{x} \leq (1 + 4\varepsilon)z^*$ where z^* is the value of an optimal solution. The algorithm makes $O(m\varepsilon^{-2}\log m)$ calls to the oracle, where m is the number of rows in A. If the exact oracle in Khandekar's algorithm is replaced by a κ-approximation algorithm, it computes a feasible solution $\hat{x} \in \mathbb{R}^n_{\geq 0}$ to (1) such that $c^T\hat{x} \leq (1 + 4\varepsilon)z^*/\kappa$. The algorithm is given as Algorithm 1 and can be thought of as the algorithmic dual of the FPTAS for multicommodity flows given in [8]. We use A_i to denote the i-th row of A. The algorithm constructs vectors $x(t) \in \mathbb{R}^n_{\geq 0}$, for $t = 0, 1, \ldots,$ until $M(t) := \min_{i\in[m]} A_i x(t)/b_i$ becomes at least $T := \frac{\ln m}{\varepsilon^2}$. Define the *active list* at time t by $L(t) := \{i \in [m] : A_i x(t-1)/b_i < T\}$. For $i \in L(t)$, define

$$p_i(t) := (1 - \varepsilon)^{A_i x(t-1)/b_i}, \tag{3}$$

and set $p_i(t) = 0$ for $i \notin L(t)$. At each time t, the algorithm calls the oracle with the vector $z_t = p(t)/\|p(t)\|_1$, and increases the variable $x_{j(t)}$ by

$$\delta(t) := \min_{i\in L(t) \text{ and } a_{i,j(t)}\neq 0} \frac{b_i}{a_{i,j(t)}}, \tag{4}$$

where $j(t)$ is the index returned by the oracle. Due to lack of space, the proof of following theorem and corollary are presented in the full paper [7].

Theorem 2. *Let $\varepsilon \in (0, \frac{1}{2}]$ and let z^* be the value of an optimum solution to (1). Procedure Covering($\mathcal{O}_\kappa$) (see Algorithm 1) terminates in at most $m\lceil\varepsilon^{-2}\ln m\rceil$ iterations with a feasible solution $\hat{x}$ of (1) of at most $m\lceil\varepsilon^{-2}\ln m\rceil$ positive components. At termination, it holds that*

$$c^T\hat{x} \leq \frac{(1 + 4\varepsilon)}{\kappa} z^*. \tag{5}$$

We observe that the proof of Theorem 2 can be modified to give:

Corollary 1. *Suppose $b = \mathbf{1}$, $c = \mathbf{1}$, and we use the following oracle $\mathcal{O}'$ instead of $\mathcal{O}$ in Algorithm 1:*

$\mathcal{O}'(A, z)$: *Given* $z \in \mathbb{R}^m_{\geq 0}$, *such that* $\mathbf{1}^T z = 1$, *the oracle finds a column* j *of* A *such that* $z^T A \mathbf{1}_j \geq 1$.

Then the algorithm terminates in at most $m\lceil \varepsilon^{-2} \ln m \rceil$ *iterations with a feasible solution* $\hat{x}$ *having at most* $m\lceil \varepsilon^{-2} \ln m \rceil$ *positive entries, such that* $\mathbf{1}^T \hat{x} \leq 1 + 4\varepsilon$.

2.2 Finding a Dominating Convex Combination

Recall that we use $\mathcal{N}$ to index the elements in $\mathcal{Q}_\mathcal{I}$. We assume the availability of an α-integrality-gap-verifier $\mathcal{F}$ for $\mathcal{Q}_\mathcal{I}$. We will use the results of the preceding section and show how to obtain for any $x^* \in \mathcal{Q}$ and any positive ε a convex composition of points in $\mathcal{Q}_\mathcal{I}$ that covers $\alpha x^*/(1+4\varepsilon)$. Our algorithm requires $O(s\varepsilon^{-2} \ln s)$ calls to the oracle, where s is the support of x^*.

Theorem 3. *Let* $\varepsilon > 0$ *be arbitrary. Given a fractional point* $x^* \in \mathcal{Q}$ *and an* α*-integrality-gap verifier* $\mathcal{F}$ *for* $\mathcal{Q}_\mathcal{I}$, *we can find a convex combination* $\bar{x}$ *of integral points in* $\mathcal{Q}_\mathcal{I}$ *such that*

$$\frac{\alpha}{1+4\varepsilon} \cdot x^* \leq \bar{x} = \sum_{i \in \mathcal{N}} \lambda_i x^i.$$

The convex decomposition has size at most $s\lceil \varepsilon^{-2} \ln s \rceil$, *where* s *is the number of positive entries of* x^*. *The algorithm makes at most* $s\lceil \varepsilon^{-2} \ln s \rceil$ *calls to the integrality-gap verifier.*

Proof. The task of finding the multipliers λ_i is naturally formulated as a covering LP, namely,

$$\begin{aligned} \min \quad & \sum_{i \in \mathcal{N}} \lambda_i \\ s.t. \quad & \sum_{i \in \mathcal{N}} \lambda_i x^i_j \geq \alpha \cdot x^*_j \quad \text{for all } j, \\ & \sum_{i \in \mathcal{N}} \lambda_i \geq 1, \quad \lambda_i \geq 0. \\ & \lambda_i \geq 0. \end{aligned} \tag{6}$$

Clearly, we can restrict our attention to the $j \in S^+ := \{j : x^*_j > 0\}$ and rewrite the constraint for $j \in S^+$ as $\sum_{i \in \mathcal{N}} \lambda_i x^i_j/(\alpha \cdot x^*_j) \geq 1$. For simplicity of notation, we assume $S^+ = [1..s]$. In the language of the preceding section, we have $m = s+1$, $n = |\mathcal{N}|$, $c = \mathbf{1}$, $b = \mathbf{1}$ and the variable $x = \lambda$. The matrix $A = (a_{j,i})$ is as follows (note that we use j for the row index and i for the column index):

$$a_{j,i} := \begin{cases} x^i_j/(\alpha x^*_j) & 1 \leq j \leq s, i \in \mathcal{N} \\ 1 & j = s+1, i \in \mathcal{N} \end{cases}$$

Thus we can apply Corollary 1 of Sect. 2.1, provided we can efficiently implement the required oracle $\mathcal{O}'$. We do so using $\mathcal{F}$.

Oracle $\mathcal{O}'$ has arguments $(A, \tilde{z})$ such that $1^T\tilde{z} = 1$. Let us conveniently write $\tilde{z} = (w, z)$, where $w \in \mathbb{R}^s_{\geq 0}$, $z \in \mathbb{R}_{\geq 0}$, and $\sum_{j=s}^{j=1} w_j + z = 1$. Oracle $\mathcal{O}'$ needs to find a column i such that $\tilde{z}^T A\mathbf{1}_i \geq 1$. In our case $\tilde{z}^T A\mathbf{1}_i = \sum_{j=1}^{s} w_j x_j^i/\alpha x_j^* + z$, and we need to find a column i for which this expression is at least one. Since z does not depend on i, we concentrate on the first term. Define

$$V_j := \begin{cases} \frac{w_j}{\alpha x_j^*} & \text{for } j \in S^+ \\ 0 & \text{otherwise.} \end{cases}$$

Call algorithm $\mathcal{F}$ with $x^* \in \mathcal{Q}$ and $V := (V_1, \ldots, V_d)$. $\mathcal{F}$ returns an integer solution $x^i \in \mathcal{Q}_\mathcal{I}$ such that

$$V^T x^i = \sum_{j \in S^+} \frac{w_j}{\alpha x_j^*} x_j^i \geq \alpha \cdot V^T x^* = \sum_{j \in S^+} w_j,$$

and hence,

$$\sum_{j \in S^+} \frac{w_j}{\alpha x_j^*} x_j^i + z \geq \sum_{j \in S^+} w_j + z = 1.$$

Thus i is the desired column of A.

It follows by Corollary 1 that Algorithm 1 finds a feasible solution $\lambda' \in \mathbb{R}^{|\mathcal{N}|}_{\geq 0}$ to the covering LP (6), and a set $\mathcal{Q}'_\mathcal{I} \subseteq \mathcal{Q}_\mathcal{I}$ of vectors (returned by $\mathcal{F}$), such that $\lambda'_i > 0$ only for $i \in \mathcal{N}'$, where $\mathcal{N}'$ is the index set returned by oracle $\mathcal{O}'$ and $|\mathcal{N}'| \leq s\lceil \varepsilon^{-2} \ln s \rceil$ also $\Lambda := \sum_{i \in \mathcal{N}'} \lambda'_i \leq (1 + 4\varepsilon)$. Scaling λ'_i by Λ, we obtain a set of multipliers $\{\lambda_i = \lambda'_i/\Lambda : \ i \in \mathcal{N}'\}$, such that $\sum_{i \in \mathcal{N}'} \lambda_i = 1$ and

$$\sum_{i \in \mathcal{N}'} \lambda_i x^i \geq \frac{\alpha}{1 + 4\varepsilon} x^*. \tag{7}$$

We may assume $x_j^i = 0$ for all $j \notin S^+$ whenever $\lambda_i > 0$; otherwise simply replace x^i by a vector in which all components not in S^+ are set to zero, by using packing property this is possible. □

2.3 From Dominating Convex Combination to Exact Convex Decomposition

We will show how to turn a dominating convex combination into an exact decomposition. The construction is general and uses only the packing property. Such a construction seems to have been observed in [15], but was not made explicit. Kraft, Fadaei, and Bichler [14] describe an alternative construction. Their construction may increase the size of the convex decomposition (= number of non-zero λ_i) by a multiplicative factor s and an additive factor s^2. In contrast, our construction increases the size only by an additive factor s.

Theorem 4. *Let $x^* \in \mathcal{Q}$ be dominated by a convex combination $\sum_{i \in \mathcal{N}} \lambda_i x^i$ of integral points in $\mathcal{Q}_\mathcal{I}$, i.e.,*

$$\sum_{i \in \mathcal{N}} \lambda_i x^i \geq x^*. \tag{8}$$

Algorithm 2. Changing a dominating convex decomposition into an exact decomposition

Require: A packing convex set $\mathcal{Q}$ and point $x^* \in \mathcal{Q}$ and a convex combination $\sum_{i \in \mathcal{N}} \lambda_i x^i$ of integral points in $\mathcal{Q}_{\mathcal{I}}$ dominating x^*.
Ensure: A convex decomposition $x^* = \sum_{i \in \mathcal{N}'} \lambda_i x^i$ with $x^i \in \mathcal{Q}_{\mathcal{I}}$.
while there is an $i \in \mathcal{N}$ and a j such that $\lambda_i x^i_j > 0$ and $\sum_{h \in \mathcal{N}} \lambda_h x^h - \lambda_i \mathbf{1}_j \geq x^*$ **do**
  replace x^i by $x^i - \mathbf{1}_j$.
end while
while $\Delta_j := \sum_{i \in \mathcal{N}} \lambda_i x^i - x^*_j > 0$ for some j **do**
  {for all $i \in \mathcal{N}$ and all j: if $\lambda_i x^i_j > 0$ then $\sum_{h \in \mathcal{N}} \lambda_h x^h - \lambda_i \mathbf{1}_j < x^*$}
  Let j be such that $\Delta_j > 0$ and let i be such that $\lambda_i x^i_j > 0$. Let $B = \{j \in S^+ : x^i_j \neq 0 \text{and} \Delta_j > 0\}$ and let $b = |B|$. Renumber the coordinates such that $B = \{1, \ldots, b\}$ and $\Delta_1 / x^i_1 \leq \ldots \leq \Delta_b / x^i_b$.
  For $\ell \in \{0, \ldots, b\}$ define a vector y^ℓ by $y^\ell_j = x^i_j$ for $j \leq \ell$ and $y^\ell_j = 0$ for $j > \ell$.
  Change the left-hand side of (8) as follows: replace λ_i by $\lambda_i - \Delta_b / x^i_b$; for $1 \leq \ell < b$, increase the coefficient of y^ℓ by $\Delta_{\ell+1} / x^i_{\ell+1} - \Delta_\ell / x^i_\ell$; and increase the coefficient of y^0 by Δ_1 / x^i_1.
end while

Then Algorithm 2 achieves equality in (8). It increases the size of the convex combination by at most s, where s is the number of positive components of x^*.

Proof. Let $\mathbf{1}_j$ be the j-th unit vector. As long as there is an $i \in \mathcal{N}$ and a j such that $\lambda_i x^i_j > 0$ and replacing x^i by $x^i - \mathbf{1}_j$ maintains feasibility, i.e., satisfies constraint (8), we perform this replacement. Note that x^i is an integer vector in $\mathcal{Q}_{\mathcal{I}}$, therefore $x^i - \mathbf{1}_j$ remains positive vector and with using packing property, it is also in $\mathcal{Q}_{\mathcal{I}}$. We may therefore assume that the set of vectors indexed by $\mathcal{N}$ satisfy a minimality condition which is for all $i \in \mathcal{N}$ and $j \in S^+$ with $\lambda_i x^i_j > 0$

$$\sum_{h \in \mathcal{N}} \lambda_h x^h_j - \lambda_i \mathbf{1}_j < x^*_j \tag{9}$$

We will establish (9) as an invariant of the second while-loop.

For $j \in S^+$, let $\Delta_j = \sum_{i \in \mathcal{N}} \lambda_i x^i_j - x^*_j$. Then $\Delta_j \geq 0$ and, by (9), for every $j \in S^+$ and $i \in \mathcal{N}$, with $\lambda_i \neq 0$ either $x^i_j = 0$ or $\Delta_j < \lambda_i \leq \lambda_i x^i_j$. If $\Delta_j = 0$ for all $j \in S^+$, we are done. Otherwise, choose j and $i \in \mathcal{N}$ such that $\Delta_j > 0$ and $x^i_j > 0$. Let $B = \{j \in S^+ : x^i_j \neq 0 \text{ and } \Delta_j > 0\}$ be the indices in the support of x^i for which Δ_j is non-zero. We will change the left-hand side of (8) such that equality holds for all indices in B. The change will not destroy an already existing equality for an index outside B and hence the number of indices for which equality holds increases by $|B|$.

Let $b = |B|$. By renumbering the coordinates, we may assume $B = \{1, \ldots, b\}$ and $\Delta_1 / x^i_1 \leq \ldots \leq \Delta_b / x^i_b$. For $j \in [b]$, we clearly have

$$\lambda_i - \frac{\Delta_j}{x^i_j} = \lambda_i - \frac{\Delta_b}{x^i_b} + \frac{\Delta_b}{x^i_b} - \frac{\Delta_{b-1}}{x^i_{b-1}} + \cdots + \frac{\Delta_{j+1}}{x^i_{j+1}} - \frac{\Delta_j}{x^i_j}.$$

Multiplying by x_j^i and adding zero a few times, we obtain

$$\lambda_i x_j^i - \Delta_j = \left(\lambda_i - \frac{\Delta_b}{x_b^i}\right) x_j^i + \sum_{\ell=j}^{b-1} \left(\frac{\Delta_{\ell+1}}{x_{\ell+1}^i} - \frac{\Delta_\ell}{x_\ell^i}\right) x_j^i + \sum_{\ell=1}^{j-1} \left(\frac{\Delta_{\ell+1}}{x_{\ell+1}^i} - \frac{\Delta_\ell}{x_\ell^i}\right) 0 + \frac{\Delta_1}{x_1^i} 0.$$

For $\ell \in \{0, \ldots, b-1\}$ define a vector y^ℓ by $y_j^\ell = x_j^i$ for $j \le \ell$ and $y_j^\ell = 0$ for $j > \ell$. Then $x_j^i = y_j^\ell$ for $\ell \ge j$ and $0 = y_j^\ell$ for $\ell < j$. Hence for all $j \le b$

$$\lambda_i x_j^i - \Delta_j = \left(\lambda_i - \frac{\Delta_b}{x_b^i}\right) x_j^i + \sum_{\ell=1}^{b-1} \left(\frac{\Delta_{\ell+1}}{x_{\ell+1}^i} - \frac{\Delta_\ell}{x_\ell^i}\right) y_j^\ell + \frac{\Delta_1}{x_1^i} y_j^0. \tag{10}$$

Note that the coefficients on the right-hand side of (10) are non-negative and sum up to λ_i. Also, by the packing property of $\mathcal{Q}$, $y^\ell \in \mathcal{Q}_\mathcal{I}$ for $0 \le \ell < b$. We now change the left-hand side of (8) as follows: we replace λ_i by $\lambda_i - \Delta_b/x_b^i$; for $1 \le \ell < b$, we increase the coefficient of y^ℓ by $\Delta_{\ell+1}/x_{\ell+1}^i - \Delta_\ell/x_\ell^i$; and we increase the coefficient of y^0 by Δ_1/x_1^i. As a result, we now have equality for all indices in B. The Δ_j for $j \notin B$ are not affected by this change.

We still need to establish that (9) holds for the vectors y^ℓ, $0 \le \ell < b$, that have a non-zero coefficient in the convex combination. Note first that $y_j^\ell > 0$ implies $j \in B$. Also (8) holds with equality for all $j \in B$. Thus (9) holds.

Consider any iteration of the second while-loop. It adds up to b vectors to the convex decomposition and decreases the number of nonzero Δ's by b. Thus the total number of vectors added to the convex decomposition is at most s. □

2.4 Fast Convex Decomposition

Proof (of Theorem 1). Theorem 3 yields a convex combination of integer points of $\mathcal{Q}_I$ dominating $\alpha x^*/1 + 4\varepsilon$. Theorem 4 turns this dominating convex combination into an exact combination. It adds up to $|S|$ additional vectors to the convex combination. The complexity bounds follow directly from the referenced theorems. □

3 Approximately Truthful-in-Expectation Fractional Mechanisms

In this section we assume that we do not want to solve the LP-relaxation of SWM exactly and there is an FPTAS for it. Then by using the FPTAS, we construct a randomized fractional mechanism, Algorithm 3, and state the following theorem. For the proof of the theorem and FPTAS algorithm see the full paper [7].

Theorem 5. *Let $\varepsilon_0 \in (0, 1/2]$, $\varepsilon = \Theta(\frac{\varepsilon_0^5}{n^4})$ and $\gamma = (1-\varepsilon)(1-\varepsilon_0)$. Given an ε-approximation algorithm for $\mathcal{Q}$, Algorithm 3 defines a fractional randomized mechanism with the following conditions:*

No positive transfer. (11)

Individually rational with probability $1 - \varepsilon_0$. (12)

$(1-\varepsilon_0)$*-truthful-in-expectation.* (13)

γ*-social efficiency, where γ depends on ε_0, and ε.* (14)

In order to present Algorithm 3, we make some assumption and define some notation. Let us assume that the problem is *separable* that means the variables can be partitioned into disjoint groups, one for each player, such that the value of an allocation for a player depends only on the variables in his group, i.e.,

$$v_i(x) = v_i(x_i),$$

where x_i is the set of variables associated with player i. Formally, any outcome $x \in \mathcal{Q} \subseteq \mathbb{R}^d$ can be written as $x = (x_1, \ldots, x_n)$ where $x_i \in \mathbb{R}^{d_i}$ and $d = d_1 + \ldots + d_n$.[3] We further assume that for each player $i \in [n]$, there is an optimal allocation $u^i \in \mathcal{Q}$ that maximizes his value for every valuation v_i, i.e.,

$$v_i(u^i) = \max_{z \in \mathcal{Q}} v_i(z), \tag{15}$$

for every $v_i \in \mathcal{V}_i$, where $\mathcal{V}_i$ denote the all possible valuations of player i. In combinatorial auction, the allocation u^i allocates all the items to player i. Let

$$L_i := \sum_{j \neq i} v_j(u^j) \quad \text{and} \quad \beta_i := \varepsilon L_i. \tag{16}$$

Note that L_i does not depend on the valuation of player i. Let $\mathcal{A}$ be an FPTAS for LP relaxation of SWM. We use $\mathcal{A}(v, \varepsilon)$ to denote the outcome of $\mathcal{A}$ on input v and ε. If ε is understood, we simply write $\mathcal{A}(v)$; $\mathcal{A}(v)$ is a fractional allocation in Q. In the following, we will apply $\mathcal{A}$ to different valuations which we denote by $v = (v_i, v_{-i})$, $\bar{v} = (\bar{v}_i, v_{-i})$, and $v' = (\mathbf{0}, v_{-i})$. Here v_i is the reported valuation of player i, $\bar{v}_i$ is his true valuation and $v'_i = \mathbf{0}$. We denote the allocation returned by $\mathcal{A}$ on input v (resp., $\bar{v}$, v') by x (resp., $\bar{x}$, x') and use the payment rule:

$$p_i(v) := \max\{p_i^{VCG}(v) - \beta_i, 0\} \tag{17}$$

where

$$p_i^{VCG}(v) := v_{-i}(x') - v_{-i}(x).$$

$v_{-i}(x) = \sum_{j \neq i} v_j(x), x = \mathcal{A}(v)$ and $x' = \mathcal{A}(0, v_{-i})$. Observe the similarity in the definition of $p_i^{VCG}(v)$ to the VCG payment rule. In both cases, the payment is defined as the difference of the total value of two allocations to the players different from i. The first allocation ignores the influence of player i ($x' = \mathcal{A}(0, v_{-i})$) and the second allocation takes it into account ($x = \mathcal{A}(v)$). Define $q_0 = (1 - \frac{\varepsilon_0}{n})^n$, $\bar{\varepsilon} = \varepsilon_0/2$, and $q_j = (1 - q_0)/n$ for $1 \leq j \leq n$. Let $\eta = \bar{\varepsilon}(1 - q_0)^2/n^3$, $\eta' = \eta/q_j$, and $\varepsilon = \eta\bar{\varepsilon}(1 - q_0)/(8n)$. Let $U_i(v)$ be the utility of player i obtained by the mechanism which has an allocation function $\mathcal{A}$ and payment rule (17). Following [5], we call player i *active* if the following two conditions hold:

$$U_i(v) + \frac{\bar{\varepsilon} q_i}{q_0} v_i(u^i) \geq \frac{q_i}{q_0} \eta' L_i, \tag{18}$$

$$v_i(u^i) \geq \eta L_i. \tag{19}$$

[3] In the combinatorial auction problem, variable x_i comprises all variables $x_{i,S}$ and the value of an allocation for player i depends only on the variables $x_{i,S}$.

Algorithm 3. The mechanism M of Theorem 5. The vectors u^i are defined as in (15) and the quantities L_i are defined in (16). The definitions of q_0, q_j, active and inactive player are given in the proof of Theorem 5.

Require: A valuation vector v, a packing convex set $\mathcal{Q}$ and an approximation scheme $\mathcal{A}$.
Ensure: An allocation $x \in \mathcal{Q}$ and a payment $p \in \mathbb{R}^n$
1: Let ε be defined as in the below.
2: Choose an index $j \in \{0, 1, \ldots, n\}$, where 0 is chosen with probability q_0 and $j \in \{1, \ldots, n\}$ is chosen with probability $q_j = (1 - q_0)/n$.
3: **if** $j = 0$ **then**
4: Use ε-approximation algorithm $\mathcal{A}$ to compute an allocation $x = (x_1, \ldots, x_n) \in \mathcal{Q}$ and compute payments with payment rule (17). For all inactive i, change x_i and p_i to zero.
5: **else**
6: For every $1 \le i \le n$, set

$$\begin{cases} x_i = u^i, p_i = \eta' L_i & \text{if } i = j \text{ and } i \text{ is active,} \\ x_i = u^i, p_i = 0 & \text{if } i = j \text{ and } i \text{ is inactive,} \\ x_i = 0, p_i = 0 & \text{if } i \neq j. \end{cases}$$

7: **end if**
8: **return** (x, p)

Now, we briefly explain Algorithm 3. Let us choose a random number $j \in \{0, 1, \ldots, n\}$ with probability q_j. If $j = 0$, we run ε-approximation algorithm $\mathcal{A}$ on v to compute allocation $x = (x_1, \ldots, x_n)$. Then we change x_i and p_i to zero for all inactive i. And if $j \neq 0$, we give optimal set u^j to j-th player and charged him with a price $\eta' L_j$ if he is active and zero otherwise. For all other players, we do not assign any item to them and do not charge them any price.

4 Approximately Truthful-in-Expectation Integral Mechanisms

In this subsection we obtain a randomized mechanism M' which returns an integral allocation. Let $\varepsilon > 0$ be arbitrary. First run Algorithm 3 to obtain x and $p(v)$. Then compute a convex decomposition of $\frac{\alpha}{1+4\varepsilon}x$, which is $\frac{\alpha}{1+4\varepsilon}x = \sum_{j \in \mathcal{N}} \lambda_j^x x^j$. Finally with probability λ_j^x (we used superscript to distinguish the convex decompositions of x) return the allocation x^j and charge i-th player, the price $p_i(v)\frac{v_i(x^j)}{v_i(x)}$, if $v_i(x) > 0$, and zero otherwise. In the following theorem we show mechanism M' is indeed an approximately truthful-in-expectation integral mechanism whose proof is appeared in the full paper [7].

Theorem 6. *Suppose that* $\varepsilon_0 \in (0, 1/2]$ *be any constant,* $\varepsilon = \Theta(\frac{\varepsilon_0^5}{n^4})$ *and* $\gamma = \alpha(1-\varepsilon)(1-\varepsilon_0)/(1+4\varepsilon)$. *Then we obtain a randomized integral mechanism satisfying Conditions (11) to (14).*

References

1. Arora, S., Hazan, E., Kale, S.: Multiplicative weights method: a meta-algorithm and its applications. Technical report, Princeton University, USA (2006)
2. Bienstock, D., Iyengar, G.: Approximating fractional packings and coverings in O(1/epsilon) iterations. SIAM J. Comput. **35**(4), 825–854 (2006)
3. Carr, R.D., Vempala, S.: Randomized metarounding. Random Struct. Algorithms **20**(3), 343–352 (2002)
4. Christodoulou, G., Elbassioni, K., Fouz, M.: Truthful mechanisms for exhibitions. In: Saberi, A. (ed.) WINE 2010. LNCS, vol. 6484, pp. 170–181. Springer, Heidelberg (2010)
5. Dughmi, S., Roughgarden, T., Vondrák, J., Yan, Q.: An approximately truthful-in-expectation mechanism for combinatorial auctions using value queries. CoRR abs/1109.1053 (2011)
6. Elbassioni, K., Jha, M.: On the power of combinatorial bidding in web display ads. In: ICIW (2015, to appear)
7. Elbassioni, K., Mehlhorn, K., Ramezani, F.: Towards more practical linear programming-based techniques for algorithmic mechanism design. In: Arxive (2015)
8. Garg, N., Könemann, J.: Faster and simpler algorithms for multicommodity flow and other fractional packing problems. In: FOCS, pp. 300–309 (1998)
9. Grigoriadis, M.D., Khachiyan, L.G.: A sublinear-time randomized approximation algorithm for matrix games. Oper. Res. Lett. **18**(2), 53–58 (1995)
10. Grötschel, M., Lovász, L., Schrijver, A.: Geometric Algorithms and Combinatorial Optimization. Springer, New York (1988)
11. Hoefer, M., Kesselheim, T., Vöcking, B.: Approximation algorithms for secondary spectrum auctions. In: SPAA, pp. 177–186 (2011)
12. Khandekar, R.: Lagrangian Relaxation Based Algorithms for convex Programming Problems. PhD thesis, Indian Institute of Technology, Delhi, India (2004)
13. Koufogiannakis, C., Young, N.E.: Beating simplex for fractional packing and covering linear programs. In: FOCS, pp. 494–504 (2007)
14. Kraft, D., Fadaei, S., Bichler, M.: Fast convex decomposition for truthful social welfare approximation. In: Liu, T.-Y., Qi, Q., Ye, Y. (eds.) WINE 2014. LNCS, vol. 8877, pp. 120–132. Springer, Heidelberg (2014)
15. Lavi, R., Swamy, C.: Truthful and near-optimal mechanism design via linear programming. In: FOCS, pp. 595–604 (2005)
16. Lavi, R., Swamy, C.: Truthful and near-optimal mechanism design via linear programming. J. ACM **58**(6), 25 (2011)
17. Nisan, N., Roughgarden, T., Tardos, E., Vazirani, V.V.: Algorithmic Game Theory. Cambridge University Press, Cambridge (2007)
18. Plotkin, S.A., Shmoys, D.B., Tardos, É.: Fast approximation algorithms for fractional packing and covering problems. In: FOCS, pp. 495–504 (1991)
19. Young, N.E.: Sequential and parallel algorithms for mixed packing and covering. In: FOCS, pp. 538–546 (2001)

Equilibria of Plurality Voting: Lazy and Truth-Biased Voters

Edith Elkind[1], Evangelos Markakis[2], Svetlana Obraztsova[3], and Piotr Skowron[4](✉)

[1] University of Oxford, Oxford, UK
[2] Athens University of Economics and Business, Athens, Greece
[3] Tel Aviv University, Tel Aviv, Israel
[4] University of Warsaw, Warsaw, Poland
p.skowron@mimuw.edu.pl

Abstract. We present a systematic study of Plurality elections with strategic voters who, in addition to having preferences over election winners, also have secondary preferences, governing their behavior when their vote cannot affect the election outcome. Specifically, we study two models that have been recently considered in the literature: *lazy* voters, who prefer to abstain when they are not pivotal, and *truth-biased* voters, who prefer to vote truthfully when they are not pivotal. For both lazy and truth-biased voters, we are interested in their behavior under different tie-breaking rules (lexicographic rule, random voter rule, random candidate rule). Two of these six combinations of secondary preferences and tie-breaking rules have been studied in prior work; for the remaining four, we characterize pure Nash equilibria (PNE) of the resulting strategic games and study the complexity of related computational problems. We then use these results to analyze the impact of different secondary preferences and tie-breaking rules on the election outcomes. Our results extend to settings where some of the voters are non-strategic.

1 Introduction

Plurality voting is a popular tool for collective decision-making in many domains, including both human societies and multiagent systems. Under this voting rule, each voter is supposed to vote for her most favorite candidate (or abstain); the winner is then the candidate that receives the highest number of votes. If several candidates have the highest score, the winner is chosen among them using a *tie-breaking rule*; popular tie-breaking rules include the *lexicographic rule*, which imposes a fixed priority order over the candidates; the *random candidate rule*, which picks one of the tied candidates uniformly at random; and the *random voter rule*, which picks the winner among the tied candidates according to the preferences of a randomly chosen voter.

In practice, voters are often *strategic*, i.e., they may vote non-truthfully if they can benefit from doing so. In that case, an election can be viewed as a game, where the voters are the players, and each player's space of actions includes voting for any candidate or abstaining. For deterministic rules (such as Plurality

M. Hoefer (Ed.): SAGT 2015, LNCS 9347, pp. 110–122, 2015.
DOI: 10.1007/978-3-662-48433-3_9

with lexicographic tie-breaking), the behavior of strategic voters is determined by their preference ordering, i.e., a ranking of the candidates, whereas for randomized rules a common approach is to specify utility functions for the voters; i.e., the voters are assumed to maximize their *expected utility* under the lottery induced by tie-breaking. The outcome of the election can then be identified with a pure Nash equilibrium (PNE) of the resulting game.

However, under Plurality and with 3 or more voters, this approach fails to provide a useful prediction of voting behavior: for each candidate c there is a PNE where c is the unique winner, irrespective of the voters' preferences. Indeed, if there are at least 3 voters, the situation where all of them vote for c is a PNE, as no voter can change the election outcome. Such equilibria may disappear if we use a more refined model of voters' preferences that captures additional aspects of their decision-making. For instance, in practice, if a voter feels that her vote is unlikely to have any effect on the outcome, she may decide to abstain from the election. Also, voters may be averse to lying about their preferences, in which case they can be expected to vote for their top candidate unless there is a clear strategic reason to vote for someone else. By taking into account these aspects of voters' preferences, we can obtain a more faithful model of their behavior.

The problem of characterizing and computing the equilibria of Plurality voting, both for "lazy" voters (i.e., ones who prefer to abstain when they are not pivotal) and for "truth-biased" voters (ones who prefer to vote truthfully when they are not pivotal), has recently received a considerable amount of attention. However, it is difficult to compare the existing results, since they rely on different tie-breaking rules. In particular, Desmedt and Elkind [6], who study lazy voters, use the random candidate tie-breaking rule, and Obraztsova et al. [17] consider truth-biased voters and the lexicographic tie-breaking rule. Thus, it is not clear whether the differences between the results in these papers can be attributed to the voters' secondary preferences, or to the tie-breaking rule.

The primary goal of our paper is to tease out the effects of different features of these models, by systematically considering all the combinations of secondary preferences and tie-breaking rules. We consider two types of secondary preferences (lazy voters and truth-biased voters) and three tie-breaking rules (the lexicographic rule, the random voter rule, and the random candidate rule); while two of these combinations have been studied earlier by [6, 17], to the best of our knowledge, the remaining four possibilities have not been considered before. For each of the new scenarios, we characterize the set of PNE for the resulting game; in doing so, we also fill in a gap in the characterization of [6] for lazy voters and random candidate tie-breaking. We then consider the problems of deciding whether a given game admits a PNE and whether a given candidate can be a co-winner/unique winner in some PNE of a given game. For all settings, we determine the computational complexity of each of these problems, classifying them as either polynomial-time solvable or NP-complete. Our characterization results enable us to analyze the impact of various features of our model on the election outcomes, and thereby evaluate the plausibility of our assumptions about voters' secondary preferences. Finally, we briefly discuss the implications of our results in the setting where some of the voters may be *principled*, i.e., always vote truthfully.

Related Work. Equilibria of Plurality voting have been investigated by a number of researchers, starting with [10]. However, most of the earlier works either consider solution concepts other than pure Nash equilibria, such as iterative elimination of dominated strategies [7,13], or assume that voters have incomplete information about each others' preferences [14]. Both types of secondary preferences (lazy voters and truth-biased voters) appear in the social choice literature, see, respectively, [2,3,19] and [8,11]. In computational social choice, truth-biased voters have been considered by Meir et al. [12] in the context of dynamics of Plurality voting; subsequently, Plurality elections with truth-biased voters have been investigated empirically by Thompson et al. [20] and theoretically by Obraztsova et al. [17]. To the best of our knowledge, the only paper to study computational aspects of Plurality voting with lazy voters is that of Desmedt and Elkind [6].

Our approach to tie-breaking is well-grounded in existing work. Lexicographic tie-breaking is standard in the computational social choice literature. The random candidate rule has been discussed by [6], and, more recently, by [15,16]. The random voter rule is used to break ties under the Schulze method [18]; the complexity of manipulation under this tie-breaking rule has been studied by [1].

2 Preliminaries

For any positive integer t, we denote the set $\{1, \ldots, t\}$ by $[t]$. We consider elections with a set of *voters* $N = [n]$ and a set of *alternatives*, or *candidates*, $C = \{c_1, \ldots c_m\}$. Each voter is associated with a *preference order*, i.e., a strict linear order over C; we denote the preference order of voter i by $\succ_i$. The list $(\succ_1, \ldots, \succ_n)$ is called a *preference profile*. For each $i \in N$, we set a_i to be the top choice of voter i, and let $\mathbf{a} = (a_1, \ldots, a_n)$. Given two disjoint sets of candidates X, Y and a preference order $\succ$, we write $X \succ Y$ if in $\succ$ all candidates from X are ranked above all candidates from Y.

We also assume that each voter $i \in N$ is endowed with a *utility function* $u_i : C \to \mathbb{N}$; $u_i(c_j)$ is the utility derived by voter i if c_j is the unique election winner. We require that $u_i(c) \neq u_i(c')$ for all $i \in N$ and all $c, c' \in C$ such that $c \neq c'$. The vector $\mathbf{u} = (u_1, \ldots, u_n)$ is called the *utility profile*. Voters' preference orders and utility functions are assumed to be consistent, i.e., for each $i \in N$ and every pair of candidates $c, c' \in C$ we have $c \succ_i c'$ if and only if $u_i(c) > u_i(c')$; when this is the case, we will also say that $\succ_i$ is *induced* by u_i. Sometimes, instead of specifying preference orders explicitly, we will specify the utility functions only, and assume that voters' preference orders are induced by their utility functions; on other occasions, it will be convenient to reason in terms of preference orders.

A *lottery* over C is a vector $\mathbf{p} = (p_1, \ldots, p_m)$ with $p_j \geq 0$ for all $j \in [m]$ and $\sum_{j \in [m]} p_j = 1$. The value p_j is the probability assigned to candidate c_j. The *expected utility* of a voter $i \in N$ from a lottery $\mathbf{p}$ is given by $\sum_{j \in [m]} u_i(c_j) p_j$.

In this work, we consider Plurality elections, where each voter $i \in N$ submits a *vote*, or *ballot*, $b_i \in C \cup \{\varnothing\}$; if $b_i = \varnothing$, voter i is said to *abstain*. The list of all votes $\mathbf{b} = (b_1, \dots, b_n)$ is also called a *ballot vector*. We say that a ballot vector is *trivial* if $b_i = \varnothing$ for all $i \in N$. Given a ballot vector $\mathbf{b}$ and a ballot b', we write $(\mathbf{b}_{-i}, b')$ to denote the ballot vector obtained from $\mathbf{b}$ by replacing b_i with b'. The *score* of an alternative c_j in an election with ballot vector $\mathbf{b}$ is given by $\text{sc}(c_j, \mathbf{b}) = |\{i \in N \mid b_i = c_j\}|$. Given a ballot vector $\mathbf{b}$, we set $M(\mathbf{b}) = \max_{c \in C} \text{sc}(c, \mathbf{b})$ and let $W(\mathbf{b}) = \{c \in C \mid \text{sc}(c, \mathbf{b}) = M(\mathbf{b})\}$, $H(\mathbf{b}) = \{c \in C \mid \text{sc}(c, \mathbf{b}) = M(\mathbf{b}) - 1\}$, $H'(\mathbf{b}) = \{c \in C \mid \text{sc}(c, \mathbf{b}) = M(\mathbf{b}) - 2\}$. These sets are useful in our analysis in the next sections. The set $W(\mathbf{b})$ is called the *winning set*. Note that if $\mathbf{b}$ is trivial then $W(\mathbf{b}) = C$. If $|W(\mathbf{b})| > 1$, the winner is selected from $W(\mathbf{b})$ according to one of the following tie-breaking rules.

(1) Under the *lexicographic rule* R^L, the winner is the candidate $c_j \in W(\mathbf{b})$ such that $j \leq k$ for all $c_k \in W(\mathbf{b})$.
(2) Under the *random candidate rule* R^C, the winner is chosen from $W(\mathbf{b})$ uniformly at random.
(3) Under the *random voter rule* R^V, we select a voter from N uniformly at random; if she has voted for a candidate in $W(\mathbf{b})$, we output this candidate, otherwise we ask this voter to report her most preferred candidate in $W(\mathbf{b})$, and output the answer. This additional elicitation step may appear difficult to implement in practice; fortunately, we can show that in equilibrium it is almost never necessary.

Thus, the outcome of an election is a lottery over C; however, for R^L this lottery is degenerate, i.e., it always assigns the entire probability mass to a single candidate. For each $X \in \{L, C, V\}$ and each ballot vector $\mathbf{b}$, let $\mathbf{p}^X(\mathbf{b})$ denote the lottery that corresponds to applying R^X to the set $W(\mathbf{b})$. From the definition of R^C, it follows that for every $c_j \in C$ it holds that if $p_j^C(\mathbf{b}) \neq 0$ then $p_j^C(\mathbf{b}) \geq \frac{1}{m}$. Similarly, for R^V, it follows that if $p_j^V(\mathbf{b}) \neq 0$ then $p_j^V(\mathbf{b}) \geq \frac{1}{n}$.

In what follows, we focus on two types of secondary preferences, namely, *lazy* voters, who prefer to abstain when their vote has no effect on the election outcome, and *truth-biased* voters, who never abstain, but prefer to vote truthfully when their vote has no effect on the election outcome. Formally, pick $\varepsilon < \min\{\frac{1}{m^2}, \frac{1}{n^2}\}$, and consider a utility profile $\mathbf{u}$ and a tie-breaking rule $R^X \in \{R^C, R^V, R^L\}$. Then

- if voter i is *lazy*, her utility in an election with ballot vector $\mathbf{b}$ under tie-breaking rule R^X is given by

$$U_i(\mathbf{b}) = \begin{cases} \sum_{j \in [m]} p_j^X(\mathbf{b}) u_i(c_j), & \text{if } b_i \in C, \\ \sum_{j \in [m]} p_j^X(\mathbf{b}) u_i(c_j) + \varepsilon, & \text{if } b_i = \varnothing. \end{cases}$$

- if voter i is *truth-biased*, her utility in an election with ballot vector $\mathbf{b}$ under tie-breaking rule R^X is given by

$$U_i(\mathbf{b}) = \begin{cases} \sum_{j \in [m]} p_j^X(\mathbf{b}) u_i(c_j), & \text{if } b_i \in C \setminus \{a_i\}, \\ \sum_{j \in [m]} p_j^X(\mathbf{b}) u_i(c_j) + \varepsilon, & \text{if } b_i = a_i, \\ -\infty, & \text{if } b_i = \varnothing. \end{cases}$$

We consider settings where all voters are of the same type, i.e., either all voters are lazy or all voters are truth-biased; we refer to these settings as *lazy* or *truth-biased*, respectively, and denote the former by $\mathcal{L}$ and the latter by $\mathcal{T}$.

We investigate all possible combinations of settings ($\mathcal{L}$, $\mathcal{T}$) and tie-breaking rules (R^L, R^C, R^V). A combination of a setting $\mathcal{S} \in \{\mathcal{L}, \mathcal{T}\}$, a tie-breaking rule $R \in \{R^L, R^C, R^V\}$ and a utility profile $\mathbf{u}$ induces a strategic game, which we will denote by $(\mathcal{S}, R, \mathbf{u})$: in this game, the players are the voters, the action space of each player is $C \cup \{\varnothing\}$, and the players' utilities $U_1, \dots, U_n$ for a vector of actions $\mathbf{b}$ are computed based on the setting and the tie-breaking rule as described above. We say that a ballot vector $\mathbf{b}$ is a *pure Nash equilibrium (PNE)* of the game $(\mathcal{S}, R, \mathbf{u})$ if $U_i(\mathbf{b}) \geq U_i(\mathbf{b}_{-i}, b')$ for every voter $i \in N$ and every $b' \in C \cup \{\varnothing\}$.

For each setting $\mathcal{S} \in \{\mathcal{L}, \mathcal{T}\}$ and each tie-breaking rule $R \in \{R^L, R^C, R^V\}$, we define three algorithmic problems, which we call $(\mathcal{S}, R)$-ExistNE, $(\mathcal{S}, R)$-TieNE, and $(\mathcal{S}, R)$-SingleNE. In each of these problems, we are given a candidate set C, $|C| = m$, a voter set N, $|N| = n$, and a utility vector $\mathbf{u} = (u_1, \dots, u_n)$, where each u_i is represented by m numbers $u_i(c_1), \dots, u_i(c_m)$; these numbers are positive integers given in binary. In $(\mathcal{S}, R)$-TieNE and $(\mathcal{S}, R)$-SingleNE we are also given the name of a target candidate $c_p \in C$. In $(\mathcal{S}, R)$-ExistNE we ask if $(\mathcal{S}, R, \mathbf{u})$ has a PNE. In $(\mathcal{S}, R)$-TieNE we ask if $(\mathcal{S}, R, \mathbf{u})$ has a PNE $\mathbf{b}$ with $|W(\mathbf{b})| > 1$ and $c_p \in W(\mathbf{b})$. In $(\mathcal{S}, R)$-SingleNE we ask if $(\mathcal{S}, R, \mathbf{u})$ has a PNE $\mathbf{b}$ with $W(\mathbf{b}) = \{c_p\}$. Each of these problems is obviously in NP, as we can simply guess an appropriate ballot vector $\mathbf{b}$ and check that it is a PNE.

In what follows, we omit some proofs due to space constraints; the omitted proofs can be found in the full version of the paper [9].

3 Lazy Voters

In this section, we study PNE in Plurality games with lazy voters. The case where the tie-breaking rule is R^C has been analyzed in detail by Desmedt and Elkind [6], albeit for a slightly different model; we complement their results by considering R^L and R^V.

We start by extending a result of [6] to all three tie-breaking rules considered here.

Proposition 1. *For every $R \in \{R^L, R^C, R^V\}$ and every utility profile $\mathbf{u}$, if a ballot vector $\mathbf{b}$ is a PNE of $(\mathcal{L}, R, \mathbf{u})$ then for every voter $i \in N$ either $b_i = \varnothing$ or $b_i \in W(\mathbf{b})$. If $|W(\mathbf{b})| = 1$, there is exactly one voter $i \in N$ with $b_i \neq \varnothing$.*

Proof. Suppose that $b_i \neq \varnothing$, $b_i \notin W(\mathbf{b})$ for some voter $i \in N$. Then if i changes her vote to $\varnothing$, the set $W(\mathbf{b})$ will not change, so i's utility would improve by ε, a contradiction with $\mathbf{b}$ being a PNE of $(\mathcal{L}, R, \mathbf{u})$. Similarly, suppose that $|W(\mathbf{b})| = 1$ and there are two voters $i, i' \in N$ with $b_i \neq \varnothing$, $b_{i'} \neq \varnothing$. It has to be the case that $b_i = b_{i'} = c_j$ for some $c_j \in C$, since otherwise $|W(\mathbf{b})| > 1$. But then if voter i changes her vote to $\varnothing$, c_j will remain the election winner, so i's utility would improve by ε, a contradiction.

Lexicographic Tie-breaking. The scenario where voters are lazy and ties are broken lexicographically turns out to be fairly easy to analyze.

Theorem 1. *For any utility profile* $\mathbf{u}$ *the game* $G = (\mathcal{L}, R^L, \mathbf{u})$ *has the following properties:*

(1) If $\mathbf{b}$ *is a PNE of* G *then* $|W(\mathbf{b})| \in \{1, m\}$*. Moreover,* $|W(\mathbf{b})| = m$ *if and only if* $\mathbf{b}$ *is the trivial ballot and all voters rank* c_1 *first.*
(2) If $\mathbf{b}$ *is a PNE of* G *then there exists at most one voter* i *with* $b_i \neq \varnothing$*.*
(3) G *admits a PNE if and only if all voters rank* c_1 *first (in which case* c_1 *is the unique PNE winner) or there exists a candidate* c_j *with* $j > 1$ *such that (i)* $\mathrm{sc}(c_j, \mathbf{a}) > 0$ *and (ii) for every* $k < j$ *it holds that all voters prefer* c_j *to* c_k*. If such a candidate exists, he is unique, and wins in all PNE of* G*.*

The following corollary is directly implied by Theorem 1.

Corollary 1. $(\mathcal{L}, R^L)$-ExistNE, $(\mathcal{L}, R^L)$-SingleNE *and* $(\mathcal{L}, R^L)$-TieNE *are in* P.

Remark 1. The reader may observe that, counterintuitively, while the lexicographic tie-breaking rule appears to favor c_1, it is impossible for c_1 to win the election unless he is ranked first by all voters. In contrast, c_2 wins the election as long as he is ranked first by at least one voter and no voter prefers c_1 to c_2. In general, the lexicographic tie-breaking rule favors lower-numbered candidates with the exception of c_1. As for c_1, his presence mostly has a destabilizing effect: if some, but not all voters rank c_1 first, no PNE exists. This phenomenon is an artifact of our treatment of the trivial ballot vector: it disappears if we assume (as [6] does) that when $\mathbf{b} = (\varnothing, \dots, \varnothing)$ the election is declared invalid and the utility of each voter is $-\infty$: under this assumption c_1 is the unique possible equilibrium winner whenever he is ranked first by at least one voter.

Randomized Tie-breaking. We now consider R^C and R^V. Desmedt and Elkind [6] give a characterization of utility profiles that admit a PNE for lazy voters and R^C. However, there is a small difference between our model and theirs regarding the trivial ballot vector, as explained in Remark 1 above. Further, their results implicitly assume that the number of voters n exceeds the number of candidates m; if this is not the case, Theorem 2 in their paper is incorrect (see Remark 2).

Thus, we will now provide a full characterization of utility profiles $\mathbf{u}$ such that $(\mathcal{L}, R^C, \mathbf{u})$ admits a PNE, and describe the corresponding equilibrium ballot profiles. Our characterization result remains essentially unchanged if we replace R^C with R^V: for almost all utility profiles $\mathbf{u}$ and ballot vectors $\mathbf{b}$ it holds that $\mathbf{b}$ is a PNE of $(\mathcal{L}, R^C, \mathbf{u})$ if and only if it is a PNE of $(\mathcal{L}, R^V, \mathbf{u})$; the only exception is the case of full consensus (all voters rank the same candidate first).

Theorem 2. *Let* $\mathbf{u} = (u_1, \dots, u_n)$ *be a utility profile over* C, $|C| = m$, *and let* $R \in \{R^C, R^V\}$*. The game* $G = (\mathcal{L}, R, \mathbf{u})$ *admits a PNE if and only if one of the following conditions holds:*

(1) all voters rank some candidate c_j first;
(2) each candidate is ranked first by at most one voter, and $\forall \ell \in N$: $\frac{1}{n}\sum_{i\in N} u_\ell(a_i) \geq \max_{i\in N\setminus\{\ell\}} u_\ell(a_i)$.
(3) there exists a set of candidates $X = \{c_{\ell_1}, \ldots, c_{\ell_k}\}$ *with* $2 \leq k \leq \min(n/2, m)$ *and a partition of the voters into* k *groups* $N_1, \ldots, N_k$ *of size* n/k *each such that for each* $j \in [k]$ *and each* $i \in N_j$ *we have* $c_{\ell_j} \succ_i c$ *for all* $c \in X \setminus \{c_{\ell_j}\}$, *and, moreover,* $\frac{1}{k}\sum_{c\in X} u_i(c) \geq \max_{c\in X\setminus\{c_{\ell_j}\}} u_i(c)$.

Further, when condition (1) holds for some $c_j \in C$ *and* $R = R^C$, *then for each* $i \in N$ *the game* G *has a PNE where* i *votes for* c_j *and all other voters abstain, whereas if* $R = R^V$, *the game* G *has a PNE where all voters abstain; if condition (2) holds, then* G *has a PNE where each voter votes for her top candidate; and if condition (3) holds for some set* X, *then* G *has a PNE where each voter votes for her favorite candidate in* X. *The game* G *has no other PNE.*

Remark 2. Desmedt and Elkind [6] claim (Theorems 1 and 2) that for R^C and lazy voters, a PNE exists if and only if the utility profile satisfies either condition (1) or (3) with the constraint $k \leq n/2$ removed. To see why this is incorrect, consider a 2-voter election over $C = \{x, y, z\}$, where the voters' utility functions are consistent with preference orders $x \succ y \succ z$ and $x \succ z \succ y$, respectively. According to [6], the vector (y, z) is a PNE. This is obviously not true: each of the voters would prefer to change her vote to x. Note, however, that the two characterizations differ only when $m \geq n$, and in practice the number of voters usually exceeds the number of candidates.

Desmedt and Elkind [6] show that checking condition (3) of Theorem 2 is NP-hard; in their proof $n > m$, and the proof does not depend on how the trivial ballot is handled. Further, their proof shows that checking whether a given candidate belongs to some such set X is also NP-hard. On the other hand, Theorem 2 shows that PNE with singleton winning sets only arise if some candidate is unanimously ranked first, and this condition is easy to check. We summarize these observations as follows.

Corollary 2. *For* $R \in \{R^C, R^V\}$, *the problems* $(\mathcal{L}, R)$-ExistNE *and* $(\mathcal{L}, R)$-TieNE *are* NP*-complete, whereas* $(\mathcal{L}, R)$-SingleNE *is in* P.

4 Truth-Biased Voters

For truth-biased voters, our exposition follows the same pattern as for lazy voters: we present some general observations, followed by a quick summary of the results for lexicographic tie-breaking, and continue by analyzing randomized tie-breaking. The following result is similar in spirit to Proposition 1.

Proposition 2. *For every* $R \in \{R^L, R^C, R^V\}$ *and every utility profile* $\mathbf{u}$, *if a ballot vector* $\mathbf{b}$ *is a PNE of* $(\mathcal{T}, R, \mathbf{u})$ *then for every voter* $i \in N$ *we have* $b_i = a_i$ *or* $b_i \in W(\mathbf{b})$.

Lexicographic Tie-breaking. Obraztsova et al. [17] characterize the PNE of the game $(\mathcal{T}, R^L, \mathbf{u})$. Their characterization is quite complex, and we will not reproduce it here. However, for the purposes of comparison with the lazy voters model, we will use the following description of *truthful* equilibria.

Proposition 3 (Obraztsova et al. [17], Theorem 1). *Consider a utility profile* $\mathbf{u}$*, let* $\mathbf{a}$ *be the respective truthful ballot vector, and let* $j = \min\{r \mid c_r \in W(\mathbf{a})\}$*. Then* $\mathbf{a}$ *is a PNE of* $(\mathcal{T}, R^L, \mathbf{u})$ *if and only if neither of the following conditions holds:*

(1) $|W(\mathbf{a})| > 1$*, and there exists a candidate* $c_k \in W(\mathbf{a})$ *and a voter* i *such that* $a_i \neq c_k$ *and* $c_k \succ_i c_j$*.*
(2) $H(\mathbf{a}) \neq \emptyset$*, and there exists a candidate* $c_k \in H(\mathbf{a})$ *and a voter* i *such that* $a_i \neq c_k$*,* $c_k \succ_i c_j$*, and* $k < j$*.*

We will also utilize a crucial property of non-truthful PNE. For this, we first need the following definition.

Definition 1. Consider a ballot vector $\mathbf{b}$, where candidate c_j is the winner under R^L. A candidate $c_k \neq c_j$ is called a *threshold candidate with respect to* $\mathbf{b}$ if either (1) $k < j$ and $\mathrm{sc}(c_k, \mathbf{b}) = \mathrm{sc}(c_j, \mathbf{b}) - 1$ or (2) $k > j$ and $\mathrm{sc}(c_k, \mathbf{b}) = \mathrm{sc}(c_j, \mathbf{b})$. We denote the set of threshold candidates with respect to $\mathbf{b}$ by $T(\mathbf{b})$.

That is, a threshold candidate is someone who could win the election if he had one additional vote. A feature of all non-truthful PNE is that there must exist at least one threshold candidate. The intuition for this is that, since voters who are not pivotal prefer to vote truthfully, in any PNE that arises under strategic voting, the winner receives just enough votes so as to beat the required threshold (as set by the threshold candidate) and not more. Formally, we have the following lemma.

Lemma 1 (Obraztsova et al. [17], Lemma 2). *Consider a utility profile* $\mathbf{u}$*, let* $\mathbf{a}$ *be the respective truthful ballot vector, and let* $\mathbf{b} \neq \mathbf{a}$ *be a non-truthful PNE of* $(\mathcal{T}, R^L, \mathbf{u})$*. Then* $T(\mathbf{b}) \neq \emptyset$*. Further,* $\mathrm{sc}(c_k, \mathbf{b}) = \mathrm{sc}(c_k, \mathbf{a})$ *for every* $c_k \in T(\mathbf{b})$*, i.e., all voters whose top choice is* c_k *vote for* c_k*.*

The existence of a threshold candidate is an important observation about the structure of non-truthful PNE, and we will use it repeatedly in the sequel. Note that the winner in $\mathbf{a}$ does not have to be a threshold candidate in a non-truthful PNE $\mathbf{b}$.

Obraztsova et al. show that, given a candidate $c_p \in C$ and a score s, it is computationally hard to decide whether the game $(\mathcal{T}, R^L, \mathbf{u})$ has a PNE $\mathbf{b}$ where c_p wins with a score of s. This problem may appear to be "harder" than $(\mathcal{T}, R^L)$-TieNE or $(\mathcal{T}, R^L)$-SingleNE, as one needs to ensure that c_p obtains a specific score; on the other hand, it does not distinguish between c_p being the unique top-scorer or being tied with other candidates and winning due to tie-breaking. We now complement this hardness result by showing that all three problems we consider are NP-hard for $\mathcal{T}$ and R^L.

Theorem 3. $(\mathcal{T}, R^L)$-SINGLENE, $(\mathcal{T}, R^L)$-EXISTNE, *and* $(\mathcal{T}, R^L)$-TIENE *are* NP*-complete.*

The proof is by reduction from MAXIMUM k-SUBSET INTERSECTION (MSI); see [9] for a formal definition of this problem. Surprisingly, the complexity of MSI was very recently posed as an open problem by Clifford and Popa [5]; subsequently, MSI was shown to be hard under Cook reductions in [21]. In our proof we first establish NP-hardness of MSI under Karp reductions, which may be of independent interest, and then show NP-hardness of our problems by constructing reductions from MSI.

Randomized Tie-breaking. It turns out that for truth-biased voters, the tie-breaking rules R^C and R^V induce identical behavior by the voters; unlike for lazy voters, this holds even if all voters rank the same candidate first.

For clarity, we present our characterization result for randomized tie-breaking in three parts. We start by considering PNE with winning sets of size at least 2; the analysis for this case turns out to be very similar to that for lazy voters.

Theorem 4. *Let* $\mathbf{u} = (u_1, \ldots, u_n)$ *be a utility profile over* C, $|C| = m$, *and let* $R \in \{R^C, R^V\}$. *The game* $G = (\mathcal{T}, R, \mathbf{u})$ *admits a PNE with a winning set of size at least* 2 *if and only if one of the following conditions holds:*

(1) each candidate is ranked first by at most one voter, and, moreover, $\frac{1}{n}\sum_{i\in N} u_\ell(a_i) \geq \max_{i\in N\setminus\{\ell\}} u_\ell(a_i)$ *for each* $\ell \in N$.
(2) there exists a set of candidates $X = \{c_{\ell_1}, \ldots, c_{\ell_k}\}$ *with* $2 \leq k \leq \min(n/2, m)$ *and a partitioning of the voters into* k *groups* $N_1, \ldots, N_k$ *of size* n/k *each such that for each* $j \in [k]$ *and each* $i \in N_j$ *we have* $c_{\ell_j} \succ_i c$ *for all* $c \in X \setminus \{c_{\ell_j}\}$, *and, moreover,* $\frac{1}{k}\sum_{c\in X} u_i(c) \geq \max_{c\in X\setminus\{c_{\ell_j}\}} u_i(c)$.

Further, if condition (1) holds, then G *has a PNE where each voter votes for her top candidate, and if condition (2) holds for some* X, *then* G *has a PNE where each voter votes for her favorite candidate in* X. *The game* G *has no other PNE.*

The case where the winning set is a singleton is surprisingly complicated. We will first characterize utility profiles that admit a truthful PNE with this property.

Theorem 5. *Let* $\mathbf{u} = (u_1, \ldots, u_n)$ *be a utility profile over* C, *let* $R \in \{R^C, R^V\}$, *and suppose that* $W(\mathbf{a}) = \{c_j\}$ *for some* $c_j \in C$. *Then* $\mathbf{a}$ *is a PNE of the game* $G = (\mathcal{T}, R, \mathbf{u})$ *if and only if for every* $i \in N$ *and every* $c_k \in H(\mathbf{a}) \setminus \{a_i\}$, *it holds that* $c_j \succ_i c_k$.

Finally, we consider elections that have non-truthful equilibria with singleton winning sets.

Theorem 6. *Let* $\mathbf{u} = (u_1, \ldots, u_n)$ *be a utility profile over* C, *let* $R \in \{R^C, R^V\}$, *and consider a ballot vector* $\mathbf{b}$ *with* $W(\mathbf{b}) = \{c_j\}$ *for some* $c_j \in C$ *and* $b_r \neq a_r$ *for some* $r \in N$. *Then* $\mathbf{b}$ *is a PNE of the game* $G = (\mathcal{T}, R, \mathbf{u})$ *if and only if all of the following conditions hold:*

(1) $b_i \in \{a_i, c_j\}$ *for all* $i \in N$;
(2) $H(\mathbf{b}) \neq \emptyset$;
(3) $c_j \succ_i c_k$ *for all* $i \in N$ *and all* $c_k \in H(\mathbf{b}) \setminus \{b_i\}$;
(4) for every candidate $c_\ell \in H'(\mathbf{b})$ *and each voter* $i \in N$ *with* $b_i = c_j$, *i prefers* c_j *to the lottery where a candidate is chosen from* $H(\mathbf{b}) \cup \{c_j, c_\ell\}$ *according to* R.

We now consider the complexity of ExistNE, TieNE, and SingleNE for truth-biased voters and randomized tie-breaking. The reader may observe that the characterization of PNE with ties in Theorem 4 is essentially identical to the one in Theorem 2. As a consequence, we immediately obtain that $(\mathcal{T}, R^C)$-TieNE and $(\mathcal{T}, R^V)$-TieNE are NP-hard. For ExistNE and SingleNE, a simple modification of the proof of Theorem 3 shows that these problems remain hard under randomized tie-breaking. These observations are summarized in the following corollary.

Corollary 3. *For* $R \in \{R^C, R^V\}$, $(\mathcal{T}, R)$-SingleNE, $(\mathcal{T}, R)$-TieNE, *and* $(\mathcal{T}, R)$-ExistNE *are* NP-*complete.*

5 Comparison

We are finally in a position to compare the different models considered in this paper.

Tie-breaking Rules. We have demonstrated that in equilibrium the two randomized tie-breaking rules (R^C and R^V) induce very similar behavior, and identical election outcomes, both for lazy and for truth-biased voters. This is quite remarkable, since under truthful voting these tie-breaking rules can result in very different lotteries. In contrast, there is a substantial difference between the randomized rules and the lexicographic rule. For instance, with lazy voters, ExistNE is NP-hard for R^C and R^V, but polynomial-time solvable for R^L. Further, R^L is, by definition, not neutral, and Theorem 1 demonstrates that candidates with smaller indices have a substantial advantage. For truth-biased voters the impact of tie-breaking rules is less clear: while we have NP-hardness results for all three rules, it appears that, in contrast with lazy voters, PNE induced by randomized tie-breaking are “simpler” than those induced by R^L.

Lazy vs. Truth-Biased Voters. Under lexicographic tie-breaking, the sets of equilibria induced by the two types of secondary preferences are incomparable: there exists a utility profile $\mathbf{u}$ such that the sets of candidates who can win in PNE of $(\mathcal{L}, R^L, \mathbf{u})$ and $(\mathcal{T}, R^L, \mathbf{u})$ are disjoint.

Example 1. Let $C = \{c_1, c_2, c_3\}$, *and consider a 4-voter election with one vote of the form* $c_2 \succ c_3 \succ c_1$, *and three votes of the form* $c_3 \succ c_2 \succ c_1$. *The only PNE of* $(\mathcal{L}, R^L, \mathbf{u})$ is $(c_2, \varnothing, \varnothing, \varnothing)$, *where* c_2 *wins, whereas the only PNE of* $(\mathcal{T}, R^L, \mathbf{u})$ is (c_2, c_3, c_3, c_3), *where* c_3 *wins.*

For randomized tie-breaking, the situation is more interesting. For concreteness, let us focus on R^C. Note first that the utility profiles for which there exist PNE with winning sets of size 2 or more are the same for both voter types. Further, if $(\mathcal{L}, R^C, \mathbf{u})$ has a PNE $\mathbf{b}$, with $|W(\mathbf{b})| = 1$ (which happens only if there is a unanimous winner), then $\mathbf{b}$ is also a PNE of $(\mathcal{T}, R^C, \mathbf{u})$. However, $(\mathcal{T}, R^C, \mathbf{u})$ may have additional PNE, including some non-truthful ones. In particular, for truth-biased voters, the presence of a strong candidate is sufficient for stability: Proposition 3 implies that if there exists a $c \in C$ such that $\text{sc}(c, \mathbf{a}) \geq \text{sc}(c', \mathbf{a}) + 2$ for all $c' \in C \setminus \{c\}$, then for any $R \in \{R^L, R^C, R^V\}$ the truthful ballot vector $\mathbf{a}$ is a PNE of $(\mathcal{T}, R, \mathbf{u})$ with $W(\mathbf{a}) = \{c\}$.

Existence of PNE. For truth-biased voters, one can argue that, when the number of voters is large relative to the number of candidates, under reasonable probabilistic models of elections, the existence of a strong candidate (as defined in the previous paragraph) is exceedingly likely. Thus, elections with truth-biased voters typically admit stable outcomes; this is corroborated by the experimental results of [20]. In contrast, for lazy voters stability is more difficult to achieve, unless there is a candidate that is unanimously ranked first: under randomized tie-breaking rules, there needs to be a very precise balance among candidates that end up being in $W(\mathbf{b})$, and under R^L the eventual winner has to Pareto-dominate all candidates that lexicographically precede him.

Quality of PNE. In all of our models, a candidate ranked last by all voters cannot be elected, in contrast to the basic game-theoretic model for Plurality voting. However, not all non-desirable outcomes are eliminated: under R^V and R^C both lazy voters and truth-biased voters can still elect a Pareto-dominated candidate with non-zero probability in PNE. This has been shown for lazy voters and R^C (Example 1 in [6]), and the same example works for truth-biased voters and R^V. A similar construction shows that a Pareto-dominated candidate may win under R^L when voters are truth-biased. In contrast, lazy voters cannot elect a Pareto-dominated candidate under R^L: Theorem 1 shows that the winner has to be ranked first by some voter.

We can also measure the quality of PNE by analyzing the Price of Anarchy (PoA) in both models. The study of PoA in the context of voting has been

Table 1. Complexity results: P stands for "polynomial-time solvable", NPc stands for "NP-complete".

	SINGLENE	TIENE	EXISTNE
$(\mathcal{L}, R^L)$	P (Corollary 1)	P (Corollary 1)	P (Corollary 1)
$(\mathcal{L}, R^C)$	P (Corollary 2)	NPc (Corollary 2)	NPc (Corollary 2)
$(\mathcal{L}, R^V)$	P (Corollary 2)	NPc (Corollary 2)	NPc (Corollary 2)
$(\mathcal{T}, R^L)$	NPc (Theorem 3)	NPc (Theorem 3)	NPc (Theorem 3)
$(\mathcal{T}, R^C)$	NPc (Corollary 3)	NPc (Corollary 3)	NPc (Corollary 3)
$(\mathcal{T}, R^V)$	NPc (Corollary 3)	NPc (Corollary 3)	NPc (Corollary 3)

recently initiated by Branzei et al. [4]. The additive version of PoA, which was considered in [4], is defined as the worst-case difference between the score of the winner under truthful voting and the truthful score of a PNE winner. It turns out that PoA can be quite high, both for lazy and for truth-biased voters: in the full version of the paper we show that PoA $= \Omega(n)$ in all of our models. Even though these results are not encouraging, PoA is only a worst-case analysis and we expect a better performance on average. Indeed, for the truth-biased model, this is supported by the experimental evaluation in [20].

6 Conclusions

We have characterized PNE of Plurality voting for several combinations of secondary preferences and tie-breaking rules. Our complexity results are summarized in Table 1.

Our results extend to the setting where some of the voters are *principled*, i.e., always vote truthfully (and never abstain). Due to space constraints, we are unable to fully describe these extensions (see [9]). Briefly, the presence of principled voters has the strongest effect on lazy voters and lexicographic tie-breaking, as illustrated by the following example, whereas for other settings the effect is less pronounced.

Example 2. Consider an election over a candidate set $C = \{c_1, \dots, c_m\}$, $m > 1$, where there are two principled voters who both vote for c_m, and two lazy voters who both rank c_m last. Then the ballot vector where both lazy voters abstain is a PNE (with winner c_m). Moreover, for every $j \in [m-1]$ the ballot vector where both lazy voters vote for c_j is a PNE as well (with winner c_j).

In the absence of principled voters, PNE for lazy voters require very precise coordination among the voters and seem to be very different from what we observe in real life. In contrast, for truth-biased voters the presence of a strong candidate implies the existence of a truthful equilibrium, which requires little coordination among the players. It is therefore tempting to conclude that truth bias has a greater explanatory power than laziness. However, we demonstrated that the presence of principled voters changes this equation. Extending our analysis to a mixture of all three voter types is perhaps the most prominent open problem suggested by our work.

Acknowledgements. The work of Elkind was partially supported by ERC-StG 639945. Markakis was supported by the European Union (European Social Fund - ESF) and Greek national funds through the Operational Program "Education and Lifelong Learning" of the National Strategic Reference Framework (NSRF) Research Funding Program: THALES, investing in knowledge society through the European Social Fund. Obraztsova's work was partially supported by RFFI grant 14-01-00156-a.

References

1. Aziz, H., Gaspers, S., Mattei, N., Narodytska, N., Walsh, T.: Ties matter: complexity of manipulation when tie-breaking with a random vote. In: AAAI 2013, pp. 74–80 (2013)
2. Battaglini, M.: Sequential voting with abstention. Games Econ. Behav. **51**, 445–463 (2005)
3. Borgers, T.: Costly voting. Am. Econ. Rev. **94**(1), 57–66 (2004)
4. Branzei, S., Caragiannis, I., Morgenstern, J., Procaccia, A.D.: How bad is selfish voting? In: AAAI 2013, pp. 138–144 (2013)
5. Clifford, R., Popa, A.: Maximum subset intersection. Inf. Process. Lett. **111**(7), 323–325 (2011)
6. Desmedt, Y., Elkind, E.: Equilibria of plurality voting with abstentions. In: ACM EC 2010, pp. 347–356 (2010)
7. Dhillon, A., Lockwood, B.: When are plurality rule voting games dominance-solvable? Games Econ. Behav. **46**, 55–75 (2004)
8. Dutta, B., Sen, A.: Nash implementation with partially honest individuals. Games Econ. Behav. **74**(1), 154–169 (2012)
9. Elkind, E., Markakis, E., Obraztsova, S., Skowron, P.: Equilibria of plurality voting: Lazy and truth-biased voters. arXivabs/1409.4132 (2014)
10. Farquharson, R.: Theory of Voting. Yale University Press, New Haven (1969)
11. Laslier, J.F., Weibull, J.W.: A strategy-proof condorcet jury theorem. Scand. J. Econ. (2012)
12. Meir, R., Polukarov, M., Rosenschein, J.S., Jennings, N.R.: Convergence to equilibria in plurality voting. In: AAAI 2010, pp. 823–828 (2010)
13. Moulin, H.: Dominance solvable voting schemes. Econometrica **47**, 1337–1351 (1979)
14. Myerson, R., Weber, R.: A theory of voting equilibria. Am. Polit. Sci. Rev. **87**(1), 102–114 (1993)
15. Obraztsova, S., Elkind, E.: On the complexity of voting manipulation under randomized tie-breaking. In: IJCAI 2011, pp. 319–324 (2011)
16. Obraztsova, S., Elkind, E., Hazon, N.: Ties matter: Complexity of voting manipulation revisited. In: AAMAS 2011, pp. 71–78 (2011)
17. Obraztsova, S., Markakis, E., Thompson, D.R.M.: Plurality voting with truth-biased agents. In: Vöcking, B. (ed.) SAGT 2013. LNCS, vol. 8146, pp. 26–37. Springer, Heidelberg (2013)
18. Schulze, M.: A new monotonic, clone-independent, reversal symmetric, and condorcet-consistent single-winner election method. Soc. Choice Welfare **36**(2), 267–303 (2011)
19. Sinopoli, F.D., Iannantuoni, G.: On the generic strategic stability of Nash equilibria if voting is costly. Econ. Theory **25**(2), 477–486 (2005)
20. Thompson, D.R.M., Lev, O., Leyton-Brown, K., Rosenschein, J.S.: Empirical analysis of plurality election equilibria. In: AAMAS 2013, pp. 391–398 (2013)
21. Xavier, E.: A note on a maximum k-subset intersection problem. Inf. Process. Lett. **112**(12), 471–472 (2012)

Auctions

The Combinatorial World (of Auctions) According to GARP

Shant Boodaghians[1(✉)] and Adrian Vetta[2]

[1] Department of Mathematics and Statistics, McGill University, Montreal, Canada
shant.boodaghians@mail.mcgill.ca
[2] Department of Mathematics and Statistics, School of Computer Science, McGill University, Montreal, Canada
vetta@math.mcgill.ca

Abstract. Revealed preference techniques are used to test whether a data set is compatible with rational behaviour. They are also incorporated as constraints in mechanism design to encourage truthful behaviour in applications such as combinatorial auctions. In the auction setting, we present an efficient combinatorial algorithm to find a virtual valuation function with the optimal (additive) rationality guarantee. Moreover, we show that there exists such a valuation function that both is individually rational and is minimum (that is, it is component-wise dominated by any other individually rational, virtual valuation function that approximately fits the data). Similarly, given upper bound constraints on the valuation function, we show how to fit the maximum virtual valuation function with the optimal additive rationality guarantee. In practice, revealed preference bidding constraints are very demanding. We explain how approximate rationality can be used to create relaxed revealed preference constraints in an auction. We then show how combinatorial methods can be used to implement these relaxed constraints. Worst/best-case welfare guarantees that result from the use of such mechanisms can be quantified via the minimum/maximum virtual valuation function.

1 Introduction

Underlying the theory of consumer demand is a standard rationality assumption: given a set of items with price vector $\boldsymbol{p}$, a consumer will demand the bundle $\boldsymbol{x}$ of maximum utility whose cost is at most her budget B. Of fundamental import, therefore, is whether or not the decision making behaviour of a real consumer is consistent with the maximization of a utility function. Samuelson [18,19] introduced *revealed preference* to provide a theoretical framework within which to analyse this question. Furthermore, this concept now lies at the heart of current empirical work in the field; see, for example, Gross [11] and Varian [23]. Specifically, Samuelson [18] conjectured that the *weak axiom of revealed preference* (WARP) was a necessary and sufficient condition for *integrability* – the ability to construct a utility function which fits observed behaviour.

However, Houtthakker [14] proved that the weak axiom was insufficient. Instead, he presented a *strong axiom of revealed preference* (SARP) and showed

M. Hoefer (Ed.): SAGT 2015, LNCS 9347, pp. 125–136, 2015.
DOI: 10.1007/978-3-662-48433-3_10

non-constructively that it was necessary and sufficient in the case where behaviour is determined via a single-valued demand function. Afriat [1] provided an extension to multi-valued demand functions – where ties are allowed – by showing that the *generalized axiom of revealed preference* (GARP) is necessary and sufficient for integrability.[1] Furthermore, Afriat's approach was constructive (producing monotonic, concave, piecewise-linear utility functions) and applied to the setting of a finite collection of observational data. This rendered his method more suitable for practical use.

In addition to its prominence in testing for rational behaviour, revealed preference has become an important tool in mechanism design. A notable area of application is auction design. For combinatorial auctions, Ausubel, Cramton and Milgrom [4] proposed bidding activity rules based upon WARP. These rules are now standard in the combinatorial clock auction, one of the two prominent auction mechanisms used to sell bandwidth. In part, the WARP bidding rules have proved successful because they are extremely difficult to game [6]. Harsha et al. [13] examine GARP-based bidding rules, and Ausubel and Baranov [3] advocate incorporating such constraints into bandwidth auctions. Based upon Afriat's theorem, these GARP-based rules imply that there always exists a utility function that is compatible with the bidding history. This gives the desirable property that a bidder in an auction will always have at least one feasible bid – a property that cannot be guaranteed under WARP.

Revealed preference also plays a key role in motivating the *generalised second price mechanism* used in adword auctions. Indeed, these position auctions have welfare maximizing solutions with respect to a revealed preference equilibrium concept; see [8,23].

1.1 Our Results

Multiple methods have been proposed to approximately measure how consistent a data set is with rational behaviour; see Gross [11] for a comparison of a sample of these approaches. In this paper, we show how a graphical viewpoint of revealed preference can be used to obtain a virtual valuation function that best fits the data set. Specifically, we show in Sect. 3 that an individually rational virtual valuation function can be obtained such that its additive deviation from rationality is exactly the minimum mean length of a cycle in a bidding graph. This additive guarantee cannot be improved upon. Furthermore, we show there exists a unique *minimum* valuation function from amongst all individually rational virtual valuation functions that optimally fit the data. Similarly, given a set of upper bound constraints, we show how to find the unique *maximum* virtual valuation that optimally fits the data, if it exists.

Imposing revealed preference bidding rules can be harsh. Indeed, Cramton [6] states that "there are good reasons to simplify and somewhat weaken the revealed preference rule". These reasons include complexity issues, common

[1] Afriat [1] gave several equivalent necessary and sufficient conditions for integrability. One of these, *cyclical consistency*, is equivalent to GARP as shown by Varian [21].

value uncertainty, the complication of budget constraints, and the fact that a bidder's assessment of its valuation function often *changes* as the auction progresses! The concept of approximate rationality, however, naturally induces a relaxed form of revealed preference rules. We examine such relaxed bidding rules in Sect. 4, show how they can be implemented combinatorially, and show how to construct the minimal and maximal valuation functions which fit the data, which may be useful for quantifying worst/best-case welfare guarantees.

2 Revealed Preference

2.1 Revealed Preference with Budgets

We first review revealed preference. We then examine its use in auction design and describe how to formulate it in terms of a bidding graph. The standard revealed preference model instigated by Samuelson [18] is as follows. We are given a set of observations $\{(B_1, \boldsymbol{p}_1, \boldsymbol{x}_1), (B_2, \boldsymbol{p}_2, \boldsymbol{x}_2), \ldots, (B_T, \boldsymbol{p}_T, \boldsymbol{x}_T)\}$. At time t, $1 \leq t \leq T$, the set of items has a price vector $\boldsymbol{p}_t$ and the consumer chooses to spend her budget B_t on the bundle $\boldsymbol{x}_t$.[2] We say that bundle $\boldsymbol{x}_t$ is (directly) revealed preferred to bundle $\boldsymbol{y}$, denoted $\boldsymbol{x}_t \succeq \boldsymbol{y}$, if $\boldsymbol{y}$ was affordable when $\boldsymbol{x}_t$ was purchased. We say that bundle $\boldsymbol{x}_t$ is strictly revealed preferred to bundle $\boldsymbol{y}$, denoted $\boldsymbol{x}_t \succ \boldsymbol{y}$, if $\boldsymbol{y}$ was (strictly) cheaper than $\boldsymbol{x}_t$ when $\boldsymbol{x}_t$ was purchased. This gives *revealed preference* (1) and *strict revealed preference* (2):

$$\boldsymbol{p}_t \cdot \boldsymbol{y} \leq \boldsymbol{p}_t \cdot \boldsymbol{x}_t \quad \Rightarrow \quad \boldsymbol{x}_t \succeq \boldsymbol{y} \quad (1) \qquad\qquad \boldsymbol{p}_t \cdot \boldsymbol{y} < \boldsymbol{p}_t \cdot \boldsymbol{x}_t \quad \Rightarrow \quad \boldsymbol{x}_t \succ \boldsymbol{y} \quad (2)$$

Furthermore, a basic assumption is that the consumer optimises a locally non-satiated utility function.[3] Consequently, at time t she will spend her entire budget, *i.e.*, $\boldsymbol{p}_t \cdot \boldsymbol{x}_t = B_t$. In the absence of ties, preference orderings give relations that are anti-symmetric and transitive. This leads to an axiomatic approach to revealed preference formulated in terms of WARP and SARP by Houthakker [14]. The *weak axiom of revealed preference* (WARP) states that the relation should be asymmetric, *i.e.* $\boldsymbol{x} \succeq \boldsymbol{y} \Rightarrow \boldsymbol{y} \not\succeq \boldsymbol{x}$. Its transitive closure, the *strong axiom of revealed preference* (SARP) states that the relation should be acyclic. Our interest lies in the general case where ties are allowed. This produces what we dub the *k-th Axiom of Revealed Preference* (KARP): Given a fixed integer k and any $\kappa \leq k$

$$\boldsymbol{x}_t = \boldsymbol{x}_{t_0} \succeq \boldsymbol{x}_{t_1} \succeq \cdots \succeq \boldsymbol{x}_{t_{\kappa-1}} \succeq \boldsymbol{x}_{t_\kappa} = \boldsymbol{y} \quad \Rightarrow \quad \boldsymbol{y} \not\succ \boldsymbol{x}_t \,. \tag{3}$$

There are two very important special cases of KARP. For $k = 1$, this is simply WARP, *i.e.* $\boldsymbol{x}_t \succeq \boldsymbol{y} \quad \Rightarrow \quad \boldsymbol{y} \not\succ \boldsymbol{x}_t$. This is just the basic property that for a preference ordering, we cannot have that $\boldsymbol{y}$ is strictly preferred to $\boldsymbol{x}_t$ if $\boldsymbol{x}_t$ is

---

[2] It is not necessary to present the model in terms of "time". We do so because this best accords with the combinatorial auction application.

[3] Local non-satiation states that for any bundle $\boldsymbol{x}$ there is a more preferred bundle arbitrarily close to $\boldsymbol{x}$. A monotonic utility function is locally non-satiated, but the converse need not hold.

preferred to $\boldsymbol{y}$. On the other hand, suppose we take k to take be arbitrarily large (or simply larger than the total number of observed bundles). Then we have the *Generalized Axiom of Revealed Preference* (GARP), the simultaneous application of KARP for each value of k. In particular, GARP encodes the property of transitivity of preference relations. Specifically, for any k, if $\boldsymbol{x}_t = \boldsymbol{x}_{t_0} \succeq \boldsymbol{x}_{t_1} \succeq \cdots \succeq \boldsymbol{x}_{t_{k-1}} \succeq \boldsymbol{x}_{t_k} = \boldsymbol{y}$ then, by transitivity, $\boldsymbol{x}_t \succeq \boldsymbol{y}$. The first axiom of revealed preference then implies that $\boldsymbol{y} \not\succ \boldsymbol{x}_t$.

The underlying importance of GARP follows from a classical result of Afriat [1]: there exists a nonsatiated, monotone, concave utility function that rationalizes the data if and only if the data satisfy GARP. Brown and Echenique [5] examine the setting of indivisible goods and Echenique et al. [7] consider the consequent computational implications.

2.2 Revealed Preference in Combinatorial Auctions

As discussed, a major application of revealed preference in mechanism design concerns combinatorial auctions. Here there are some important distinctions from the standard revealed preference model presented in Sect. 2.1. First, consumers are assumed to have quasilinear utility functions that are linear in money. Thus, they seek to maximise profit. Second, the standard assumption is that bidders have *no* budgetary constraints. For example, if profitable opportunities arise that require large investments then these can be obtained from perfect capital markets. (This assumption is slightly unrealistic; Harsha et al. [13] show how to implement a budgeted revealed preference model for combinatorial auctions). Third, the observations $(\boldsymbol{p}_t, \boldsymbol{x}_t)$, for each $1 \le t \le T$, are typically not purchases but are bids made over a collection of auction rounds. When offered a set of prices at time t the consumer bids for bundle $\boldsymbol{x}_t$.

So what would a model of revealed preference be in this combinatorial auction setting? Suppose the bidder has an arbitrarily large budget B. In particular, prices will never be so high that she cannot afford to buy every item. Second, to model quasilinear utility functions, we treat money as a good. Specifically, given a bundle of items $\boldsymbol{x} = (x_1, \ldots, x_n)$ and an amount x_0 of money we denote by $\hat{\boldsymbol{x}} = (x_0, x_1, \ldots, x_n)$ the concatenation of x_0 and $\boldsymbol{x}$. If $\boldsymbol{p} = (p_1, \ldots, p_n)$ is the price vector for the the non-monetary items, then $\hat{\boldsymbol{p}} = (1, p_1, \ldots, p_n)$ gives the prices of all items including money.

In this $n+1$ dimensional setting, let us select bundle $\hat{\boldsymbol{x}}_t$ at time t. As the budget B is arbitrarily large, we can certainly afford the bundle $\boldsymbol{x}_s$ at this time. But we may not be able to afford bundle $\hat{\boldsymbol{x}}_s$, as then we must also pay for the monetary component at a cost of $B - \boldsymbol{p}_s \cdot \boldsymbol{x}_s$. However, we can afford the bundle $\boldsymbol{x}_s$ plus an amount $B - \boldsymbol{p}_t \cdot \boldsymbol{x}_s$ of money. Applying revealed preference to $\{\hat{\boldsymbol{x}}, \hat{\boldsymbol{p}}\}$, we have revealed that $\hat{\boldsymbol{x}}_t = (B - \boldsymbol{p}_t \cdot \boldsymbol{x}_t, \boldsymbol{x}_t) \succeq (B - \boldsymbol{p}_t \cdot \boldsymbol{x}_s, \boldsymbol{x}_s)$. Hence, by quasilinearity, subtracting the monetary component from both sides, we have,

$$(0, \boldsymbol{x}_t) \succeq ((B - \boldsymbol{p}_t \cdot \boldsymbol{x}_s) - (B - \boldsymbol{p}_t \cdot \boldsymbol{x}_t), \boldsymbol{x}_s) = (\boldsymbol{p}_t \cdot \boldsymbol{x}_t - \boldsymbol{p}_t \cdot \boldsymbol{x}_s, \boldsymbol{x}_s),$$

equivalently, $v(\boldsymbol{x}_t) \ge v(\boldsymbol{x}_s) + \boldsymbol{p}_t \cdot \boldsymbol{x}_t - \boldsymbol{p}_t \cdot \boldsymbol{x}_s$. Rearranging, we have $v(\boldsymbol{x}_t) - \boldsymbol{p}_t \boldsymbol{x}_t \ge v(\boldsymbol{x}_s) - \boldsymbol{p}_t \boldsymbol{x}_s$. Symmetrically, we get $v(\boldsymbol{x}_s) - \boldsymbol{p}_s \boldsymbol{x}_s \ge v(\boldsymbol{x}_t) - \boldsymbol{p}_s \boldsymbol{x}_t$, and combining both, we get

$$\boldsymbol{p}_t - \boldsymbol{p}_s) \cdot \boldsymbol{x}_s \geq (\boldsymbol{p}_t - \boldsymbol{p}_s) \cdot \boldsymbol{x}_t \,, \tag{4}$$

which is the bidding rule based upon WARP as defined by Ausubel, Cramton and Milgrom [4]. This can be derived directly from the assumption of quasilinear utility, as done in the full version of this paper. We can now extend this bidding rule to incorporate indirect comparisons in a similar fashion to the extension from WARP to SARP via transitivity. This produces a GARP-based bidding rule. To wit, suppose we bid for the money-less bundle $\boldsymbol{x}_i$ at time t_i, for all $0 \leq i \leq k$, where $1 \leq t_i \leq T$. Thus we have revealed that

$$(0, \boldsymbol{x}_i) \;\succeq\; ((B - \boldsymbol{p}_i \cdot \boldsymbol{x}_{i+1}) - (B - \boldsymbol{p}_i \cdot \boldsymbol{x}_i), \boldsymbol{x}_{i+1}) \;=\; (\boldsymbol{p}_i \cdot \boldsymbol{x}_i - \boldsymbol{p}_i \cdot \boldsymbol{x}_{i+1}, \boldsymbol{x}_{i+1})$$

This induces the inequality $v(\boldsymbol{x}_i) - \boldsymbol{p}_i \cdot \boldsymbol{x}_i \geq v(\boldsymbol{x}_{i+1}) - \boldsymbol{p}_i \cdot \boldsymbol{x}_{i+1}$. Summing over all i, we obtain $\sum_{i=0}^k (v(\boldsymbol{x}_i) - \boldsymbol{p}_i \cdot \boldsymbol{x}_i) \geq \sum_{i=0}^k (v(\boldsymbol{x}_{i+1}) - \boldsymbol{p}_i \cdot \boldsymbol{x}_{i+1})$, where the sum in the subscripts are taken modulo k. Rearranging now gives the combinatorial auction KARP-based bidding activity rule:

$$(\boldsymbol{p}_k - \boldsymbol{p}_0) \cdot \boldsymbol{x}_0 \;\geq\; \sum\nolimits_{i=1}^{k} (\boldsymbol{p}_i - \boldsymbol{p}_{i-1}) \cdot \boldsymbol{x}_i \,. \tag{5}$$

For k arbitrarily large, this gives the GARP-based bidding rule. In order to qualitatively analyze the consequences of imposing KARP-based activity rules, it is informative to now provide a graphical interpretation of the these rules.

2.3 A Graphical View of Revealed Preference

Given the set of price-bid pairings $\{(\boldsymbol{p}_t, \boldsymbol{x}_t) : 1 \leq t \leq T\}$, we create a directed graph $G = (V, A)$, called the *bidding graph*, to which we will assign arc lengths ℓ. There is a vertex in V for each possible bundle – that is, there are 2^n bundles in an n-item auction. For each observed bid $\boldsymbol{x}_t$, $1 \leq t \leq T$, there is an arc $(\boldsymbol{x}_t, \boldsymbol{y})$ for each bundle $\boldsymbol{y} \in V$. In order to define the length $\ell_{\boldsymbol{x}_t,\boldsymbol{y}}$ of an arc $(\boldsymbol{x}_t, \boldsymbol{y})$, note that Inequality (4) applied to $\boldsymbol{x}_s = \boldsymbol{y}$ gives $v(\boldsymbol{y}) \;\leq\; v(\boldsymbol{x}_t) + \boldsymbol{p}_t \cdot (\boldsymbol{y} - \boldsymbol{x}_t)$, otherwise we would prefer bundle $\boldsymbol{y}$ at time t. For the arc length, we would like to simply set $\ell_{\boldsymbol{x}_t,\boldsymbol{y}} = \boldsymbol{p}_t \cdot (\boldsymbol{y} - \boldsymbol{x}_t)$. Observe, however, that the bundle $\boldsymbol{x}_t$ may be chosen in more than one time period. That is, possibly $\boldsymbol{x}_t = \boldsymbol{x}_{t'}$, for some $t \neq t'$. Therefore the bidding graph is, in fact, a multigraph. It suffices, though, to represent only the most stringent constraints imposed by the bidding behaviour. Thus, we obtain a simple graph by setting $\ell_{\boldsymbol{x}_t,\boldsymbol{y}} = \min_{t'} \{\boldsymbol{p}_{t'} \cdot (\boldsymbol{y} - \boldsymbol{x}_t) : \boldsymbol{x}_{t'} = \boldsymbol{x}_t\}$. Now the WARP-based bidding rule (4) of Ausubel et al. [4] is equivalent to $(\boldsymbol{p}_t - \boldsymbol{p}_s) \cdot \boldsymbol{x}_s - (\boldsymbol{p}_t - \boldsymbol{p}_s) \cdot \boldsymbol{x}_t \geq 0$. But

$$\begin{aligned}\ell_{\boldsymbol{x}_s,\boldsymbol{x}_t} + \ell_{\boldsymbol{x}_t,\boldsymbol{x}_s} &= \min_{s'} \{\boldsymbol{p}_{s'} \cdot (\boldsymbol{x}_t - \boldsymbol{x}_s) : \boldsymbol{x}_{s'} = \boldsymbol{x}_s\} + \min_{t'} \{\boldsymbol{p}_{t'} \cdot (\boldsymbol{x}_s - \boldsymbol{x}_t) : \boldsymbol{x}_{t'} = \boldsymbol{x}_t\} \\ &\leq \boldsymbol{p}_s \cdot (\boldsymbol{x}_t - \boldsymbol{x}_s) + \boldsymbol{p}_t \cdot (\boldsymbol{x}_s - \boldsymbol{x}_t) = (\boldsymbol{p}_t - \boldsymbol{p}_s) \cdot \boldsymbol{x}_s - (\boldsymbol{p}_t - \boldsymbol{p}_s) \cdot \boldsymbol{x}_t \,.\end{aligned}$$

It is then easy to see that the bidding constraint (4) is violated if and only if the bidding graph contains no negative digons (cycles of length two). Furthermore, we can interpret KARP and GARP is a similar fashion. Hence, the k-th axiom of revealed preference is equivalent to requiring that the bidding graph not contain any negative cycles of cardinality at most $k + 1$, and GARP is equivalent to requiring no negative cycles at all. Thus, we can formalize the preference axioms in terms of the lengths of negative cycles in a directed graph. We remark that a cyclic view of revealed preference is briefly outlined by Vohra [25]. For us, this cyclic formulation has important consequences in testing for the extent of bidding deviations from the axioms. We will quantify this exactly in Sect. 3. Before doing so, though, we remark that the focus on cycles also has important computational consequences.

First, recall that the bidding graph G contains an exponential number of vertices, one for every subset of the items. Of course, it is not practical to work with such a graph. Observe, however, that a bundle $\boldsymbol{y} \notin \{\boldsymbol{x}_1, \boldsymbol{x}_2 \ldots, \boldsymbol{x}_T\}$ has zero out-degree in G. Consequently, $\boldsymbol{y}$ cannot be contained in any cycle. Thus, it will suffice to consider only the subgraph induced by the bids $\{\boldsymbol{x}_1, \boldsymbol{x}_2 \ldots, \boldsymbol{x}_T\}$. In a combinatorial auction there is typically one bid per time period and the number of periods is quite small.[4] Hence, the induced subgraph of the bidding graph that we actually need is of a very manageable size.

Second, one way to implement a bidding rule is via a mathematical program; see, for example, Harsha et al. [13]. The cyclic interpretation of a bidding rule has two major advantages: we can test the rule very quickly by searching for negative cycles in a graph. For example, we can test for negative cycles of length at most $k + 1$ either by fast matrix multiplication or directly by looking for shortest paths of length k using the Bellman-Ford algorithm in $O(T^3)$ time. Another major advantage is that a bidder can interpret the consequence of a prospective new bid dynamically by consideration of the bidding graph. This is extremely important in practice. In contrast, bidding rules that require using an optimization solver as a black-box are very opaque to bidders.

3 Approximate Virtual Valuation Functions

For combinatorial auctions, Afriat's result that GARP is necessary and sufficient for rationalisability can be reformulated as:

Theorem 3.1. *A valuation function which rationalises bidding behaviour exists if and only if the bidding graph has no negative cycle.*

This is a simple corollary of Theorem 3.2 below; see also [25]. From an economic perspective, however, what is most important is not whether agents are perfectly rational but "whether optimization is a reasonable way to describe some behavior" [22].[5] It is then important to study the consequences of approximately rational behaviour, see, for example, Akerlof and Yellen [2]. First, though, is it possible to quantify the degree to which agents are rational? Gross [11] examines assorted methods to test the degree of rationality. Notable amongst them is the

[4] For example, in a bandwidth auction there are at most a few hundred rounds.

[5] Indeed, several schools of thought in the field of bounded rationality argue that people utilize simple (but often effective) heuristics rather than attempt to optimize; see, for example, [10].

Afriat Efficiency Index [1,22]. Here the condition required to imply a preference is strengthened multiplicatively. Specifically, $\boldsymbol{x}_t \succeq \boldsymbol{y}$ only if $\boldsymbol{p}_t \cdot \boldsymbol{y} \leq \lambda \cdot \boldsymbol{p}_t \cdot \boldsymbol{x}_t$ where $\lambda < 1$. We examine this index with respect to the bidding graph in Sect. 4.3. For combinatorial auctions, a variant of this constraint was examined experimentally by Harsha et al. [13].

Here we show how to quantify exactly the degree of rationality present in the data via a parameter of the bidding graph. Moreover, we are able to go beyond multiplicative guarantees and obtain stronger additive bounds. To wit, we say that $\hat{v}$ is an *ϵ-approximate virtual valuation function* if, for all t and for any bundle $\boldsymbol{y}$, $\hat{v}(\boldsymbol{x}_t) - \boldsymbol{p}_t \cdot \boldsymbol{x}_t \geq \hat{v}(\boldsymbol{y}) - \boldsymbol{p}_t \cdot \boldsymbol{y} - \epsilon$. Note that if $\epsilon = 0$, then the bidder is optimizing with respect to a virtual valuation function, *i.e.* is rational. We remark that the term *virtual* reflects the fact that $\hat{v}$ need not be the real valuation function (if one exists) of the bidder, but if it is then the bidding is termed *truthful*.

3.1 Minimum Mean Cycles and Approximate Virtual Valuations

We now examine exactly when a bidding strategy is approximately rational. It turns out that the key to understanding approximate deviations from rationality is the minimum mean cycle in the bidding graph. Given a cycle C in G, its mean length is $\mu(C) = \frac{\sum_{a \in C} \ell_a}{|C|}$. We denote by $\mu(G) = \min_C \mu(C)$ the *minimum mean length* of a cycle in G, and we say that C^* is a *minimum mean cycle* if $C^* \in \operatorname{argmin}_C \mu(C)$. We can find a minimum mean cycle in polynomial time using the classical techniques of Karp [15].

Theorem 3.2. *An ϵ-approximate valuation function which (approximately) rationalises bidding behaviour exists if and only if the bidding graph has minimum mean cycle $\mu(G) \geq -\epsilon$.*

Proof. From the bidding graph G we create an auxiliary directed graph $\hat{G} = (\hat{V}, \hat{A})$ with vertex set $\hat{V} = \{\boldsymbol{x}_1, \boldsymbol{x}_2, \ldots, \boldsymbol{x}_T\}$. The arc set is complete with arc lengths $\hat{\ell}_{\boldsymbol{x}_s, \boldsymbol{x}_t} = \ell_{\boldsymbol{x}_s, \boldsymbol{x}_t} - \mu(G)$. Observe that, by construction, every cycle in $\hat{G}$ is of non-negative length. It follows that we may obtain shortest path distances $\hat{d}$ from any arbitrary root vertex r. Thus, for any arc $(\boldsymbol{x}_t, \boldsymbol{y})$, we have $\hat{d}(\boldsymbol{y}) \leq \hat{d}(\boldsymbol{x}_t) + \hat{\ell}_{\boldsymbol{x}_t, \boldsymbol{y}} = \hat{d}(\boldsymbol{x}_t) + \ell_{\boldsymbol{x}_t, \boldsymbol{y}} - \mu(G) \leq \hat{d}(\boldsymbol{x}_t) + \boldsymbol{p}_t \cdot (\boldsymbol{y} - \boldsymbol{x}_t) - \mu(G)$. So, if we set $\hat{v}(\boldsymbol{x}) = \hat{d}(\boldsymbol{x})$, for each $\boldsymbol{x}$, then $\hat{v}(\boldsymbol{x}_t) - \boldsymbol{p}_t \cdot \boldsymbol{x}_t \geq \hat{v}(\boldsymbol{y}) - \boldsymbol{p}_t \cdot \boldsymbol{y} + \mu(G)$. for all t. Therefore, by definition of ϵ-approximate bidding, we have that $\hat{v}$ is a $(-\mu)$-approximate virtual valuation function.

Conversely, let $\hat{v}$ be an ϵ-approximate virtual valuation function which rationalises the graph, and take some cycle C of minimum mean length in the bidding graph. Suppose for a contradiction that $\mu(C) < -\epsilon$. By ϵ-approximability, we have $\hat{v}(\boldsymbol{x}_s) - \boldsymbol{p}_s \cdot \boldsymbol{x}_s \geq \hat{v}(\boldsymbol{x}_t) - \boldsymbol{p}_s \cdot \boldsymbol{x}_t - \epsilon$. But $\ell_{\boldsymbol{x}_s, \boldsymbol{x}_t} \geq \boldsymbol{p}_s \cdot (\boldsymbol{x}_t - \boldsymbol{x}_s)$. Therefore $\ell_{\boldsymbol{x}_s \boldsymbol{x}_t} \geq \hat{v}(\boldsymbol{x}_t) - \hat{v}(\boldsymbol{x}_s) - \epsilon$. Summing over every arc in the cycle we obtain $\ell(C) = \sum_{(\boldsymbol{x}, \boldsymbol{y}) \in C} \ell_{\boldsymbol{x}\boldsymbol{y}} \geq \sum_{(\boldsymbol{x}, \boldsymbol{y}) \in C} (\hat{v}(\boldsymbol{y}) - \hat{v}(\boldsymbol{x}) - \epsilon) = -|C| \cdot \epsilon$. Thus $\mu(C) \geq -\epsilon$, giving the desired contradiction. □

3.2 Individually Rational Virtual Valuation Functions

Theorem 3.2 shows how to obtain a virtual valuation function with the best possible additive approximation guarantee: any valuation rationalising the bidding graph G must allow for an additive approximation of at least $-\mu(G)$. However, there is a problem. Such a valuation function may not actually be compatible with the data; specifically, it may not be individually rational. For *individual rationality*, we require, for each time t, that $\hat{v}(\boldsymbol{x}_t) - \boldsymbol{p}_t \cdot \boldsymbol{x}_t \geq 0$. But individually rationality is (almost certainly) violated for the the root node r since we have $\hat{v}(\boldsymbol{x}_r) = 0$. It is possible to obtain an individually rational, approximate, virtual valuation function simply by taking the $\hat{v}$ from Theorem 3.2 and adding a huge constant to value of each package. This operation, of course, is entirely unnatural and the resulting valuation function is of little practical value.

The Minimum Individually Rational Virtual Valuation Function. We say that $v()$ is the *minimum individually rational, ϵ-approximate virtual valuation function* if $v(\boldsymbol{x}_t) \leq \omega(\boldsymbol{x}_t)$ for each $1 \leq t \leq T$, for any other individually rational, ϵ-approximate virtual valuation function $\omega()$. This leads to the questions: (i) Does such a valuation function exist? and (ii) Can it be obtained efficiently? The answer to both these questions is yes.

Theorem 3.3. *The minimum individually rational, μ-approximate virtual valuation function exists and can be found in polynomial time.*

Proof. We create an auxiliary directed graph H from $\hat{G}$ by adding a sink vertex $\boldsymbol{z}$. We add an arc $(\boldsymbol{x}_t, \boldsymbol{z})$ of length $-\boldsymbol{p}_t \cdot \boldsymbol{x}_t$, for each $1 \leq t \leq T$, allowing for repeated arcs. Because $\hat{G}$ contains no negative cycle, neither does H. Therefore, there exist shortest path distances in H. Denote by $\hat{d}()$ the shortest path distance from vertex $\boldsymbol{x}_t$ *to* $\boldsymbol{z}$ in H. We claim that setting $v(\boldsymbol{x}_t) = -\hat{d}(\boldsymbol{x}_t)$ gives the minimum individually rational, μ-approximate virtual valuation function.

To begin, let's verify that $v()$ is an individually rational, μ-approximate virtual valuation function. First, we require that $v()$ is individually rational. Now the direct path consisting of the arc $(\boldsymbol{x}_t, \boldsymbol{z})$ is at least as long as the shortest path from $\boldsymbol{x}_t$ to $\boldsymbol{z}$. Thus, $-\boldsymbol{p}_t \cdot \boldsymbol{x}_t \geq \hat{d}(\boldsymbol{x}_t)$. Individual rationality then follows as $v(\boldsymbol{x}_t) = -\hat{d}(\boldsymbol{x}_t) \geq \boldsymbol{p}_t \cdot \boldsymbol{x}_t$. Second we need to show that $v()$ is μ-approximate. Consider a pair $\{\boldsymbol{x}_s, \boldsymbol{x}_t\}$. The shortest path conditions imply that $-v(\boldsymbol{x}_s) = \hat{d}(\boldsymbol{x}_s) \leq \hat{\ell}_{st} + \hat{d}(\boldsymbol{x}_t) = (\ell_{st} - \mu) + \hat{d}(\boldsymbol{x}_t) = (\ell_{st} - \mu) - v(\boldsymbol{x}_t)$. Here the inequality follows from the shortest path conditions on $\hat{d}()$. Therefore, by definition of ℓ_{st},$v(\boldsymbol{x}_t) \leq v(\boldsymbol{x}_s) + \ell_{st} - \mu = v(\boldsymbol{x}_s) + \min_{s'} \{\boldsymbol{p}_{s'} \cdot (\boldsymbol{x}_t - \boldsymbol{x}_s) : \boldsymbol{x}_{s'} = \boldsymbol{x}_s\} - \mu \leq v(\boldsymbol{x}_s) + \boldsymbol{p}_s \cdot (\boldsymbol{x}_t - \boldsymbol{x}_s) - \mu$. Hence, $v()$ is μ-approximate as desired.

Finally we require that $v()$ is minimum individually rational. So, take any other individually rational, μ-approximate virtual valuation $\omega()$. We must show that $v(\boldsymbol{x}_t) \leq \omega(\boldsymbol{x}_t)$ for every bundle $\boldsymbol{x}_t$. Now consider the shortest path tree T in H corresponding to $\hat{d}()$. If $(\boldsymbol{x}_t, \boldsymbol{z})$ is an arc in T (and at least one such arc exists) then $-\boldsymbol{p}_t \cdot \boldsymbol{x}_t = \hat{d}(\boldsymbol{x}_t)$. Thus $v(\boldsymbol{x}_t) - \boldsymbol{p}_t \cdot \boldsymbol{x}_t = (-\boldsymbol{p}_t \cdot \boldsymbol{x}_t) - \hat{d}(\boldsymbol{x}_t) = 0 \leq \omega(\boldsymbol{x}_t) - \boldsymbol{p}_t \cdot \boldsymbol{x}_t$. Here the inequality follows by the individual rationality of $\omega()$.

Thus $v(\boldsymbol{x}_t) \leq \omega(\boldsymbol{x}_t)$. Now suppose that $v(\boldsymbol{x}_s) > \omega(\boldsymbol{x}_s)$ for some $\boldsymbol{x}_s$. We may take $\boldsymbol{x}_s$ to be the closest vertex to the root $\boldsymbol{z}$ in T with this property. We have seen that $\boldsymbol{x}_s$ cannot be a child of $\boldsymbol{z}$. So let $(\boldsymbol{x}_s, \boldsymbol{x}_t)$ be an arc in T. As $\boldsymbol{x}_t$ is closer to the root than $\boldsymbol{x}_s$, we know $v(\boldsymbol{x}_t) \leq \omega(\boldsymbol{x}_t)$. Then, as T is a shortest path tree, we have $\hat{d}(\boldsymbol{x}_s) = \hat{\ell}_{st} + \hat{d}(\boldsymbol{x}_t)$. Consequently $-v(\boldsymbol{x}_s) = \hat{\ell}_{st} - v(\boldsymbol{x}_t)$, and so $\omega(\boldsymbol{x}_t) \geq v(\boldsymbol{x}_t) = \hat{\ell}_{st} + v(\boldsymbol{x}_s) > \hat{\ell}_{st} + \omega(\boldsymbol{x}_s)$. But then $\omega(\boldsymbol{x}_t) > \omega(\boldsymbol{x}_s) + \ell_{st} - \mu = \omega(\boldsymbol{x}_s) + \min_{s'} \{\boldsymbol{p}_{s'} \cdot (\boldsymbol{x}_t - \boldsymbol{x}_s) : \boldsymbol{x}_{s'} = \boldsymbol{x}_s\} - \mu$. It follows that there is at least one time period when $\boldsymbol{x}_s$ was selected in violation of the μ-optimality of $\omega()$. So $v()$ is a minimum individually rational, μ-approximate virtual valuation function. □

The Maximum (Individually Rational) Virtual Valuation Function. The minimum individually rational virtual valuation function allows us to obtain worst-case social welfare guarantees when revealed preference is used in mechanism design, see Sect. 4. For the best-case welfare guarantees, we are interested in finding the *maximum* virtual valuation function. In general, this need not exist as we may add an arbitrary constant to each bundle valuation given by the minimum individually rational virtual valuation function. But, it does exist provided we have an upper bound on the valuation of at least one bundle. This is often the case. For example in a combinatorial auction if a bidder drops out of the auction at time $t+1$, then $\boldsymbol{p}_{t+1} \cdot \boldsymbol{x}_t$ is an upper bound on the value of bundle $\boldsymbol{x}_t$. Furthermore, in practice, bidders (and the auctioneer) often have (over)-estimates of the maximum possible value of some bundles. So suppose we are given a set I and constraints of the form $v(\boldsymbol{x}_i) \leq \beta_i$ for each $i \in I$. Then there is a *unique* maximum μ-approximate virtual valuation function. Due to space constraints the proof of this result and all that follow have been omitted from this extended abstract; they can be found in the full version.

Theorem 3.4. *Given a set of constraints, the maximum μ-approximate virtual valuation function exists and can be found in polynomial time.*

4 Revealed Preference Auction Bidding Rules

So far, we have focused upon how to test the degree of rationality reflected in a data set. Specifically, we saw in Theorem 3.2 that the minimum mean length of a cycle, $\mu(G)$, gives an exact and optimal goodness of fit measure for rationality. Furthermore, Theorem 3.3 explained how to quickly obtain the minimum individually rational valuation function that best fits the data.

Recall, however, that revealed preference is also used as tool in mechanism design. In particular, we saw in Sect. 2.2 how revealed preference is used to impose bidding constraints in combinatorial auctions. We will now show how to apply the combinatorial arguments we have developed to create other relaxed revealed preference constraints.

4.1 Relaxed Revealed Preference Bidding Rules

Consider a combinatorial auction at time (round) t where our prior price-bundle bidding pairs are $\{(\boldsymbol{p}_1, \boldsymbol{x}_1), (\boldsymbol{p}_2, \boldsymbol{x}_2), \ldots, (\boldsymbol{p}_{t-1}, \boldsymbol{x}_{t-1})\}$. By Inequality (4) in Sect. 2.2, rational bidding at time t implies that $v(\boldsymbol{x}_t) - \boldsymbol{p}_t \cdot \boldsymbol{x}_t \geq v(\boldsymbol{x}_s) - \boldsymbol{p}_t \cdot \boldsymbol{x}_s$, for all $s < t$. Moreover, a necessary condition is then that $(\boldsymbol{p}_t - \boldsymbol{p}_s) \cdot \boldsymbol{x}_s \geq (\boldsymbol{p}_t - \boldsymbol{p}_s) \cdot \boldsymbol{x}_t$ and this can easily be checked by searching for negative length digons in the bidding graph induced by the first t bids. If such a cycle is found then the bid $(\boldsymbol{p}_t, \boldsymbol{x}_t)$ is not permitted by the auction mechanism.

The non-permittal of bids is clearly an extreme measure, and one that can lead to the exclusion of bidders from the auction even when they still have bids they wish to make. In this respect, it may be desirable for the mechanism to use a relaxed set of revealed preference bidding rules. The natural approach is to insist not upon strictly rational bidders but rather just upon approximately rational bidders. Specifically, the auction mechanism may (dynamically) select a desired degree ϵ of rationality. This requires $v(\boldsymbol{x}_t) - \boldsymbol{p}_t \cdot \boldsymbol{x}_t \geq v(\boldsymbol{x}_s) - \boldsymbol{p}_t \cdot \boldsymbol{x}_s - \epsilon$, for all $s < t$. A necessary condition then is $(\boldsymbol{p}_t - \boldsymbol{p}_s) \cdot \boldsymbol{x}_s \geq (\boldsymbol{p}_t - \boldsymbol{p}_s) \cdot \boldsymbol{x}_t - 2\epsilon$, and we can test this *relaxed* WARP-based bidding rule by insisting that every digon has mean length at least $-\epsilon$. Similarly, the *relaxed* KARP-based bidding rule is

$$(\boldsymbol{p}_k - \boldsymbol{p}_0) \cdot \boldsymbol{x}_0 \;\geq\; \sum_{i=1}^{k} (\boldsymbol{p}_i - \boldsymbol{p}_{i-1}) \cdot \boldsymbol{x}_i - (k+1) \cdot \epsilon \qquad (6)$$

The *relaxed* GARP-based bidding rule applies the relaxed KARP-based bidding rule for every choice of k. The imposition of the relaxed GARP-based bidding rule ensures approximate rationality.

Theorem 4.1. *A set of price-bid pairings $\{(\boldsymbol{p}_t, \boldsymbol{x}_t) : 1 \leq t \leq T\}$ has a corresponding ϵ-approximate individually rational virtual valuation function if and only if it satisfies the relaxed* GARP*-based bidding rule.*

4.2 Relaxed KARP-Based Bidding Rules

Theorem 4.1 tells us that imposing the relaxed GARP-based bidding rule ensures approximate rationality. But, in practice, even WARP-based bidding rules are often confusing to real bidders. There is likely therefore to be some resistance to the idea of imposing the whole gamut of GARP-based bidding rules. We believe that this combinatorial view of revealed preference, where the bidding rules can be tested via cycle examination, will eradicate some of the confusion. However, for simplicity, there is some worth in quantitatively examining the consequences of imposing a weaker relaxed KARP-based bidding rule rather than the GARP-based bidding rule. To test for the relaxed KARP-based bidding rules, we simply have to examine cycles of length at most $k+1$. Now suppose the KARP-based bidding rules are satisfied. By finding the $\mu(G)$ in the bidding graph we can still obtain the best-fit additive approximation guarantee, but we no longer have that this guarantee is ϵ. We can still, though, prove a strong additive approximation guarantee even for small values of k. To do this we need the following result.

Theorem 4.2. *Given a complete directed graph G with arc lengths ℓ. If every cycle of cardinality at most $k+1$ has non-negative length then the minimum mean length of a cycle is at least $-\frac{\ell^{\max}}{k}$, where $\ell^{\max} = \max_{e \in E(G)} |\ell_e|$.*

This result is important as it allows us to bound the degree of rationality that must arise whenever we impose the relaxed KARP-based bidding rule.

Corollary 4.1. *Given a set of price-bid pairings $\{(\boldsymbol{p}_t, \boldsymbol{x}_t) : 1 \leq t \leq T\}$ that satisfy the relaxed* KARP*-based bidding rule. Then there is a $(\frac{b^{\max}}{k} + \epsilon)$-approximate individually rational virtual valuation function, where $b^{\max}$ is the maximum bid made by the bidder during the auction.*

One may ask whether the additive approximation guarantee in Corollary 4.1 can be improved. The answer is no; Theorem 4.2 is tight.

Lemma 4.1. *There is a graph G where each cycle of cardinality at most $k+1$ has non-negative length and the minimum mean length of a cycle is $-\ell^{\max}/k$.*

4.3 Alternate Bidding Rules

Interestingly other bidding rules used in practice or proposed in the literature can be viewed in the graphical framework. For example, bid withdrawals correspond to vertex deletion in the bidding graph, whilst budget constraints and the Afriat Efficiency Index can be formulated in terms of arc-deletion. We introduce here the application to budget constraints; we discuss other applications in the full version of the paper.

Recall that, in our bidding graph, each arc encodes information about the head node relative to the tail node. However, this information is irrelevant if the head node is not affordable at the time where the arc was recorded. This simply means we must remove the arc $(\boldsymbol{x}_t, \boldsymbol{x}_s)$ if $\boldsymbol{x}_s$ was not affordable at time t. A μ-approximate individual rational virtual valuation that is compatible with the budget constraints then exists if and only if the resultant bidding graph has minimum mean cycle length at least $-\mu$.

In a budgeted combinatorial auction, we may assume a fixed budget $B_t = B$ for all t. Harsha et al. [13] then explain how to implement budgeted revealed preference in a combinatorial auction. Their method applies to the case when the fixed budget B is unknown to the auction mechanism. To do this, upper and lower bounds on feasible budgets are maintained dynamically via a linear program. It is also straightforward to do this combinatorially using the bidding graph; as the auction proceeds, the price of any purchased bundle is a lower bound on the budget B. Furthermore, for any value of B, we may ignore arcs $(\boldsymbol{x}_t, \boldsymbol{x}_s)$ such that $\boldsymbol{p}_t \boldsymbol{x}_s > B$. There is a greatest value of B for which, after ignoring arcs, G has no negative cycles: this is an upper bound on B. A bid is not permitted if these two bounds contradict one another.

References

1. Afriat, S.: The construction of a utility function from expenditure data. Int. Econ. Rev. **8**, 67–77 (1967)
2. Akerlof, G., Yellen, J.: Can small deviations from rationality make significant differences to economic equilibria? Am. Econ. Rev. **75**(4), 708–720 (1985)
3. Ausubel, L., Baranov, O.: Market design and the evolution of the combinatorial clock auction. Am. Econ. Rev. **104**(5), 446–451 (2014)
4. Ausubel, L., Cramton, P., Milgrom, P.: The clock-proxy auction: a practical combinatorial auction design. In: Cramton, P., Shoham, Y., Steinberg, R. (eds.) Combinatorial Auctions, pp. 115–138. MIT Press, Cambridge (2006)
5. Brown, D., Echenique, F.: Supermodularity and preferences. J. Econ. Theor. **144**(3), 1004–1014 (2009)
6. Cramton, P.: Spectrum auction design. Rev. Ind. Organ. **42**(2), 161–190 (2013)
7. Echenique, F., Golovin, D., Wierman, A.: A revealed preference approach to computational complexity in economics. In: Proceedings of EC, pp 101–110 (2011)
8. Edelman, B., Ostrovsky, M., Schwarz, M.: Internet advertising and the generalized second-price auction: selling billions of dollars worth of keywords. Am. Econ. Rev. **97**(1), 242–259 (2007)
9. Fostel, A., Scarf, H., Todd, M.: Two new proofs of Afriat's theorem. Econ. Theor. **24**, 211–219 (2004)
10. Gigerenzer, G., Selten, R. (eds.): Bounded Rationality: the Adaptive Toolbox. MIT Press, Cambridge (2001)
11. Gross, J.: Testing data for consistency with revealed preference. Rev. Econ. Stat. **77**(4), 701–710 (1995)
12. Gul, F., Stacchetti, E.: Walrasian equilibrium with gross substitutes. J. Econ. Theor. **87**, 95–124 (1999)
13. Harsha, P., Barnhart, C., Parkes, D., Zhang, H.: Strong activity rules for iterative combinatorial auctions. Comput. Oper. Res. **37**(7), 1271–1284 (2010)
14. Houthakker, H.: Revealed preference and the utility function. Economica New Ser. **17**(66), 159–174 (1950)
15. Karp, R.: A characterization of the minimum cycle mean in a digraph. Discrete Math. **23**(3), 309–311 (1978)
16. Kelso, A., Crawford, P.: Job matching, coalition formation, and gross substitutes. Econometrica **50**(6), 1483–1504 (1982)
17. Milgrom, P.: Putting auction theory to work: the simultaneous ascending auction. J. Polit. Econ. **108**, 245–272 (2000)
18. Samuelson, P.: A note on the pure theory of consumer's behavior. Economica **5**(17), 61–71 (1938)
19. Samuelson, P.: Consumption theory in terms of revealed preference. Economica **15**(60), 243–253 (1948)
20. Varian, H.: Revealed preference. In: Szenberg, M., Ramrattand, L., Gottesman, A. (eds.) Samulesonian Economics and the 21st Century, pp. 99–115. Oxford University Press, New York (2005)
21. Varian, H.: The nonparametric approach to demand analysis. Econometrica **50**(4), 945–973 (1982)
22. Varian, H.: Goodness-of-fit in optimizing models. J. Econometrics **46**, 125–140 (1990)
23. Varian, H.: Position auctions. Int. J. Ind. Organ. **25**(6), 1163–1178 (2007)
24. Varian, H.: Revealed preference and its applications. Working paper (2011)
25. Vohra, R.: Mechanism Design: A Linear Programming Approach. Cambridge University Press, Cambridge (2011)

Auction Design with a Revenue Target

Paul W. Goldberg[1] and Bo Tang[2](✉)

[1] Oxford University, Oxford, UK
[2] University of Liverpool, Liverpool, UK
bo.tang@liv.ac.uk

Abstract. In many fund-raising situations, a revenue target is specified. This suggests that the fund-raiser is interested in maximizing the probability to achieve this revenue target, rather than in maximizing the expected revenue. We study this topic from the perspective of Bayesian mechanism design, in a setting where a seller has a certain good that he can supply at no cost, and there are buyers whose joint valuation for the good comes from some given prior distribution. We present an algorithm to find the optimal truthful auction for two buyers with independent valuations via a direct characterization of the optimal auction. In contrast, we show the problem is NP-hard when the number of buyers is arbitrary or the distributions are correlated. Both negative results can be modified to show NP-hardness of designing auctions for risk-averse sellers.

Our main results address the design of *simple* auctions for many buyers, again in the context of a revenue target. For *Sequential Posted Price Auctions*, we provide a FPTAS to compute the optimal posted prices for a given sequence of buyers. For *Monopoly Price Auctions*, we apply the results of [8] on sparse covers of distributions to obtain a PTAS in a setting where the seller has a constraint on discriminatory pricing, consisting of a fixed set of prices he may use.

1 Introduction

There is a considerable literature on the algorithmic challenge of designing auctions that maximise the expected revenue obtained from a set of buyers. In this paper we consider a related objective where instead of maximising the expected revenue, the auctioneer has been given some revenue target T, and wishes to maximise the probability of raising at least T. This objective gives rise to new and interesting algorithmic challenges, and has some plausible real-world motivations, discussed below.

We work in the classical Bayesian setting of a collection of buyers whose valuations (prices they are willing to pay) for items being sold, are assumed to be drawn from some known prior distribution D. We are interested in designing mechanisms that are incentive compatible and individually rational. D in combination with a mechanism M results in a distribution over the revenue R obtained. A standard objective is to choose M to maximise the expected

P.W. Goldberg — Supported by EPSRC under grant EP/K01000X/1.

M. Hoefer (Ed.): SAGT 2015, LNCS 9347, pp. 137–149, 2015.
DOI: 10.1007/978-3-662-48433-3_11

value of R. A more general setting assumes a non-decreasing "utility of money" function u, and aims to maximize the expectation of $u(R)$. In this revenue-target setting, u is a shifted Heaviside function, equal to 0 for $R < T$ and 1 for $R \geq T$. Certain concave functions u have been used to model risk aversion, however the functions u considered here are not concave.

In this paper we focus on the "digital goods" setting, where the seller can supply unlimited copies of some good, at no cost. We also assume that the buyers have unit demand, so that a buyer's type is represented by a probability distribution over his valuation for a copy of the item. This special case is a simplified model of the fund-raising situations mentioned below. In the context of digital goods and unit demands, maximisation of the *expected* revenue can be decomposed into revenue-maximisation from each buyer independently. In contrast, when we switch to a revenue target, we find that the deal offered to a buyer should depend on the outcomes of the deals offered to other buyers.

This revenue-target setting is motivated by various real-world scenarios. Charitable fund-raising typically identify a target revenue to be raised. Similarly, in Internet crowd-sourcing platforms that support fund-raising for business start-ups (Kickstarter, Indiegogo, RocketHub etc.), it is typical to aim for some amount of money, and if that target is not reached, the would-be investors get their money back. (Our model doesn't properly capture this situation; we mention it to emphasise the importance of revenue targets in practice.) While a fund-raising effort is not the same thing as an auction, to some extent it can be modelled as one: an approach to a donor (or investor) corresponds to an attempt to sell an item to a would-be buyer. In cases where goods are sold at auction, it may be more desirable to raise a particular amount of money than to maximise the expected revenue. For example, in a bankruptcy situation, the administrator may wish to sell a collection of items so as to prioritise repaying the top-tier creditors. And while the FCC spectrum auction wants to raise as much money as possible, it is also required to cover its costs.

1.1 Our Results

We consider the problem parametrised by the number of buyers n, and the support size m of their value distributions. With multiple buyers, it is #P-complete to compute the exact success probability (probability to achieve revenue target T) for a given auction (Proposition 2). Given this obstacle, in Sect. 3 we consider a basic case of two buyers having uncorrelated valuations. We exhibit a polynomial-time algorithm to exactly compute the optimal truthful auction that maximises the probability to achieve T, given as input any discrete prior distributions. We do this via a structural characterisation of auctions that optimise the probability of achieving a given revenue target. This characterisation totally differs from the one maximising expected revenue and allows us to restrict to auctions with a geometric property that makes the problem tractable.

We show contrasting hardness results for correlated valuations or n buyers with independent distributions. Specifically, it is shown to be NP-complete to compute the optimal auction for three buyers having correlated valuations and

NP-hard for n buyers with independent distributions. Note that, in the latter case, a truthful auction may not necessarily be succinctly representable. We overcome the obstacle via proving the hardness for a class of succinct auctions and showing there exists a truthful auction with good performance if and only if there exists a good succinct auction in the constructed instance.

Our main algorithmic results are in Sect. 4, for two prevalent auctions following the trend of designing *simple* auctions. The first one is the *Sequential Posted Price Auction* introduced by Chawla et al. [4] to approximate the expected revenue in multi-dimensional Bayesian mechanism design. In this auction, the seller offers a take-it-or-leave-it price to each buyer sequentially. Given a sequence of buyers, we are able to provide a *fully polynomial-time approximation scheme* (FPTAS) to compute an approximately optimal sequential posted price auction that maximizes the success probability with an additive error. Second, we consider the *Monopoly Price Auction* where the seller offers take-it-or-leave-it prices to buyers simultaneously. This type of auctions was studied in [13] for selling goods with limited supply. We apply results of [8] on sparse covers of Poisson binomial distributions to obtain a PTAS when the seller has a limitation on discriminatory pricing, i.e., is only allowed to use few distinct prices.

1.2 Related Work

There has been a long line of research on maximizing expected revenue in Bayesian mechanism design starting from the seminal work by Myerson [15]. Recently, Cai et al [3] developed a general framework reducing revenue maximization to social welfare maximization. They also applied the framework to optimize certain non-linear functions [2]. However, the mechanisms they derived are randomized and Bayesian truthful, not deterministic truthful mechanisms studied in this paper.

Another line of research studied auction design for risk-averse sellers that can be regarded as maximizing a concave function of the revenue (cf. [17]). Sundararajan and Yan [18] studied the auction design problem for a risk averse seller and gave robust mechanisms (without knowledge of the concave function) which achieve constant approximations when buyers' distributions are independent. The approximation ratio has been improved to $e/(e-1)$ by Bhalgat et al. [1] by using the knowledge of concave functions. Our work complements their results by providing some corresponding intractability results.

We mention several negative results on revenue maximization in deterministic mechanism design. Diakonikolas et al. [10] showed that it is NP-hard to maximize revenue given a welfare constraint. Chen et al. [6] proved that it is NP-hard to maximize revenue in a multi-dimensional setting with a single unit-demand buyer when the valuations of items are independently distributed. For correlated buyers, Papadimitriou and Pierrakos [16] proved that it is NP-hard to approximate the optimal expected revenue for a single-item auction. However, in digital goods setting, the revenue maximizing auction can be constructed easily by computing the optimal price for each bidder separately based on their distributions conditioned on others' bids.

The study of digital goods auctions was initiated by Goldberg et al. [11]. Recently, Chen et al. [5] derived the optimal competitive auction with the benchmark defined to measure worst-case over all buyer profiles. In contrast, our benchmark measure is the average cases based on the prior distribution. Another related concept is "profit extractor" (see Sect. 6.2.4 in [12]) which is a decision problem the profit maximization in the prior-free setting.

Threshold probability maximization is a classical objective in stochastic optimization and has been studied for several combinatorial optimization problems (cf. [14] and references therein). However no incentive issues were considered before when optimizing this objective. The technique we apply to approximate the optimal monopoly price auction is based on [8]. These results have been shown helpful in computing Nash Equilibria [9] and learning sums of random variables [7]. But to our knowledge, this paper is their first application in auction design.

2 Preliminaries

Auction Setting. We study an auction environment where a seller wants to sell copies of an item to n bidders. Each bidder/buyer i is interested in a single copy of the item and values it at a privately known value v_i. A valuation profile $\mathbf{v}$ is the vector of all bidders' valuations, i.e. $\mathbf{v} = (v_1, \ldots, v_n)$. We consider a deterministic single-round sealed-bid auction where each bidder submits a bid b_i to express how much he is willing to pay for the item. After soliciting submitted bids $\mathbf{b} = (b_1, \ldots, b_n)$, the seller must decide whether each bidder i wins an item and how much he needs to pay. Bidder i's utility is the difference between his value v_i and his payment if he wins a item; otherwise he pays 0 and gets utility 0 to guarantee *individual rationality*, that is, no bidders will get a negative utility in the auction.

We assume every bidder in the auction is rational and aims to maximize his own utility by choosing the best bidding strategy. An auction is said to be truthful if for each bidder i, bidding his true valuation (i.e. $b_i = v_i$) is a dominant strategy no matter what the other bidders bid. It is known that truthful auctions can be characterized by *bid-independent* auctions where for each bidder i, the auction computes a threshold price p_i that does not depend on b_i but may depend on the bids of the other bidders $\mathbf{b}_{-i} = (b_1, \ldots, b_{i-1}, b_{i+1}, \ldots, b_n)$. In other words, there exists a pricing function for bidder i such that $p_i = f_i(\mathbf{b}_{-i})$ and i wins the item iff $b_i \geq p_i$ and his payment is p_i if he wins. So it suffices to consider bid-independent auctions when designing truthful auctions.

Thus any truthful or bid-independent auction A can be represented by n pricing functions $(f_1, \ldots, f_n)$ where f_i is the pricing function for bidder i which maps other bidders' valuations $\mathbf{v}_{-i}$ to the threshold price p_i. For convenience, we use $x_i(\mathbf{v})$ to denote the allocation rule of the auction, i.e. $x_i(\mathbf{v}) = 1$ if i wins an item when the valuation profile is $\mathbf{v}$; otherwise $x_i(\mathbf{v}) = 0$. Hence, the revenue of A on profile $\mathbf{v}$ is $R^A(\mathbf{v}) = \sum_{i \in [n]} x_i(\mathbf{v}) f_i(\mathbf{v}_{-i})$ where $[n]$ denotes the set $\{1, \ldots, n\}$. We also use $R_i^A(\mathbf{v})$ to denote the revenue of the auction A from bidder i, i.e., $R_i^A(\mathbf{v}) = x_i(\mathbf{v})f_i(\mathbf{v}_{-i})$. We will omit A from the notation if the auction is clear from the context.

Representation of Prior Distribution. We assume the seller has prior knowledge of the bidders' valuations, which is represented by a distribution on the valuation profile $\mathbf{v}$. In particular, we use D to denote the distribution on the valuation profile and V to denote the support of D. We denote the probability that the valuation profile is $\mathbf{v}$ by $\Pr[\mathbf{v}]$ for all $\mathbf{v} \in V$. Obviously, the distribution D can be represented in the size of V (denoted by $|V|$ or $|D|$) by explicitly describing $\Pr[\mathbf{v}]$ for all $\mathbf{v} \in V$. We also use $V_i = \{v_i^1, \ldots, v_i^{m_i}\}$ to denote the set of all possible value of v_i in D, where m_i is $|V_i|$ and $v_i^1 < v_i^2 < \cdots < v_i^{m_i}$. For convenience, we define $v_i^0 = 0$ and assume $0 \in V_i$.

We say the bidders' valuations are independently distributed if D is a product distribution, i.e. $D = \times_{i\in[n]} D_i$ where D_i is the distribution on buyer i's valuations; otherwise they are correlated. For convenience, we say the bidders are independent (or correlated) according to whether their valuations are independently distributed. For independent bidders, D can be represented using space $O(n \cdot m)$ where $m = \max_i m_i$.

We consider a seller with revenue target T and his utility is 1 if the revenue raised in the auction is at least T; otherwise his utility is 0. Given an instance $\mathcal{I} = (D, T)$ with the profile distribution D and revenue target T, the seller's utility in an auction A is $\Pr_{\mathbf{v}\sim D}[R^A(\mathbf{v}) \geq T]$. We also call this value the *performance* of auction A on instance $\mathcal{I}$. So an auction is an optimal truthful auction for an instance $\mathcal{I}$ if no truthful auction can outperform A on the instance $\mathcal{I}$. Similarly, we say A is c-additive approximately optimal if no truthful auction can perform better than the performance of A plus a parameter c. It is without loss of generality to assume the range of pricing function for bidder i is V_i as shown in the following proposition. The intuition is that rounding prices up to the next valuation of the agent will not decrease the revenue of the auction.

Proposition 1. *For any distribution profile D and truthful auction A, there exists another truthful auction A' such that the range of pricing functions for bidder i in A' is V_i for all $i \in [n]$ and $R^{A'}(\mathbf{v}) \geq R^A(\mathbf{v})$ for all profiles $\mathbf{v}$.*

Simple Auctions. We consider two types of simple auctions called monopoly price auctions and sequential posted price auctions. A *monopoly price auction* is a truthful auction with pricing functions $(f_1, \ldots, f_n)$ where each function f_i depends only on the prior distribution D and not on the other bids $\mathbf{b}_{-i}$. We say an auction is a *sequential posted price auction* with respect to an order σ if f_i may depend on D together with the bids of buyers who precede i in σ, i.e. $(b_1, \ldots, b_{i-1})$ if buyers are indexed according to σ. The following proposition shows the hardness of evaluating the performance of a given monopoly price auction. This is proved via a reduction from counting the solutions of KNAPSACK.

Proposition 2. *Given a monopoly price auction for independent bidders, it is #P-complete to compute the probability of achieving a revenue target.*

3 Optimal Truthful Auction for Two Independent Bidders

Recall that any truthful auction for two bidders can be represented by two pricing functions f_1 and f_2. By Proposition 1, we only need to consider $f_1 : V_2 \to V_1$ which maps bidder 2's valuations to bidder 1's threshold prices and $f_2 : V_1 \to V_2$. First of all, we show that the general problem reduces to a restricted version where bidders' distributions have support $\{0, \dots, m\} \times \{0, \dots, m\}$ and the target revenue is m, for some positive integer m. The intuition is mapping values of one agent to indices and mapping values of the second agent to intervals of $T - v_1$.

Lemma 3. *Given any instance $\mathcal{I} = (D, T)$ with an independent profile distribution $D = D_1 \times D_2$ (D_i having support V_i) and a target revenue T, there exists an integer $m \leq \min\{|V_1|, |V_2|\} + 1$ and another instance $\mathcal{I}' = (D', T')$ such that*

(a) $D' = D'_1 \times D'_2$ has the support $\{0, \dots, m\} \times \{0, \dots, m\}$ and $T' = m$
(b) Given an instance $\mathcal{I}$, the instance $\mathcal{I}'$ can be found in time linear in m
(c) Given any optimal truthful auction for $\mathcal{I}'$, it is possible to construct an optimal truthful auction for $\mathcal{I}$ in time linear in m.

For the case with two independent bidders, we assume $V_1 = V_2 = \{0, \dots, m\}$ and $T = m$. We also use q_1^i and q_2^j to denote probabilities $\Pr[v_1 = i]$ and $\Pr[v_2 = j]$ respectively and $R(i, j)$ to be the revenue from the profile (i, j). Regarding pricing functions, we can assume $f_1(0) = m$ and $f_2(0) = m$, since otherwise we can increase $f_1(0)$ or $f_2(0)$ to m without loss of the objective. In the following lemmas, we show that there exists an optimal auction with several nice properties. The first one is monotonicity of f_1 and f_2. Intuitively, the lemma says once one bidder's valuation increases, the seller will get more revenue from this bidder and set a lower price for the other bidder as a consequence.

Lemma 4. *There exists an optimal truthful auction for two independent bidders such that the pricing functions are monotonically non-increasing.*

By Lemma 3 we assume the valuations of both bidders are in $\{0, \dots, m\}$ and the target revenue is m. So for any profile $\mathbf{v}$ such that $v_1 < m$ and $v_2 < m$, the seller must sell items to both bidders to achieve the target revenue. Based on this observation, we are able to show another property of f_1 and f_2.

Lemma 5. *There exists an optimal truthful auction $A = (f_1, f_2)$ for two independent bidders such that f_1 is non-increasing and for any $i \in \{0, \dots, m\}$,*

$$f_2(i) = \begin{cases} m & \text{if } \forall j \in \{0, \dots, m\},\ i < f_1(j) \\ j & \text{if } \exists j \in \{0, \dots, m\},\ f_1(j) \leq i < f_1(j-1) \\ f_2(m-1) & \text{if } \forall j \in \{0, \dots, m\},\ i \geq f_1(j), \text{ i.e.i} = \text{m since } f_1(0) = \text{m} \end{cases}$$

Intuitively, the optimal auction described in the above lemma divides all profiles into four areas. In area one, the auction allocates nothing and in area two it sells both items. In area three (or four), the auction only sells a single copy with a price m to bidder 1 (or bidder 2). In addition, as shown in Fig. 1, the values of f_2 in this auction only depend on f_1. Thus, in order to design the optimal auction, we only need to find the optimal f_1, then a suitable f_2 follows by Lemma 5. Before characterizing the optimal f_1, we introduce some new notations. Given a non-increasing function f_1, let $J \subseteq [m]$ be the set of indices such that $f_2(j) < f_2(j-1)$. We denote the set J by $\{j_1, j_2, \ldots, j_{|J|}\}$ with an increasing order, i.e. $j_\ell < j_{\ell+1}$. Let $i_\ell = f_1(j_\ell)$ as illustrated in Fig. 2. We also define $i_0 = j_{|J|+1} = m+1$ for simplicity. Then for all $\ell = 1, \ldots, |J|$ and $j_\ell \leq j < j_{\ell+1}$, $f_1(j) = i_\ell$ by the definition of j_ℓ. In addition, for all $\ell = 1, \ldots, |J|$ and $i_\ell \leq i < i_{\ell-1}$, $f_2(i) = j_\ell$ by Lemma 5. This is because j_ℓ is the j such that $f_1(j) \leq i < f_1(j-1)$. Then we can prove the following lemma.

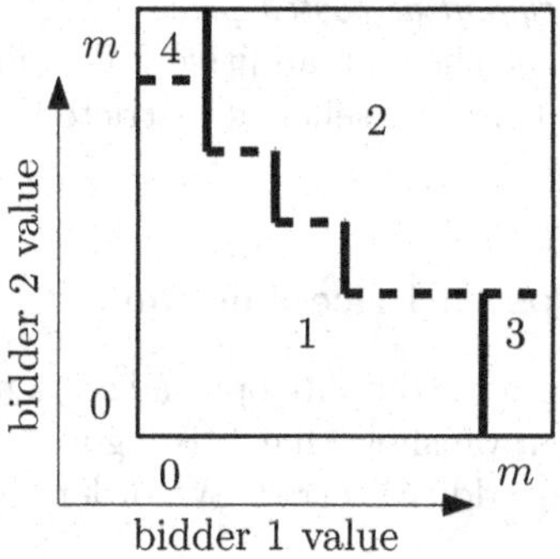

Fig. 1. Illustration of the computation of f_2 for a given f_1 based on Lemma 5. Again, the vertical bold lines are f_1 and the horizontal dashed lines are the resulting f_2. We also mark the four areas mentioned in the text.

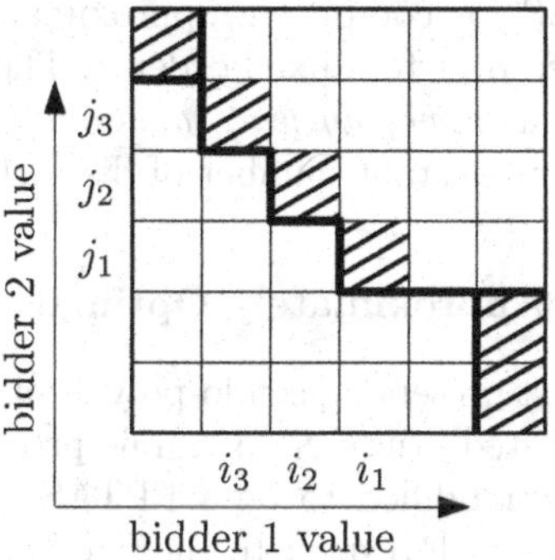

Fig. 2. Illustration of the definition of the set J, the values j_ℓ and i_ℓ when the pricing functions f_1 and f_2 are given as vertical and horizontal bold lines respectively. The shawed squares illustrate the profiles with revenue at least m.

Lemma 6. *There exists an optimal auction $A = (f_1, f_2)$ such that $i_\ell + j_\ell = m$ for all $\ell = 1, \ldots, |J|$ where i_ℓ and j_ℓ are defined by f_1 as above.*

By the above lemma, we can characterize the optimal auction by only using the set J, i.e. the values of $\{j_1, \ldots, j_{|J|}\}$. Given the set J, we can compute f_1 and f_2 by Lemmas 6 and 5 respectively. Based on this characterization, we are able to show the main theorem in this section.

Theorem 7. *Given a distribution $D = D_1 \times D_2$ for two independent bidders and a target revenue for the seller, an optimal truthful auction can be found in time $O(m^3)$ where $m = \min\{|D_1|, |D_2|\}$.*

We have contrasting NP-hardness for more general cases. Both results can be modified for the cases with risk-averse sellers.

Theorem 8. *It is NP-complete to compute an optimal auction for three correlated bidders, having a joint prior distribution presented as a set of probabilities on a finite set of support points.*

Theorem 9. *It is NP-hard to compute the optimal auction for n independent bidders even when each bidder has only two possible valuations, i.e.* $|V_i| = 2$.

4 Near-Optimal Simple Auctions for Independent Bidders

In this section, we study the following simple auctions for sellers with a target revenue when the bidders are independent. In Sect. 4.1, we present an additive FPTAS for computing approximately optimal *sequential posted price* auctions with respect to a fixed order σ. Then in Sect. 4.2 we show an additive PTAS for optimal *monopoly price auctions*, in a setting where the seller is restricted to using a constant number of distinct prices.

4.1 Approximately Optimal Sequential Posted Price Auction

We first present a pseudo-polynomial time algorithm to compute optimal sequential posted prices via dynamic programming. Then we show that this algorithm can be modified to be a FPTAS with respect to additive error. We order the bidders with respect to the fixed order σ.

Recall that in a sequential posted price mechanism, the seller offers take-it-or-leave-it prices to the buyers sequentially with respect to a given order σ and the computation of the price for buyer i is based on the results of all buyers preceding i, together with the valuation distributions. Note that the optimal sequential posted price for any sequence of buyers, performs at least as well as the optimal monopoly price auction. In contrast with the objective of expected revenue maximization, our objective of a target revenue means that the price offered to bidder i may depend on the revenue gained from the first $i-1$ bidders. This allows us to solve the problem by the following dynamic programming. Let $Q[i,r]$ be the maximal probability to achieve revenue r by selling items to buyers from i to n. By Proposition 1, it is sufficient to consider the case that $p_i \in V_i$ where V_i is the support of buyer i's valuation distribution. It is easy to see $Q[i,r] = 1$ if $r \leq 0$ and $Q[i,r] = 0$ if $i > n$ and $r > 0$. For the other cases when $i \leq n$ and $r > 0$ we have

$$Q[i,r] = \max_{p_i \in V_i}\{Q[i+1, r-p_i] \cdot \Pr[v_i \geq p_i] + Q[i+1, r] \cdot (1 - \Pr[v_i \geq p_i])\}.$$

Thus the maximal probability to achieve target revenue T from all buyers is $Q[1,T]$. Note that solving the above dynamic programming gives a pseudo-polynomial time algorithm for the problem. Actually, we can get an additive FPTAS by rounding the dynamic programming properly.

Theorem 10. *There exists an additive FPTAS for computing approximately optimal sequential posted price auctions with respect to a fixed order of the buyers. In particular, given $\epsilon \in (0,1)$, an instance $\mathcal{I} = (D, T)$ with n independent buyers and a buyer sequence σ, an ϵ-additive approximately optimal sequential posted price auction with respect to σ can be computed in time $O(m^2 n^2 \log n \cdot 1/\epsilon \log(1/\epsilon))$ where m is the maximal support size, i.e. $\max_{i \in [n]}\{|D_i|\}$.*

4.2 Approximately Optimal Monopoly Price Auction

In this section, we present a PTAS for computing the optimal monopoly price auction when the seller is restricted to a given constant-sized set of distinct prices, and for each buyer has to select one of those prices for that buyer. Recall that in a monopoly price auction, the seller offers those take-it-or-leave-it prices to the buyers simultaneously, and the prices are only based on the valuation distributions. Our PTAS uses results of [8] on Poisson Binomial Distributions. First of all, we review the definitions and results. For any two random variables X and Y supported on a finite set A, their total variation distance is defined as

$$d_{\mathrm{TV}}(X, Y) = \frac{1}{2} \sum_{a \in A} |\Pr[X = a] - \Pr[Y = a]|.$$

We use the following result in the proof of Theorems 14 and 15.

Lemma 11 (Lemma 2 in [8]). *Let $X_1, \ldots, X_n$ be mutually independent random variables, and let $Y_1, \ldots, Y_n$ be mutually independent random variables. Then*

$$d_{\mathrm{TV}}(\sum_{i=1}^{n} X_i, \sum_{i=1}^{n} Y_i) \leq \sum_{i=1}^{n} d_{\mathrm{TV}}(X_i, Y_i).$$

A distribution is said to be a *Poisson Binomial Distribution* (PBD) of order n if it is a discrete probability distribution consisting of the sum of n independent indicator random variables. The distribution is parameterized by a vector $(r_i)_{i=1}^n \in [0,1]^n$ of probabilities and is denoted by $\mathrm{PBD}(r_1, \ldots, r_n)$. Let $\mathcal{S}_n$ be the set of all PBDs of order n. We review a construction of an efficient and proper ϵ-cover for $\mathcal{S}_n$.

Theorem 12 (Theorem 1 in [8]). *For all $n, \epsilon > 0$, there exists a set $\mathcal{S}_{n,\epsilon} \subset \mathcal{S}_n$ such that*

1. *$\mathcal{S}_{n,\epsilon}$ is an ϵ-cover of $\mathcal{S}_n$ in total variation distance; that is, for all $D \in \mathcal{S}_n$, there exists some $D' \in \mathcal{S}_{n,\epsilon}$ such that $d_{\mathrm{TV}}(D, D') \leq \epsilon$,*
2. *$|\mathcal{S}_{n,\epsilon}| \leq n^2 + n \cdot (\frac{1}{\epsilon})^{O(\log^2 1/\epsilon)}$,*
3. *$\mathcal{S}_{n,\epsilon}$ can be computed in time $O(n^2 \log n) + O(n \log n) \cdot (\frac{1}{\epsilon})^{O(\log^2 1/\epsilon)}$.*

Moreover, all distributions $\mathrm{PBD}(r_1, \ldots, r_n) \in \mathcal{S}_{n,\epsilon}$ in the cover satisfy at least one of the following properties, for some positive integer $t = t(\epsilon) = O(1/\epsilon)$.

- *(t-sparse form) there is some $\ell \leq t^3$ such that, for all $i \leq \ell$, $r_i \in \{\frac{1}{t^2}, \frac{2}{t^2}, \ldots, \frac{t^2-1}{t^2}\}$ and for all $i > \ell$, $r_i \in \{0, 1\}$; or*

- *((n,t)-Binomial form) there is some $\ell \in [n]$ and $q \in \{\frac{1}{n}, \frac{2}{n}, \ldots, \frac{n}{n}\}$ such that, for all $i \leq \ell$, $r_i = q$ and for all $i > \ell$, $r_i = 0$; moreover ℓ and q satisfy $\ell q \geq t^2$ and $\ell q(1-q) \geq t^2 - t - 1$.*

In words, every PBD can be approximated by either a sparse PBD or a binomial distribution. Moreover, the following theorem tells us that if the first $O(\log 1/\epsilon)$ moments of two PBDs are the same, then the total variation distance between them is at most ϵ.

Theorem 13 (Theorem 3 in [8]). *Let $\mathcal{P} := (p_i)_{i=1}^n \in [0,1/2]^n$ and $\mathcal{Q} := (q_i)_{i=1}^n \in [0,1/2]^n$ be two collections of probability values. Let also $\mathcal{X} := (X_i)_{i=1}^n$ and $\mathcal{Y} := (Y_i)_{i=1}^n$ be two collections of mutually independent indicators with $\mathrm{E}[X_i] = p_i$ and $\mathrm{E}[Y_i] = q_i$, for all $i \in [n]$. If for some $d \in [n]$ the following condition is satisfied: $\sum_{i=1}^n p_i^\ell = \sum_{i=1}^n q_i^\ell$ for all $\ell = 1, \ldots, d$, then $d_{\mathrm{TV}}(\sum_i X_i, \sum_i Y_i) \leq 13(d+1)^{1/4} 2^{-(d+1)/2}$.*

It is easy to see that Theorem 13 holds if we replace $[0,1/2]$ with $[1/2,1]$. Moreover, by setting $d = O(\log 1/\epsilon)$, this bound becomes at most ϵ. Theorem 12 shows that there exists an efficient cover for the set of all PBDs. However, we cannot directly apply this theorem to our problem, since (given prices and prior distributions of a problem instance) the set of associated PBDs (call it S) is a proper subset of $\mathcal{S}_n$, and we need to find a cover that consists of a subset of S. Theorem 14 is intended to overcome this obstacle. Given n finite sets $W_1, \ldots, W_n$ where $W_i \subset [0,1]$ for all $i \in [n]$, let $W = \times_{i=1}^n W_i$, and let $\mathcal{S}_n(W)$ denote the set of all PBDs such that the probability of the indicator i is in W_i for all $i \in [n]$. That is $\mathcal{S}_n(W) = \{\mathrm{PBD}(r_1, \ldots, r_n) | (r_i)_{i=1}^n \in W\}$.

Theorem 14. *For all $n, \epsilon > 0$ and any n finite subsets of $[0,1]$, $W_1, \ldots, W_n$ let $W = \times_{i=1}^n W_i$. Then there exists a set $\mathcal{S}_{n,\epsilon}(W) \subset \mathcal{S}_n(W)$ such that*

1. *$\mathcal{S}_{n,\epsilon}(W)$ is an ϵ-cover of $\mathcal{S}_n(W)$ in total variation distance; that is, for all $D \in \mathcal{S}_n(W)$, there exists some $D' \in \mathcal{S}_{n,\epsilon}(W)$ such that $d_{\mathrm{TV}}(D, D') \leq \epsilon$,*
2. *$\mathcal{S}_{n,\epsilon}(W)$ can be computed in time $(\frac{n}{\epsilon})^{O(\log^2 1/\epsilon)}$ and has size at most $(\frac{n}{\epsilon})^{O(\log^2 1/\epsilon)}$.*

Given the above theorem, we can obtain an additive PTAS for computing approximately optimal monopoly price auctions, given a fixed set of allowed prices.

Theorem 15. *There exists an additive PTAS for computing approximately optimal monopoly price auctions when the seller is restricted to a fixed number of distinct prices. In particular, given $\epsilon \in (0,1)$, an instance with n independent bidders and k distinct prices the seller may use, an ϵ-additive approximately optimal monopoly price auction can be computed in time $(\frac{nk}{\epsilon})^{O(k \log^2 1/\epsilon)}$.*

Proof. We use $a_1, \ldots, a_k$ to denote the k distinct prices the seller may use. Given a monopoly price auction with price vector $(p_1, p_2, \ldots, p_n)$, we use an indicator random variable H_{ij} to indicate that the seller gets revenue a_j from buyer i, that is $H_{ij} = 1$ iff $p_i = a_j$ and $v_i \geq a_j$. Let $H_j = \sum_{i \in [n]} H_{ij}$ and

$H = \sum_{j \in [k]} a_j H_j$. Note that H is the random variable for the total revenue raised in this auction. Since the H_{ij} are indicator random variables, the H_j are Poisson Binomial random variables due to the independence among bidders. So H can be viewed as a weighted sum of k Poisson Binomial random variables. Let r_{ij} denote the probability of getting revenue exactly a_j from buyer i. Then the distribution of H_j is PBD$(r_{1j}, \cdots, r_{nj})$. The distribution of H can be represented by the vector $\mathbf{r} = (r_{ij})_{i\in[n], j\in[k]}$. Let W_i be the set of all possible $(r_{i1}, \ldots, r_{in})$ such that $r_{ij} = \Pr[v_i \geq a_j]$ if the seller use price a_j for bidder i and $r_{ij} = 0$ otherwise . It is clear that the set $W = \times_{i\in[n]} W_i$ is the set of all probability vector $\mathbf{r}$ corresponding to a feasible pricing vector $\mathbf{p}$.

Note that for any two random variables X, Y and any value T, $|\Pr[X \geq T] - \Pr[Y \geq T]| \leq d_{\mathrm{TV}}(X, Y)$. So if there exists an ϵ-cover for the set of all possible distribution of H parameterized by $\mathbf{r} \in W$, we can explore the pricing rules in the cover instead of all possible pricing rules to find a sequence of monopoly prices which approximately maximize $\Pr[H \geq T]$. In order to get such a cover, we need to modify the dynamic programming used in the proof of Theorem 14 to be k-dimensional. The moment profile $(\mu^1, \ldots, \mu^k, \nu^1, \ldots, \nu^k)$ is defined as $\mu^j = (\mu^j_1, \ldots, \mu^j_d), \nu^j = (\nu^j_1, \ldots, \nu^j_d)$ and $\mu^j_\ell, \nu^j_\ell \in \{0, (\frac{\epsilon}{nk})^\ell, 2(\frac{\epsilon}{nk})^\ell, \ldots, n\}$ for all $\ell \in [d]$ and $j \in [k]$. By a similar argument to Theorem 14 and Lemma 11, all the possible moment profiles is already an ϵ-cover. Define $A[i, \mu^1, \ldots, \mu^k, \nu^1, \ldots, \nu^k]$ to be the indicator such that it is equal to 1 iff there exists $\mathbf{r}_1 \in W_1, \ldots, \mathbf{r}_i \in W_i$ such that for all $j \in [k]$ and $\ell \in [d]$, $\sum_{i' \leq i: r'_{i'j} \in [0,1/2]} (r'_{i'j})^\ell = \mu^j_\ell$ and $\sum_{i' \leq i: r'_{i'j} \in (1/2,1]} (r'_{i'j})^\ell = \nu^j_\ell$ where $\mathbf{r}'$ is a $\frac{\epsilon}{nk}$-rounding of $\mathbf{r}$ such that r'_{ij} is a multiple of $\frac{\epsilon}{nk}$ and $r_{ij} - \frac{\epsilon}{nk} < r'_{ij} \leq r_{ij}$ for all $i \in [n]$ and $j \in [k]$.

Similarly to the proof of Theorem 14, A can be computed by the following dynamic programming. Inductively, to compute layer $i + 1$, we consider all the non-zero entries of layer i and for every such non-zero entry and every possible prices a_j, we find which entry of layer $i + 1$ we would transition to if we choose $p_i = a_j$, i.e. $r_{ij} = \Pr[v_i \geq a_j]$ and $r_{ij'} = 0$ for all $j' \neq j$. It is easy to see the overall running time to compute A is $(\frac{nk}{\epsilon})^{O(k \log^2 1/\epsilon)}$. In addition, we can find the corresponding monopoly prices for any distribution in this cover by tracing the pointers in the computation of A. Therefore, we can enumerate all possible pricing rules in this cover with size at most $(\frac{nk}{\epsilon})^{O(k \log^2 1/\epsilon)}$ to find the optimal pricing which maximize $\Pr[H \geq T]$.

The final step is to compute $\Pr[H \geq T]$ given a price vector $\mathbf{p}$. By Theorem 12, we know any PBD can be approximated by a sparse PBD or a binomial distribution. For the given price vector, we can get the corresponding H_j for all $j \in [k]$. We use Theorem 12 to compute H'_j from H_j such that H'_j is either a k/ϵ-sparse PBD or a binomial distribution and $d_{\mathrm{TV}}(H'_j, H_j) \leq \epsilon/k$ for all $j \in [k]$. Then we compute $\Pr[H'_j = T_j]$ for any value $T_j \in [0, \ldots, n]$ and $j \in [k]$. This computation can be done efficiently since H'_j is either a k/ϵ-sparse PBD or a binomial distribution. By Lemma 11, we have $d_{\mathrm{TV}}(H', H) \leq \epsilon$ where $H' = \sum_j a_j H_j$. Finally we compute $\Pr[H' \geq T] = \sum_{(T_j)_j: \sum_j a_j T_j \geq T} \prod_j \Pr[H'_j = T_j]$ by enumerating all possible $T_1, \ldots, T_k$. Since the distance between H and H' is at most ϵ, we have $\Pr[H \geq T] \geq \Pr[H' \geq T] - \epsilon$. Combine all these together, we get the additive PTAS with running time $(\frac{nk}{\epsilon})^{O(k \log^2 1/\epsilon)}$. □

5 Conclusion

We see several promising directions for future work. For independent buyers, a direct open problem is to generalize our characterization to three or more buyers. That may be achievable via an induction on the number of buyers, characterizing the optimal auction for three buyers by using the case with two buyers as a substructure. Another direction is to approximate the optimal auction via designing simple auctions. We find several examples to show the lower bounds (see full version for more details) but the upper bound is still open. Finally, we point out an interesting problem of computing optimal monopoly prices without the limitation on distinct prices.

References

1. Bhalgat, A., Chakraborty, T., Khanna, S.: Mechanism Design for a Risk Averse Seller. In: Goldberg, P.W. (ed.) WINE 2012. LNCS, vol. 7695, pp. 198–211. Springer, Heidelberg (2012)
2. Cai, Y., Daskalakis, C., Weinberg, S.: Understanding incentives: Mechanism design becomes algorithm design, In: FOCS 2013, pp. 618–627. IEEE, October 2013
3. Cai, Y., Daskalakis, C., Weinberg, S.M.: Optimal multi-dimensional mechanism design: Reducing revenue to welfare maximization. In: FOCS 2012, pp. 130–139. IEEE Computer Society, Washington (2012)
4. Chawla, S., Hartline, J.D., Malec, D.L., Sivan, B.: Multi-parameter mechanism design and sequential posted pricing. In: STOC 2010, New York, NY, pp. 311–320 (2010)
5. Chen, N., Gravin, N., Lu, P.: Optimal competitive auctions. In: STOC 2014, pp. 253–262. ACM, New York (2014)
6. Chen, X., Diakonikolas, I., Paparas, D., Sun, X., Yannakakis, M.: The complexity of optimal multidimensional pricing. In: SODA 2014, pp. 1319–1328. SIAM (2014)
7. Daskalakis, C., Diakonikolas, I., Servedio, R.A.: Learning poisson binomial distributions. In: STOC 2012, pp. 709–728. ACM, New York (2012)
8. Daskalakis, C., Papadimitriou, C.: Sparse covers for sums of indicators. Probab. Theory Relat. Fields **162**, 679–705 (2014)
9. Daskalakis, C., Papadimitriou, C.H.: Approximate Nash equilibria in anonymous games. J. Econ. Theory **156**, 207–245 (2015)
10. Diakonikolas, I., Papadimitriou, C., Pierrakos, G., Singer, Y.: Efficiency-revenue trade-offs in auctions. In: Czumaj, A., Mehlhorn, K., Pitts, A., Wattenhofer, R. (eds.) ICALP 2012, Part II. LNCS, vol. 7392, pp. 488–499. Springer, Heidelberg (2012)
11. Goldberg, A.V., Hartline, J.D., Wright, A.: Competitive auctions and digital goods. In: SODA 2001, pp. 735–744. SIAM, Philadelphia (2001)
12. Hartline, J.D.: Mechanism design and approximation. Book draft, October 2013

13. Hartline, J.D., Roughgarden, T.: Simple versus optimal mechanisms. SIGecom Exch. **5:8**(1), 1–5:3 (2009)
14. Li, J., Yuan, W.: Stochastic combinatorial optimization via poisson approximation, In: STOC 2013, pp. 971–980. ACM, New York (2013)
15. Myerson, R.B.: Optimal auction design. Math. Oper. Res. **6**(1), 58–73 (1981)
16. Papadimitriou, C.H., Pierrakos, G.: On optimal single-item auctions. In: STOC 2011, pp. 119–128. ACM, New York (2011)
17. Rothschild, M., Stiglitz, J.E.: Increasing risk: I. A definition. J. Econ. Theory **2**(3), 225–243 (1970)
18. Sundararajan, M., Yan, Q.: Robust mechanisms for risk-averse sellers. In: EC 2010, pp. 139–148. ACM, New York (2010)

Algorithmic Signaling of Features in Auction Design

Shaddin Dughmi[1], Nicole Immorlica[2], Ryan O'Donnell[3], and Li-Yang Tan[4](✉)

[1] University of Southern California, Los Angeles, USA
[2] Microsoft Research, Cambridge, USA
[3] Carnegie Mellon University, Pittsburgh, USA
[4] Simons Institute, UC, Berkeley, USA
liyang@ttic.edu

Abstract. In many markets, products are highly complex with an extremely large set of features. In advertising auctions, for example, an impression, i.e., a viewer on a web page, has numerous features describing the viewer's demographics, browsing history, temporal aspects, etc. In these markets, an auctioneer must select a few key features to signal to bidders. These features should be selected such that the bidder with the highest value for the product can construct a bid so as to win the auction. We present an efficient algorithmic solution for this problem in a setting where the product's features are drawn independently from a known distribution, the bidders' values for a product are additive over their known values for the features of the product, and the number of features is exponentially larger than the number of bidders and the number of signals. Our approach involves solving a novel optimization problem regarding the expectation of a sum of independent random vectors that may be of independent interest. We complement our positive result with a hardness result for the problem when features are arbitrarily correlated. This result is based on the conjectured hardness of learning k-juntas, a central open problem in learning theory.

1 Introduction

Much of the computer science literature on auction design assumes bidders have full knowledge of their own values. However, in many markets, this assumption

S. Dughmi—Supported in part by NSF CAREER Award CCF-1350900. Part of this work performed while the author was visiting Microsoft Research New England.

R. O'Donnell—Supported by NSF grants CCF-0747250 and CCF-1116594. Part of this work performed while the author was visiting Microsoft Research New England. Part of this work performed at the Boğaziçi University Computer Engineering Department, supported by Marie Curie International Incoming Fellowship project number 626373.

L.Y. Tan—Supported by NSF grants CCF-1115703 and CCF-1319788. Part of this research was done while visiting Carnegie Mellon University.

M. Hoefer (Ed.): SAGT 2015, LNCS 9347, pp. 150–162, 2015.
DOI: 10.1007/978-3-662-48433-3_12

is quite unrealistic in part because the item for sale is not fully observable by the bidders. In used car auctions, for example, the cars for sale are each unique items with a long list of features – make, model, year, mileage, color, etc. Time and communication constraints make it impractical for the auctioneer to provide bidders with a full description of each car. Similarly, in advertising auctions, the impressions for sale correspond to searchers, again with a long list of features – gender, age, income, zip code, search history, etc. Again it is impractical for the auctioneer to communicate all these features for each search, let alone track them all. This raises a natural question: *which features should an auctioneer signal to bidders?*

We study this question in the context of a single item auction. The item is parameterized by a large feature vector drawn from some known distribution. A bidder's value for an item is a function of its features. The goal is to signal a small subset of features to bidders such that the welfare[1] generated by the resulting auction is maximized. Trivial brute-force search can solve this problem in time $O(nk \cdot m^k)$ where n is the number of players, k is the number of allowed signals, and m is the number of features. Throughout this paper, we think of the number of bidders and allowable signals as small, whereas the number of features is exponentially larger, and thus seek running times at most linear in m.

We wish to focus attention on the algorithmic problem of selecting features, and so we make several simplifying assumptions. First we assume bidders' values are additively separable across features. This assumption is a reasonable approximation to valuations in many settings and is also a good first step in understanding general substitutable valuations. Second, as is common in much of the computer science literature on signaling [2,8,9,15], we assume bidders' values for features are known to the auctioneer. This information could be available to the auctioneer through historical data, and is also a first step in designing systems for the more common Bayesian setting.[2]

Even with these simplifying assumptions, we obtain strong negative results for the problem of finding a welfare-maximizing set of signals. We do this by relating the feature selection problem to the problem of *learning k-juntas* (i.e. m-variable boolean functions that depend only on $k \ll m$ of its coordinates) with respect to the uniform distribution[3]. Introduced by Blum in 1994 [3,7], the junta problem is a clean abstraction of learning in the presence of irrelevant information, and represents a necessary first step towards the notorious problems of learning polynomial-size decision trees and DNF formulas. Progress on the problem has been slow despite significant interest — the current best algorithm is due to G. Valiant and runs in time $O(m^{0.6k})$ [17], a polynomial improvement

[1] The welfare of a single item auction is the value of the winning bidder.

[2] Clearly, if the auctioneer knowns the values of the bidders, he can maximize welfare by simply assigning the item to the highest-value bidder, circumventing the auction altogether. We assert that even if the auctioneer has this information, market constraints require the use of a second-price auction format as is the case in, e.g., ad auctions.

[3] See Sect. 3 for a definition.

over brute-force search in time $O(m^k)$, and it is a generally accepted assumption that is no $m^{o(k)}$-time algorithm for the problem. (Indeed, it is known that the broad class *statistical query* learning algorithms require both time and sample complexity $m^{\Omega(k)}$ for the junta problem [6]). Assuming that the junta problem does in fact require time $m^{\Omega(k)}$, we show there is no $m^{o(k)}$-time algorithm that can find an $(1/n + \epsilon)$-approximately optimal set of signals.

On the positive side, we consider a setting where each feature is selected independently from a (not necessarily identical) distribution, and takes on only a constant number of values. In this case, we give an $(1-\epsilon)$-approximate algorithm that runs in time $O(m) + 2^{O(k \log(k/\epsilon))}$ for all fixed values of n. This algorithm solves a general optimization problem of potentially independent interest: for any norm $\|\cdot\|$ on $\mathbb{R}^n$, given $\theta \in \mathbb{R}^n$ and m independent mean-zero vector-valued random variables $\mathbf{X}_1, \ldots, \mathbf{X}_m$, find a subset $S \subseteq [m]$ of cardinality k that approximately maximizes $\mathbf{E}[\|\theta + \sum_{i \in S} \mathbf{X}_i\|]$. Prior to our work there were no non-trivial algorithms even when $n = 1$ — given m *real-valued* random variables $\mathbf{X}_1, \ldots, \mathbf{X}_m$ and $\theta \in \mathbb{R}$, find a k-subset $S \subseteq [m]$ that approximately maximizes $\mathbf{E}[|\theta + \sum_{i \in S} \mathbf{X}_i|]$ — and even under the assumption that all the $\mathbf{X}_i$'s are two-valued.

Related Work. Understanding the structure of optimal signaling schemes is a classical question in economics [14], and has recently generated great interest within the computer science community [2,8,9,13,15]. One line of prior work [2,9,15] studies unconstrained signaling schemes that maximize revenue. In such unconstrained settings, full information revelation is guaranteed to optimize welfare. Other prior work [8], more closely related to the current paper, studies constrained signaling schemes and seeks to maximize welfare. That work considered two settings: one in which goods were represented by high dimensional feature vectors and one where the goods were arbitrary and had to be partitioned into classes. The former setting is closely related to ours, but in the prior work the signaling schemes were arbitrary bounded-length bit strings. In this paper, we constrain our signaling schemes to announce subsets of features, an arguably more natural scheme for which the techniques of the prior work cannot be applied. For an overview of the junta problem and its role in learning theory see [4,16] and the references therein. As mentioned above its hardness is a generally accepted assumption in learning theory, and indeed it is commonly used as hardness primitive to establish the intractability of various other learning problems (e.g. [1,10–12]).

2 Preliminaries

We consider a setting in which there is a set of possible items Ω for sale, where each $\omega \in \Omega$ is summarized by an m-dimensional vector of *features* — formally, $\Omega = \prod_{j=1}^{m} \Omega_j$, where Ω_j is the set of possible values of the j'th feature. We assume that an item is drawn according to a distribution $\lambda \in \Delta_\Omega$. There is a set of n players, each of whom is equipped with a *valuation* function $v_i : \Omega \rightarrow \mathbb{R}_+$ mapping items to the real numbers. We restrict attention to *linearly separable* valuation functions, of the form $v_i(\omega) = \sum_{j=1}^{m} v_{ij}(\omega_j)$, for functions $v_{ij} : \Omega_j \rightarrow \mathbb{R}_+$.

We assume that the features of the item being sold are a-priori unknown to the players, who learn them through a *signaling scheme* mapping an items to messages, known as *signals*. In this paper, we restict attention to signaling schemes which simply fix a set $S \subseteq [m]$ of feature indices of a given size $|S| = k$, and announces $\omega_S = \{(j, \omega_j) : j \in S\}$. After players learn this partial information, some protocol — typically an auction — is run to assign the item to one of the players. We focus on auctions, such as the second-price auction, which assign the item to the player with the highest posterior expected value for the item given the features revealed. In this case the expected *social welfare*, i.e. the expected value of the winning player, can be written as follows.

$$\text{welfare}(S) = \mathbf{E}[\max_{i=1}^{n} \mathbf{E}[v_i(\omega)|\omega_S]]$$

where both expectations are over $\omega \sim \lambda$. Using v_{ij} as shorthand for the random variable $v_{ij}(\omega_j)$, we can rewrite the above expression as follows.

$$\text{welfare}(S) = \mathbf{E}\left[\max_{i=1}^{n}\left(\sum_{j\in S} v_{ij} + \sum_{j\notin S}\mathbf{E}[v_{ij}|\omega_S]\right)\right]$$

In the special case in which the features are independently distributed, this reduces to

$$\begin{aligned}\text{welfare}(S) &= \mathbf{E}\left[\max_{i=1}^{n}\left(\sum_{j\in S} v_{ij} + \sum_{j\notin S}\mathbf{E}[v_{ij}]\right)\right] \\ &= \mathbf{E}\left[\max_{i=1}^{n}\left(\sum_{j\in S}(v_{ij} - \mathbf{E}[v_{ij}]) + \sum_{j=1}^{m}\mathbf{E}[v_{ij}]\right)\right] \\ &= \mathbf{E}\left[\left\|\sum_{j\in S}(\boldsymbol{v}_j - \mathbf{E}[\boldsymbol{v}_j]) + \sum_{j=1}^{m}\mathbf{E}[\boldsymbol{v}_j]\right\|_\infty\right]\end{aligned}$$

when $\boldsymbol{v}_j$ denotes the n-dimensional random vector $(v_{1j}, v_{2j}, \ldots, v_{nj})$. Note that the vectors $\boldsymbol{v}_1, \ldots, \boldsymbol{v}_m$ are independent when the features are independently distributed.

We adopt the perspective of an auctioneer seeking to optimize his choice of signaling scheme, with the goal of maximizing the expected welfare. This is nontrivial when $0 < k < m$, and we focus on the algorithmic question of finding the best set of features $S \in \binom{[m]}{k}$. We consider this question when the distribution λ is represented explicitly. The sets $\Omega_1, \ldots, \Omega_m$ are given explicitly, as are the functions $\{v_{ij}\}_{i=1}^{n}$. In the general (correlated) case, λ is described explicitly by a list of items $\Omega' \subseteq \Omega$ with associated probabilities $\{p(\omega) : \omega \in \Omega'\}$ summing to 1 — all other items in Ω assumed to have probability 0. In the independent case, the marginal distribution of each feature j is given explicitly by the associated probabilities $\{p_j(\mu) : \mu \in \Omega_j\}$. We also consider the oracle model whereby only oracle access is given to λ; however, uniform convergence arguments reduce the algorithmic task of signaling in the oracle model to that in the explicit model, up to an arbitrarily small additive error term. In fact, our hardness result is proved in the oracle model, and thus translates to the explicit model.

3 Hardness for General Distributions

We now prove that, in general, no nontrivial approximation is possible for the feature signaling problem when the features are arbitrarily correlated. Our starting point is the conjectured hardness of a special case of the k-junta learning problem. A k-junta on m variables is a boolean function $f : \{-1, 1\}^m \to \{-1, 1\}$ which depends on only k bits of its input. When the bits $S \subseteq [m]$ determining f are unknown, and a learner is given access to sample access to evaluations $(x, f(x))$ of f on bit strings $x \in \{-1, 1\}^m$ drawn uniformly at random, it is widely believed that no algorithm can recover S in polynomial time/samples. In fact, this is believed true even for k-junta functions which compute the majority function on $k/2$ of the input bits, and the parity function on another $k/2$ bits, and then xor the results (these are listed explicitly as candidate hard functions in Blum's surveys on the junta problem [4,5]) — those functions are "balanced" in the sense we describe below.

Definition 1. *A boolean function $f : \{-1, 1\}^m \to \{-1, 1\}$ is c-balanced if the following holds for every $T \subseteq [m]$ with $|T| \leq c$, and $y \in \{-1, 1\}^c$.*

$$\Pr[f(x) = 1 | x_T = y] = \frac{1}{2},$$

where x_T denotes the projection of x onto the coordinates in T, and the probability is over x drawn uniformly from $\{-1, 1\}^m$.

Definition 2. *We say a randomized algorithm (ϵ,δ)-weakly learns a k-junta f if it outputs $S \subseteq [m]$ with $|S| \leq k$ such that, with probability at least $1 - \delta$,*

$$advantage(S) := \mathop{\mathbf{E}}_{x_S} \left[\left| \Pr_x[f(x) = 1 | x_S] - \frac{1}{2} \right| \right] \geq \epsilon$$

where x is uniformly distributed on $\{-1, 1\}^m$.

We use the following commonly believed conjecture.

Conjecture 1 (see e.g. [4,5]). There are functions $k = k(m) = o(m)$ and $c = c(m) = \Theta(k)$ such that c-balanced k-juntas on m variables can not be (ϵ,δ)-weakly learned in time $m^{o(k)}$ under the uniform distribution, for any pair of constants $\epsilon, \delta > 0$.

The above conjecture implies the following corollary.

Corollary 1. *Assuming Conjecture 1, there are functions $k = k(m) = o(m)$ and $c = c(m) = \Theta(k)$ such that no* $\mathrm{poly}(m^{o(k)}, \log \frac{1}{\delta})$*-time learning algorithm, given sample access to a c-balanced k-Junta f on m variables, outputs with probability $1 - \delta$ a set of variables S of size $O(k)$ intersecting more than c of the relevant variables of f.*

Proof. We assume that such an algorithm $\mathcal{A}$, with runtime $m^{o(k)}$ and arbitrarily small failure probability $\delta = \exp(-\Omega(k))$, exists. To simplify the proof, we assume $\mathcal{A}$ recovers a set of size $2k$ which includes $k/2$ relevant variables of a $k/2$-balanced Junta f, though the choice of constants is unimportant. We now show how to weakly learn f in time $m^{o(k)}$, and with constant success probability, violating Conjecture 1.

We learn the relevant variables $S^* \in \binom{m}{k}$ of f as follows: first, run $\mathcal{A}$ to recover $S \subseteq [m]$ with $|S| = 2k$ and $|S \cap S^*| \geq k/2$. Then, for each possible setting z of the bits S (of which there are 2^{2k}) recurse on the function $f_{S,z}$ — often referred to as a *restriction* of f — which simply replaces the portion of its input at indices S with z and then evaluates f. Note that $f_{S,z}$ remains $k/2$-balanced, and hence also $k/4$-balanced, yet is now a $k/2$-Junta on m variables. Assuming the recursive calls succeed, between them they return the set $S^* \setminus S$. To complete S^*, it then suffices to try all 2^{2k} subsets of S.

In the event all invokations of $\mathcal{A}$ in the recursion tree are successful, correctness follows by induction. It remains to bound the runtime. Note that each recursive call halves the number of variables of the Junta. Therefore, the number of recursive calls equals $2^{2k} + 2^{2k} \cdot 2^k + 2^{2k} \cdot 2^k \cdot 2^{k/2} + \ldots \leq \log 2k \cdot 2^{4k} \leq 2^{5k} = m^{o(k)}$. By essentially the same analysis, the runtime of the algorithm is also $2^{5k} \leq m^{o(k)}$. The success probability is at least $1 - \delta$ raised to a power equal to the number of calls of $\mathcal{A}$, which is a constant when $\delta = \exp(-\Omega(k))$ is sufficiently small. □

3.1 Warmup: Two Players

As a warmup, we prove our impossibility result for 2 players assuming Conjecture 1. Note that we do not need the balance assumption for the 2-player special case.

Theorem 1. *Assuming Conjecture 1, there is no $m^{o(k)}$-time $(\frac{1}{2} + \epsilon)$-approximation algorithm for the feature signaling problem with two players in the sample oracle model, for any constant $\epsilon > 0$. This holds for Monte Carlo approximation algorithms having a constant success probability.*

Proof. Given sample access to an m-bit k-Junta f, with $k = o(m)$, we construct an instance of the feature signaling problem in the sample oracle model as follows. We let $\Omega = \{-1, 0, 1\}^{2m}$, and consider two players Alice and Bob. Both players have no value for features 1 through m — i.e. $v_{ij}(.) = 0$ for $i \in \{A, B\}$ and $1 \leq j \leq m$. For the remaining features $j \in [m+1, 2m]$, Alice has value 1 if $\omega_j = 1$ and 0 otherwise, and Bob has value 1 if $\omega_j = -1$ and 0 otherwise.

The distribution λ is constructed as follows. The first m features of $\omega \sim \lambda$, which we denote by x, are uniformly distributed in $\{-1, 1\}^m$. The last m features, which we denote by y, are all set to 0, except for a single feature j^* chosen uniformly at random, which is set to $f(x)$.

Note that if f is a k-Junta determined by the bits $S^* \subseteq [m]$ with $|S^*| = k$, then $welfare(S^*) = 1$ as those bits uniquely determine which of Alice or Bob values the item being sold. To complete the proof, we now show that if $T \subseteq [2m]$ is a set of k features satisfying $welfare(T) \geq \frac{1}{2} + \epsilon$, then $S = T \cap [m]$ is a solution to the k-Junta problem with $advantage(S) = \Omega(\epsilon)$. Indeed:

$$\begin{aligned}
advantage(S) &= \mathop{\mathbf{E}}_{x_S}\left[\left|\Pr_x[f(x) = 1|x_S] - \frac{1}{2}\right|\right] \\
&= \mathop{\mathbf{E}}_{x_S}\left[\max\left(\Pr_x[f(x) = 1|x_S], \Pr_x[f(x) = -1|x_S]\right)\right] - \frac{1}{2} \\
&\geq \mathop{\mathbf{E}}_{\omega_T}\left[\max\left(\Pr_x[f(x) = 1|\omega_T], \Pr_x[f(x) = -1|\omega_T]\right)\right] - \frac{|T \setminus [m]|}{m} - \frac{1}{2} \\
&\geq \text{welfare}(T) - \frac{k}{m} - \frac{1}{2}
\end{aligned}$$

where the next to last inequality is a consequence of the fact that, with probability at least $1 - \frac{|T \setminus [m]|}{m}$, the feature j^* is not in T and therefore ω_T provides no information on $f(x)$ beyond x_S. □

3.2 n Players

Next, we show that the feature signaling problem is hard to approximate to within any constant independent of the number of players, assuming Conjecture 1. Specifically, for n players where n is a constant independent of m, we show that it is hard to approximate the feature signaling problem to within any constant exceeding $1/n$, and this holds for both the oracle and explicit representation models.

Theorem 2. *Assuming Conjecture 1, there is no $m^{o(k)}$-time, $(\frac{1}{n} + \epsilon)$-approximation algorithm for the feature signaling problem with n players in the sample oracle model, for any constant $\epsilon > 0$. This holds for Monte Carlo approximation algorithms having a constant success probability.*

Proof. Our reduction for n players generalizes that for 2 players. Specifically, Given sample access to an c-balanced k'-Junta $f : \{-1,1\}^m \to \{0,1\}$, with $k' = k/\log n$ and $c = \theta(k')$, we construct an instance of the k-feature signaling problem in the sample oracle model as follows. We let $\Omega = \{-1,1\}^{m \log n} \times \{0,1,\ldots,n\}^m$, and consider players $[n] = \{1,\ldots,n\}$. All players have no value for features 1 through $m \log n$. For the remaining features $j \in [m \log n + 1, m \log n + m]$, player i has value 1 if $\omega_j = i$ and 0 otherwise.

The distribution λ is constructed as follows. The first $m \log n$ features of $\omega \sim \lambda$, which we denote by x, are uniformly distributed in $\{-1,1\}^{m \log n}$. We partition x into sub-vectors $x_1, \ldots, x_{\log n}$, of length m each. The last m

features, which we denote by y, are all set to 0, except for a single feature i^* chosen uniformly at random, which is set to the integer encoded by the bit-string $f(x_1)f(x_2)\dots f(x_{\log n})$.

Note that since f is determined by some bits $S^* \subseteq [m]$ with $|S^*| = k'$, signaling the $k = k' \log n$ bits corresponding to the S^*th indices of each sub-vector x_i yields a welfare of 1, since those bits uniquely determine the player who values the item. We now show that any signaling algorithm with nontrivial performance must violate Corollary 1.

Indeed, consider any set T of k features computed by some algorithm for the feature signaling problem which runs in $m^{o(k)} = m^{o(k')}$ time. By Corollary 1 and the fact that $|T| = k' \log n = O(k')$, on some inputs T will not contain more than c relevant features from any sub-vector among $x_1, \dots, x_{\log n}$. By the balance property, such a set T affords no information regarding the player who values the item for sale beyond that afforded by the features $T \setminus [m \log n]$. A similar analysis to that of Theorem 1 shows that the advantage of the signaling scheme which reveals T over one which randomly assigns the item to one of the n players is at most the probability that $i^* \in T$, which is at most $\frac{k \log n}{m} = o(1)$, as needed. □

Finally, we note that any Monte Carlo algorithm with constant success probability can be boosted to one with exponentially small (in k) failure probability, as needed to violate Corollary 1.

Corollary 2. *Assuming Conjecture 1, there is no $m^{o(k)}$-time $(\frac{1}{n} + \epsilon)$-approximation algorithm for the feature signaling problem with n players in the explicit model, for any constant $\epsilon > 0$. This holds for Monte Carlo approximation algorithms having a constant success probability.*

4 An Approximation Algorithm for Independent Distributions

We cast the algorithmic task of feature selection as the following optimization problem. The inputs are $\theta \in \mathbb{R}^n$, $k \in [m]$, and independent t-valued n-dimensional random vectors $\mathbf{X}_1, \dots, \mathbf{X}_m$ with $\mathbf{E}[\mathbf{X}_i] = 0$ for all $i \in [m]$. We will assume that each $\mathbf{X}_i$ is specified as $\{(p_1, v_1), \dots, (p_t, v_t)\}$ where $\Pr[\mathbf{X}_i = v_j] = p_j$ and $\sum_{j=1}^{t} p_j = 1$, and that basic arithmetic can be done in constant time (e.g. we can compute $p_i + p_j$ in constant time, and $\|v_i\|_\infty$ in $O(n)$ time). Given $S \subseteq [m]$ we write

$$\mathsf{value}(S) = \mathbf{E}\left[\left\|\theta + \sum_{j \in S} \mathbf{X}_j\right\|_\infty\right], \tag{1}$$

and define

$$S^* = \underset{|S|=k}{\operatorname{argmax}} \{\mathsf{value}(S)\}, \qquad \mathsf{opt} = \mathsf{value}(S^*). \tag{2}$$

For $0 < \epsilon \leq \frac{1}{2}$, we say that a subset $S \subseteq [m]$ with $|S| \leq k$ is ϵ-optimal if $\mathsf{value}(S) \geq (1 - \epsilon)\mathsf{opt}$; the algorithmic task is to find an ϵ-optimal k-subset $S \subseteq [m]$ efficiently.

To see that this does in fact capture the feature selection problem where each feature is selected independently, we recall the expression for welfare(S) given in (1). Setting

$$\mathbf{X}_j := \boldsymbol{v}_j - \mathbf{E}[\boldsymbol{v}_j] \quad \forall j \in [m] \quad \text{and} \quad \theta := \sum_{j=1}^{m} \mathbf{E}[\boldsymbol{v}_j],$$

and noting that the $\mathbf{X}_j$'s do indeed satisfy $\mathbf{E}[\mathbf{X}_j] = 0$, we have that for all $S \subseteq [m]$,

$$\begin{aligned} \mathsf{value}(S) &= \mathbf{E}\left[\left\|\theta + \sum_{j\in S}\mathbf{X}_j\right\|_\infty\right] = \mathbf{E}\left[\left\|\sum_{j\in S}(\boldsymbol{v}_j - \mathbf{E}[\boldsymbol{v}_j]) + \sum_{j=1}^{m}\mathbf{E}[\boldsymbol{v}_j]\right\|_\infty\right] \\ &= \text{welfare}(S). \end{aligned}$$

Note that for any $S \subseteq [m]$ of cardinality k, the quantity $\mathsf{value}(S)$ can be computed exactly in time $O(nk \cdot t^k)$. Hence the naive algorithm which computes $\mathsf{value}(S)$ for all $\binom{m}{k}$ possible k-subsets S runs in time $O(nk \cdot (mt)^k)$ and finds S^* achieving $\mathsf{value}(S^*) = \mathsf{opt}$. As mentioned in the introduction, we will be primarily interested in the setting where the number of players n is constant, as is the number of values each feature takes, and so this runtime can be written as $m^{O(k)}$. To the best of our knowledge, prior to our work there were no known improvements to this trivial algorithm even when $n = 1$ — *given* m real-valued *random variables* $\mathbf{X}_1, \ldots, \mathbf{X}_m$ *and* $\theta \in \mathbb{R}$, *find a* k*-subset* $S \subseteq [m]$ *that approximately maximizes* $\mathbf{E}[|\theta + \sum_{i\in S}\mathbf{X}_i|]$ — and even under the assumption that all the $\mathbf{X}_i$'s are two-valued (i.e. $t = 2$).

We give an algorithm that finds an ϵ-optimal set S of cardinality k, running in time $O(m) + 2^{O(k\log(k/\epsilon))}$ for all fixed values of n and t. (In particular, this is poly(m) for all $\epsilon \geq 1/\text{polylog}(m)$ and $k \ll \frac{\log m}{\log\log m}$.)

Theorem 3. *There is an algorithm* $\mathcal{A}$ *which, given as input* $0 < \epsilon \leq \frac{1}{2}$, $k \in [m]$, $\theta \in \mathbb{R}^n$, *and independent* t*-valued* d*-dimensional random vectors* $\mathbf{X}_1, \ldots, \mathbf{X}_m$ *with* $\mathbf{E}[\mathbf{X}_i] = 0$ *for all* $i \in [m]$, *runs in time* $O(mnt) + \text{poly}(kt/\epsilon)^{knt}$ *and outputs a* k*-subset* $S \subseteq [m]$ *satisfying* $\mathsf{value}(S) \geq (1-\epsilon)\mathsf{opt}$.

The techniques we develop to establish Theorem 3 are fairly general and robust. Indeed, we obtain Theorem 3 as a special case of our most general result which we now state. Given an arbitrary norm $\|\cdot\|$ on $\mathbb{R}^n$, we may define $\mathsf{value}(\cdot)$ and opt with respect to $\|\cdot\|$ instead of $\|\cdot\|_\infty$, and hence also an analogous optimization problem of finding an ϵ-optimal k-subset. Our most general result is an efficient algorithm for this abstract optimization problem for any norm $\|\cdot\|$ on $\mathbb{R}^n$:

Theorem 4. *Fix a norm* $\|\cdot\|$ *on* $\mathbb{R}^n$. *Given* $\epsilon > 0$ *and* $k \in [m]$, *let* $\mathcal{N} = \mathcal{N}(\epsilon, k)$ *be an* (ϵ/k)*-net within the ball* $\{v \in \mathbb{R}^n : \|v\| \leq k^2/\epsilon\}$ *with the property that for every vector* v *in the ball, its closest point in* $\mathcal{N}$ *can be found in time* r. *There is* *an algorithm $\mathcal{A}$ which, given as input $\epsilon > 0$, $k \in [m]$, $\theta \in \mathbb{R}^n$, and independent t-valued n-dimensional random vectors $\mathbf{X}_1, \ldots, \mathbf{X}_m$ with $\mathbf{E}[\mathbf{X}_i] = 0$ for all $i \in [m]$, runs in time $O(mt(r+n) + nk \cdot (4\ell t)^k)$ where $\ell = (|\mathcal{N}|k^3 t/\epsilon)^{O(t)}$, and outputs a k-subset $S \subseteq [m]$ satisfying $\mathsf{value}(S) \geq (1-\epsilon)\mathsf{opt}$, where $\mathsf{value}(\cdot)$ and opt are defined with respect to $\|\cdot\|$.*

To see that Theorem 3 does in fact follow from Theorem 4, we note that for all $B, \delta > 0$, the grid points $\mathcal{N} = \big\{(\lambda_1\delta, \ldots, \lambda_n\delta) \colon \lambda_i \in \{0, 1, \ldots, \lfloor B/\delta \rfloor\}\big\}$ is a δ-net of size $(\lfloor B/\delta \rfloor + 1)^n$ within the ball $\{v \in \mathbb{R}^n \colon \|v\|_\infty \leq B\}$. Furthermore, it is clear that given any vector v in the ball, its closest vector within $\mathcal{N}$ can be computed in time $O(n)$. The remainder of this section will be devoted to proving Theorem 4. The following simple fact will be useful for us:

Fact 1. *Let $\mathbf{X}_1$ and $\mathbf{X}_2$ be independent random vectors where $\mathbf{E}[\mathbf{X}_1] = 0$. Then $\mathbf{E}[\|\mathbf{X}_1 + \mathbf{X}_2\|] \geq \mathbf{E}[\|\mathbf{X}_2\|]$. Consequently, if $S' \supseteq S$ then $\mathsf{value}(S') \geq \mathsf{value}(S)$ (and in particular, it is equivalent to maximize over all $|S| \leq k$ in the definition of opt in (2)).*

Proof. The inequality holds pointwise for every possible outcome $\theta \in \mathbb{R}^d$ of $\mathbf{X}_2$ since $\|\theta\| = \|\mathbf{E}[\mathbf{X}_1 + \theta]\| \leq \mathbf{E}[\|\mathbf{X}_1 + \theta\|]$. □

Overview of proof. We assume for the sake of scaling that $\max(\|\theta\|, \max_i \mathbf{E}[\|\mathbf{X}_i\|]) = 1$, and hence $\mathsf{opt} \geq 1$ by Fact 1. Thus to find an ϵ-optimal set S, it suffices to find one achieving value at least $\mathsf{opt} - \epsilon$; for notational simplicity, we will only achieve value at least $\mathsf{opt} - O(\epsilon)$. The main idea is to modify the random vectors $\mathbf{X}_1, \ldots, \mathbf{X}_m$ in such a way that changes $\mathsf{value}(S)$ by at most an additive $\pm O(\epsilon)$ for all k-subsets $S \subseteq [m]$, and yet results in a total of only ℓ distinct random variables $\mathbf{Y}_1, \ldots, \mathbf{Y}_\ell$ where ℓ is independent of m (i.e. many $\mathbf{X}_i$'s are modified to become the same $\mathbf{Y}_j$). If for each $j \in [\ell]$ we let M_j denote the number of $\mathbf{X}_i$'s that are modified to become $\mathbf{Y}_j$, this reduces the problem of finding an $O(\epsilon)$-optimal k-subset $S \subseteq [m]$ to that of finding $\lambda \in \mathbb{Z}^\ell$ that maximizes

$$\mathbf{E}\left[\left\|\theta + \sum_{i=1}^{\ell} \lambda_i \mathbf{Y}_i\right\|\right] \tag{3}$$

$$\text{subject to } \lambda_i \in \{0, 1, \ldots, M_i\} \text{ and } \sum_{i=1}^{\ell} \lambda_i = k. \tag{4}$$

Since there are at most $4^k\binom{\ell}{k}$ many $\lambda \in \mathbb{Z}^\ell$ satisfying (4), and for each such λ the quantity (3) can be computed in time $O(nk \cdot t^k)$, the optimal λ can be found in time $O(nk \cdot (4\ell t)^k)$.

4.1 Transforming the $\mathbf{X}_i$'s

Given numbers $a, b \in \mathbb{R}$ and $\epsilon > 0$, we write $a \overset{\epsilon}{\approx} b$ as shorthand for $|a - b| \leq \epsilon$. By the triangle inequality, if $a \overset{\epsilon_1}{\approx} b$ and $b \overset{\epsilon_2}{\approx} c$ then $a \overset{\epsilon_1+\epsilon_2}{\approx} c$.

Definition 3. *Given a parameter $B > 1$ we say that an n-dimensional random vector $\mathbf{X}_i$ is B-*bounded *if $\|\mathbf{X}_i\| \leq B$ with probability 1.*

We begin with the following proposition which states that the $\mathbf{X}_i$'s can be modified so that all of them are B-bounded; we defer its proof to the full version of this paper.

Proposition 1. *Fix a parameter $B > 1$ and assume $\mathbf{X}_i$ is not B-bounded. Then there is a B-bounded random vector $\mathbf{X}'_i$ such that*

$$\mathbf{E}[\|\mathbf{X}'_i + \mathbf{Y}\|] \overset{4(k+1)/B}{\approx} \mathbf{E}[\|\mathbf{X}_i + \mathbf{Y}\|]$$

for all random vectors $\mathbf{Y}$ that are independent of $\mathbf{X}_i$, $\mathbf{X}'_i$ and satisfy $\mathbf{E}[\|\mathbf{Y}\|] \leq k$. Furthermore, $\mathbf{X}'_i$ can be defined from $\mathbf{X}_i$ in time $O(nt)$.

As a corollary of Proposition 1, for all k-subsets $S \subseteq [m]$ containing i we have

$$\mathsf{value}(S) \overset{4(k+1)/B}{\approx} \mathbf{E}\left[\left\|\theta + \mathbf{X}'_i + \sum_{j \in S \setminus \{i\}} \mathbf{X}_j\right\|\right].$$

In words, replacing $\mathbf{X}_i$ by $\mathbf{X}'_i$ in $\mathbf{X}_1, \ldots, \mathbf{X}_n$ changes $\mathsf{value}(S)$ by at most an additive $\pm O(k/B)$ for all k-subsets $S \subseteq [m]$. Consequently, by the union bound, we may make *all* of $\mathbf{X}_1, \ldots, \mathbf{X}_n$ B-bounded and change $\mathsf{value}(S)$ by at most an additive $\pm O(k^2/B)$.

We will need a simple numerical lemma for our next modification; we defer its proof to the full version of this paper.

Lemma 2. *Let $p_1, \ldots, p_t \in (0, 1)$ where $\sum_{j=1}^{t} p_i = 1$, and $0 < \eta \leq 1$ where $1/\eta \in \mathbb{Z}$. There exist nonnegative integer multiples $p'_1, \ldots, p'_t$ of η also summing to 1 and satisfying $|p'_j - p_j| < \eta$ for all j.*

Proposition 2. *Fix parameters $B > 1$, $\delta > 0$, and $0 < \eta \leq 1$, where $1/\eta \in \mathbb{Z}$. Let $\mathcal{N}$ denote a δ-net within the ball $\{v \in \mathbb{R}^n : \|v\| \leq B\}$, and assume that for every vector v in the ball, its closest vector in the δ-net $\mathcal{N}$ can be computed in time r. Then for any B-bounded n-dimensional random vector $\mathbf{X}_i$, there is a random vector $\mathbf{X}'_i$, dependent on $\mathbf{X}_i$, such that:*

1. *all outcomes for $\mathbf{X}'_i$ are in $\mathcal{N}$;*
2. *all outcomes for $\mathbf{X}'_i$ occur with probability equal to an integer multiple of η;*
3. *$\mathbf{E}[\|\mathbf{X}'_i - \mathbf{X}_i\|] \leq \delta + 2Bt\eta$.*

Furthermore, $\mathbf{X}'_i$ can be defined from $\mathbf{X}_i$ in time $O(rt)$.

Proof. Let $\mathbf{X}_i \equiv \{(p_1, v_1), \ldots, (p_t, v_t)\}$ (i.e. $\Pr[\mathbf{X}_i = v_j] = p_j$ and $\sum_{j=1}^{t} p_j = 1$). We first consider $\mathbf{X}^*_i = \{(p_1, v^*_1), \ldots, (p_t, v^*_t)\}$, where v^*_j is the vector in $\mathcal{N}$ closest to v_j, coupled to $\mathbf{X}_i$ in such a way that $\Pr[\mathbf{X}^*_i = v^*_j \mid \mathbf{X}_i = v_j] = 1$. Since

$$\mathbf{E}[\|\mathbf{X}^*_i - \mathbf{X}_i\|] = \sum_{j=1}^{t} p_j \cdot \|v^*_j - v_j\| \leq \delta$$

and all outcomes of $\mathbf{X}_i^*$ are in $\mathcal{N}$ (i.e. satisfying (1)), it remains to show how to achieve (2) while incurring error at most $2Bt\eta$ in (3). By Lemma 2 there exist nonnegative integer multiples $p_1', \ldots, p_t'$ of η, summing to 1 and satisfying $|p_j' - p_j| < \eta$ for all j (and it is straightforward to verify that $p_1', \ldots, p_t'$ can be computed from $p_1, \ldots, p_t$ and η in time $O(t)$). We can then define $\mathbf{X}_i'$ by $\Pr[\mathbf{X}_i' = v_j^*] = p_j'$, coupled to $\mathbf{X}^*$ in such a way that $\Pr[\mathbf{X}_i^* = \mathbf{X}_i' = v_j^*] = \min(p_j, p_j')$ for all $j \in [t]$. It is clear then that

$$\Pr[\mathbf{X}_i' \neq \mathbf{X}_i^*] \leq \sum_{j=1}^{t} |p_j' - p_j| \leq t\eta,$$

and that whenever $\mathbf{X}_i' \neq \mathbf{X}_i^*$ we at least have $\|\mathbf{X}_i' - \mathbf{X}_i^*\| \leq 2B$ by the B-boundedness of $\mathbf{X}_i^*$ and $\mathbf{X}_i$. The lemma follows.

Proof of Theorem 4. Applying Proposition 1 with $B := k^2/\epsilon$, we may assume that $\mathbf{X}_1, \ldots, \mathbf{X}_m$ are all B-bounded — this modification can be carried out in time $O(mnt)$, and changes $\mathsf{value}(S)$ by at most an additive $\pm O(\epsilon)$ for all k-subsets $S \subseteq [m]$. Next, applying Proposition 2 with $\delta := \epsilon/k$ and η any number in $[\epsilon/(2kBt), \epsilon/(kBt)]$ such that $1/\eta \in \mathbb{Z}$, there is an algorithm which runs in time $O(mrt)$ (i.e. $O(rt)$ for each $\mathbf{X}_i$) and outputs $\mathbf{X}_1', \ldots, \mathbf{X}_m'$ satisfying

$$\begin{aligned}\left|\mathsf{value}(S) - \mathbf{E}\left[\left\|\theta + \sum_{i \in S} \mathbf{X}_i'\right\|_\infty\right]\right| &\leq \mathbf{E}\left[\left\|\sum_{i \in S} \mathbf{X}_i' - \mathbf{X}_i\right\|\right] \\ &\leq \sum_{i \in S} \mathbf{E}[\|\mathbf{X}_i' - \mathbf{X}_i\|] \;\leq\; k(\delta + 2Bt\eta) = O(\epsilon)\end{aligned}$$

for all k-subsets $S \subseteq [m]$. Furthermore, by Proposition 2 each $\mathbf{X}_i'$ is of the form $\{(p_1, v_1), \ldots, (p_t, v_t)\}$ where every p_i is an integer multiple of η, and every v_i is in $\mathcal{N}$. It follows that there are in fact at most

$$\ell \leq \binom{|\mathcal{N}| \cdot \eta^{-1}}{t} = (|\mathcal{N}| k^3 t/\epsilon)^{O(t)}$$

many distinct random variables $\mathbf{Y}_1, \ldots, \mathbf{Y}_\ell$ in the multiset $\{\mathbf{X}_1', \ldots, \mathbf{X}_m'\}$. Letting M_i denote the multiplicity of $\mathbf{Y}_i$ in $\{\mathbf{X}_1', \ldots, \mathbf{X}_n'\}$, we have reduced the problem of finding an $O(\epsilon)$-optimal k-subset $S \subseteq [m]$ to that of finding $\lambda \in \mathbb{Z}^\ell$ that maximizes

$$\mathbf{E}\left[\left\|\theta + \sum_{i=1}^{\ell} \lambda_i \mathbf{Y}_i\right\|\right] \tag{5}$$

$$\text{subject to } \lambda_i \in \{0, 1, \ldots, M_i\} \text{ and } \sum_{i=1}^{\ell} \lambda_i = k. \tag{6}$$

Since there are at most $4^k \binom{\ell}{k}$ many $\lambda \in \mathbb{Z}^\ell$ satisfying (6), and for each such λ the quantity (5) can be computed in time $O(nk \cdot t^k)$, the optimal λ can be found in time $O(nk \cdot (4\ell t)^k)$.

References

1. Alekhnovich, M., Braverman, M., Feldman, V., Klivans, A.R., Pitassi, T.: Learnability and automatizability. In: FOCS, pp. 621–630 (2004)
2. Alon, N., Feldman, M., Gamzu, I., Tennenholtz, M.: The asymmetric matrix partition problem. In: Chen, Y., Immorlica, N. (eds.) WINE 2013. LNCS, vol. 8289, pp. 1–14. Springer, Heidelberg (2013)
3. Blum, A.: Relevant examples and relevant features: Thoughts from computational learning theory. In: AAAI Fall Symposium on 'Relevance' (1994)
4. Blum, A.: Open problem: Learning a function of r relevant variables. In: Proceedings of COLT, pp. 731–733 (2003)
5. Blum, A.: Tutorial on Machine Learning Theory given at FOCS 2003 (2003). http://www.cs.cmu.edu/avrim/Talks/FOCS03/
6. Blum, A., Furst, M., Jackson, J., Kearns, M., Mansour, Y., Rudich, S.: Weakly learning DNF and characterizing statistical query learning using Fourier analysis. In: Proceedings of STOC, pp 253–262 (1994)
7. Blum, A., Langley, P.: Selection of relevant features and examples in machine learning. Artif. Intel. **97**(1–2), 245–271 (1997)
8. Dughmi, S., Immorlica, N., Roth, A.: Constrained signaling in auction design. In: ACM Symposium on Discrete Algorithms (SODA) (2014)
9. Emek, Y., Feldman, M., Gamzu, I., Paes-Leme, R., Tennenholtz, M.: Signaling schemes for revenue maximization. In: ACM Conference on Electronic Commerce (EC) (2012)
10. Feldman, V., Lee, H., Servedio, R.: Lower bounds and hardness amplification for learning shallow monotone formulas. J. Mach. Learn. Res. - COLT Proceedings **19**, 273–292 (2011)
11. Feldman, V., Gopalan, P., Khot, S., Ponnuswami, A.K.: On agnostic learning of parities, monomials, and halfspaces. SIAM J. Comput. **39**(2), 606–645 (2009)
12. Feldman, V., Kothari, P., Vondrák, J.: Nearly tight bounds on ℓ_1 approximation of self-bounding functions. CoRR, abs/1404.4702 (2014)
13. Guo, M., Deligkas, A.: Revenue maximization via hiding item attributes. In: Proceedings of the Twenty-Third International Joint Conference on Artificial Intelligence, pp. 157–163. AAAI Press (2013)
14. Milgrom, P., Weber, R.J.: A theory of auctions and competitive bidding. Econometrica **50**, 1089–1122 (1982)
15. Miltersen, P.B., Sheffet, O.: Send mixed signals: earn more, work less. In: Proceedings of the 13th ACM Conference on Electronic Commerce, pp. 234–247. ACM (2012)
16. Mossel, E., O'Donnell, R., Servedio, R.: Learning functions of k relevant variables. J. Comput. Syst. Sci., **69(3)**, 421–434 (2004). Previously published as "Learning juntas"
17. Valiant, G.: Finding correlations in subquadratic time, with applications to learning parities and juntas. In: 2012 IEEE 53rd Annual Symposium on Foundations of Computer Science (FOCS), pp. 11–20. IEEE (2012)

Networking

On the Efficiency of the Proportional Allocation Mechanism for Divisible Resources

George Christodoulou, Alkmini Sgouritsa, and Bo Tang(✉)

University of Liverpool, Liverpool, UK
Bo.Tang@liverpool.ac.uk

Abstract. We study the efficiency of the *proportional allocation mechanism*, that is widely used to allocate divisible resources. Each agent submits a bid for each divisible resource and receives a fraction proportional to her bids. We quantify the inefficiency of Nash equilibria by studying the Price of Anarchy (PoA) of the induced game under complete and incomplete information. When agents' valuations are concave, we show that the Bayesian Nash equilibria can be arbitrarily inefficient, in contrast to the well-known 4/3 bound for pure equilibria [12]. Next, we upper bound the PoA over Bayesian equilibria by 2 when agents' valuations are subadditive, generalizing and strengthening previous bounds on lattice submodular valuations. Furthermore, we show that this bound is tight and cannot be improved by any *simple* mechanism. Then we switch to settings with budget constraints, and we show an improved upper bound on the PoA over coarse-correlated equilibria. Finally, we prove that the PoA is *exactly* 2 for pure equilibria in the polyhedral environment.

1 Introduction

Allocating network resources, like bandwidth, among agents is a canonical problem in the network optimization literature. A traditional model for this problem was proposed by Kelly [14], where allocating these infinitely divisible resources is treated as a market with prices. More precisely, agents in the system submit bids on resources to express their willingness to pay. After soliciting the bids, the system manager prices each resource with an amount equal to the sum of bids on it. Then the agents buy portions of resources proportional to their bids by paying the corresponding prices. This mechanism is known as the *proportional allocation mechanism* or Kelly mechanism in the literature.

The proportional allocation mechanism is widely used in network pricing and has been implemented for allocating computing resources in several distributed systems [5]. In practice, each agent has different interests for different subsets and fractions of the resources. This can be expressed via a *valuation* function of the resource allocation vector, that is typically private knowledge to each agent. Thus, agents may bid strategically to maximize their own utilities, i.e., the difference

G. Christodoulou—This author was supported by EPSRC grants P/M008118/1 and EP/K01000X/1.

M. Hoefer (Ed.): SAGT 2015, LNCS 9347, pp. 165–177, 2015.
DOI: 10.1007/978-3-662-48433-3_13

between their valuations and payments. Johari and Tsitsiklis [12] observed that this strategic bidding in the proportional allocation mechanism leads to inefficient allocations, that do not maximize social welfare. On the other hand, they showed that this efficiency loss is bounded when agents' valuations are concave. More specifically, they proved that the proportional allocation game admits a *unique pure* equilibrium with Price of Anarchy (PoA) [15] at most 4/3.

An essential assumption used by Johari and Tsitsiklis is that agents have complete information of each other's valuations. However, in many realistic scenarios, the agents are only partially informed. A standard way to model incomplete information is by using the Bayesian framework, where the agents' valuations are drawn independently from some publicly known distribution, that in a sense, represents the agents' beliefs. A natural question is whether the efficiency loss is still bounded in the Bayesian setting. We give a negative answer to this question by showing that the PoA over Bayesian equilibria is at least $\sqrt{m}/2$ where m is the number of resources. This result complements the current study by Caragiannis and Voudouris [2], where the PoA of single-resource proportional allocation games is shown to be at most 2 in the Bayesian setting.

Non-concave valuation functions were studied by Syrgkanis and Tardos [20] for both complete and incomplete information games. They showed that, when agents' valuations are lattice-submodular, the PoA for coarse correlated and Bayesian Nash equilibria is at most 3.73, by applying their general smoothness framework. In this paper, we study subadditive valuations [8] that is a superclass of lattice submodular functions. We prove that the PoA over Bayesian Nash equilibria is at most 2. Moreover, we show optimality of the proportional allocation mechanism, by showing that this bound is tight and cannot be improved by any *simple* mechanism, as defined in the recent framework of Roughgarden [19].

Next, we switch to the setting where agents are constrained by budgets, that represent the maximum payment they can afford. We prove that the PoA of the proportional allocation mechanism is at most $1 + \phi \approx 2.618$, where ϕ is the golden ratio. The previously best known bound was 2.78 and for a single resource due to [2]. Finally, we consider the polyhedral environment that was previously studied by Nguyen and Tardos in [16], where they proved that pure equilibria are at least 75 % efficient with concave valuations. We prove that the PoA is exactly 2 for agents with subadditive valuations.

Related Work. The efficiency of the proportional allocation mechanism has been extensively studied in the literature of network resource allocation. Besides the work mentioned above, Johari and Tsitsiklis [13] studied a more general class of scale-free mechanisms and proved that the proportional allocation mechanism achieves the best PoA in this class. Zhang [21] and Feldman et al. [10] studied the efficiency and fairness of the proportional allocation mechanism, when agents aim at maximizing non quasi-linear utilities subject to budget constraints. Correa, Schulz and Stier-Moses [6] showed a relationship in the efficiency loss between proportional allocation mechanism and non-atomic selfish routing for not necessarily concave valuation functions.

There is a line of research studying the PoA of simple auctions for selling indivisible goods (see [1,3,11,20]). Recently, Feldman et al. [9] showed tighter upper bounds for simultaneous first and second price auctions when the agents have subadditive valuations. Christodoulou et al. [4] showed matching lower bounds for simultaneous first price auctions, and Roughgarden [19] proved general lower bounds for the PoA of all simple auctions, by using the corresponding computational or communication lower bounds of the underlying allocation problem.

2 Preliminaries

There are n *agents* who compete for m *divisible resources* with *unit* supply. Every agent $i \in [n]$ has a valuation function $v_i : [0,1]^m \to \mathbb{R}_+$, where $[n]$ denotes the set $\{1, 2, \ldots, n\}$. The valuations are normalized as $v_i(\mathbf{0}) = 0$, and monotonically non-decreasing, that is, for every $\mathbf{x}, \mathbf{x}' \in [0,1]^m$, where $\mathbf{x} = (x_j)_j, \mathbf{x}' = (x'_j)_j$ and $\forall j \in [m]\ x_j \leq x'_j$, we have $v_i(\mathbf{x}) \leq v_i(\mathbf{x}')$. Let $\mathbf{x} + \mathbf{y}$ be the componentwise sum of two vectors $\mathbf{x}$ and $\mathbf{y}$.

Definition 1. *A function $v : [0,1]^m \to \mathbb{R}_{\geq 0}$ is subadditive if, for all $\mathbf{x}, \mathbf{y} \in [0,1]^m$, such that $\mathbf{x} + \mathbf{y} \in [0,1]^m$, it is $v(\mathbf{x} + \mathbf{y}) \leq v(\mathbf{x}) + v(\mathbf{y})$.*

Remark. Lattice submodular functions used in [20] are subadditive. In the case of a single variable (single resource), any concave function is subadditive; more precisely, concave functions are equivalent to lattice submodular functions in this case. However, concave functions of many variables may not be subadditive [18].

In the *Bayesian* setting, the valuation of each agent i is drawn from a set of possible valuations V_i, according to some known probability distribution D_i. We assume that D_i's are independent, but not necessarily identical over the agents.

A mechanism can be represented by a tuple $(\mathbf{x}, \mathbf{q})$, where $\mathbf{x}$ specifies the allocation of resources and $\mathbf{q}$ specifies the agents' payments. In the mechanism, every agent i submits a non-negative bid b_{ij} for each resource j. The proportional allocation mechanism determines the allocation $x_i = (x_{ij})_j$ and payment q_i, for each agent i, as follows: $x_{ij} = \frac{b_{ij}}{\sum_{k \in [n]} b_{kj}}$, $q_i = \sum_{j \in [m]} b_{ij}$. When all agents bid 0, the allocation can be defined arbitrarily, but consistently.

Nash Equilibrium. We denote by $\mathbf{b} = (b_1, \ldots, b_n)$ the strategy profile of all agents, where $b_i = (b_{i1}, \ldots, b_{im})$ denotes the pure bids of agent i for the m resources. By $\mathbf{b}_{-i} = (b_1, \ldots, b_{i-1}, b_{i+1}, \ldots, b_n)$ we denote the strategies of all agents except for i. Any *mixed, correlated, coarse correlated or Bayesian strategy* B_i of agent i is a probability distribution over b_i. For any strategy profile $\mathbf{b}$, $\mathbf{x}(\mathbf{b})$ denotes the allocation and $\mathbf{q}(\mathbf{b})$ the payments under the strategy profile $\mathbf{b}$. The *utility* u_i of agent i is defined as the difference between her valuation for the received allocation and her payment: $u_i(\mathbf{x}(\mathbf{b}), \mathbf{q}(\mathbf{b})) = u_i(\mathbf{b}) = v_i(x_i(\mathbf{b})) - q_i(\mathbf{b})$.

Definition 2. *A bidding profile* $\mathbf{B}$ *forms the following equilibrium if for every agent i and all bids b'_i:*

Pure Nash equilibrium: $\mathbf{B} = \mathbf{b}$, $u_i(\mathbf{b}) \geq u_i(b'_i, \mathbf{b}_{-i})$.
Mixed Nash equilibrium: $\mathbf{B} = \times_i B_i$, $\mathbb{E}_{\mathbf{b} \sim \mathbf{B}}[u_i(\mathbf{b})] \geq \mathbb{E}_{\mathbf{b} \sim \mathbf{B}}[u_i(b'_i, \mathbf{b}_{-i})]$.

Correlated equilibrium: $\mathbf{B} = (B_i)_i$, $\mathbb{E}_{\mathbf{b}\sim\mathbf{B}}[u_i(\mathbf{b})|b_i] \geq \mathbb{E}_{\mathbf{b}\sim\mathbf{B}}[u_i(b'_i, \mathbf{b}_{-i})|b_i]$.
Coarse correlated equilibrium: $\mathbf{B} = (B_i)_i$, $\mathbb{E}_{\mathbf{b}\sim\mathbf{B}}[u_i(\mathbf{b})] \geq \mathbb{E}_{\mathbf{b}\sim\mathbf{B}}[u_i(b'_i, \mathbf{b}_{-i})]$.
Bayesian Nash equilibrium: $\mathbf{B}(\mathbf{v}) = \times_i B_i(v_i)$, $\mathbb{E}_{\mathbf{v}_{-i},\mathbf{b}}[u_i(\mathbf{b})] \geq \mathbb{E}_{\mathbf{v}_{-i},\mathbf{b}}[u_i(b'_i, \mathbf{b}_{-i})]$.

The first four classes of equilibria are in increasing order of inclusion. Moreover, any mixed Nash equilibrium is also a Bayesian Nash equilibrium.

Price of Anarchy (PoA). Our global objective is to maximize the sum of the agents' valuations for their received allocations, i.e., to maximize the *social welfare* $\mathrm{SW}(\mathbf{x}) = \sum_{i\in[n]} v_i(x_i)$. Given the valuations, $\mathbf{v}$, of all agents, there exists an optimal allocation $\mathbf{o}^{\mathbf{v}} = \mathbf{o} = (o_1, \ldots, o_n)$, such that $\mathrm{SW}(\mathbf{o}) = \max_{\mathbf{x}} \mathrm{SW}(\mathbf{x})$. By $o_i = (o_{i1}, \ldots, o_{im})$ we denote the optimal allocation to agent i. For simplicity, we use $\mathrm{SW}(\mathbf{b})$ and $v_i(\mathbf{b})$ instead of $\mathrm{SW}(\mathbf{x}(\mathbf{b}))$ and $v_i(x_i(\mathbf{b}))$, whenever the allocation rule $\mathbf{x}$ is clear from the context. We also use shorter notation for expectations, e.g. we use $\mathbb{E}_{\mathbf{v}}$ instead of $\mathbb{E}_{\mathbf{v}\sim\mathbf{D}}$, $\mathbb{E}[u_i(\mathbf{b})]$ instead of $\mathbb{E}_{\mathbf{b}\sim\mathbf{B}}[u_i(\mathbf{b})]$ and $u(\mathbf{B})$ for $\mathbb{E}_{\mathbf{b}\sim\mathbf{B}}[u(\mathbf{b})]$ whenever $\mathbf{D}$ and $\mathbf{B}$ are clear from the context.

Definition 3. *Let $\mathcal{I}([n],[m],\mathbf{v})$ be the set of all instances, i.e., $\mathcal{I}([n],[m],\mathbf{v})$ includes the instances for every set of agents and resources and any possible valuations that the agents might have for the resources. We define the pure, mixed, correlated, coarse correlated and Bayesian Price of Anarchy, PoA, as*

$$PoA = \max_{I\in\mathcal{I}} \max_{\mathbf{B}\in\mathcal{E}(I)} \frac{\mathbb{E}_{\mathbf{v}}[\mathrm{SW}(\mathbf{o})]}{\mathbb{E}_{\mathbf{v},\mathbf{b}\sim\mathbf{B}}[\mathrm{SW}(\mathbf{b})]},$$

where $\mathcal{E}(I)$ is the set of pure Nash, mixed Nash, correlated, coarse correlated or Bayesian Nash equilibria for the specific instance $I \in \mathcal{I}$, respectively[1].

Budget Constraints. We also consider the setting where agents are budget-constrained. That is, the payment of each agent i cannot be higher than c_i, where c_i is a non-negative value denoting agent i's budget. Following [2,20], we use *Effective Welfare* as the benchmark: $\mathrm{EW}(\mathbf{x}) = \sum_i \min\{v_i(x_i), c_i\}$. In addition, for any *randomized* allocation $\mathbf{x}$, the expected effective welfare is defined as: $\mathbb{E}_{\mathbf{x}}[\mathrm{EW}(\mathbf{x})] = \sum_i \min\{\mathbb{E}_{\mathbf{x}}[v_i(x_i)], c_i\}$.

3 Concave Valuations

In this section, we show that for concave valuations on multiple resources, Bayesian equilibria can be arbitrarily inefficient. More precisely, we prove that the Bayesian PoA is $\Omega(\sqrt{m})$ in contrast to the constant bound for pure equilibria [12]. Therefore, there is a big gap between complete and incomplete information settings. We state our main theorem in this section as follows.

Theorem 4. *When valuations are concave, the PoA of the proportional allocation mechanism for Bayesian equilibria is at least $\frac{\sqrt{m}}{2}$.*

[1] The expectation over $\mathbf{v}$ is only needed for the definition of Bayesian PoA.

Proof. We consider an instance with m resources and 2 agents with the following concave valuations. $v_1(\mathbf{x}) = \min_j\{x_j\}$ and $v_2(\mathbf{x})$ is drawn from a distribution D_2, such that some resource $j \in [m]$ is chosen uniformly at random and then $v_2(\mathbf{x}) = x_j/\sqrt{m}$. Let $\delta = 1/(\sqrt{m}+1)^2$. We claim that $\mathbf{b}(\mathbf{v}) = (b_1, b_2(v_2))$ is a pure Bayesian Nash equilibrium, where $\forall j \in [m]$, $b_{1j} = \sqrt{\delta/m} - \delta$ and, if $j \in [m]$ is the resource chosen by D_2, $b_{2j}(v_2) = \delta$ and for all $j' \neq j$ $b_{2j'} = 0$.

Under this bidding profile, agent 1 bids the same value for all resources, and agent 2 only bids positive value for a single resource associated with her valuation. Suppose that agent 2 has positive valuation for resource j, i.e., $v_2(\mathbf{x}) = x_j/\sqrt{m}$. Then the rest $m-1$ resources are allocated to agent 1 and agents are competing for resource j. Bidder 2 has no reason to bid positively for any other resource. If she bids any value b'_{2j} for resource j, her utility would be $u_2(\mathbf{b}_1, b'_{2j}) = \frac{1}{\sqrt{m}}\frac{b'_{2j}}{b_{1j}+b'_{2j}} - b'_{2j}$, which is maximized for $b'_{2j} = \sqrt{\frac{b_{1j}}{\sqrt{m}}} - b_{1j}$. For $b_{1j} = \sqrt{\delta/m} - \delta$, the utility of agent 2 is maximized for $b'_{2j} = 1/(\sqrt{m}+1)^2 = \delta$ by simple calculations.

Since $v_1(\mathbf{x})$ equals the minimum of $\mathbf{x}$'s components, agent 1's valuation is completely determined by the allocation of resource j. So the expected utility of agent 1 under $\mathbf{b}$ is $\mathbb{E}_{v_2}[u_1(\mathbf{b})] = \frac{\sqrt{\delta/m}-\delta}{\sqrt{\delta/m}-\delta+\delta} - m(\sqrt{\delta/m}-\delta) = (1-\sqrt{m\delta})^2 = \frac{1}{(\sqrt{m}+1)^2} = \delta$. Suppose now that agent 1 deviates to $b'_1 = (b'_{11}, \ldots, b'_{1m})$.

$$\begin{aligned}
\mathbb{E}_{v_2}[u_1(b'_1, b_2)] &= \frac{1}{m}\sum_j \frac{b'_{1j}}{b'_{1j}+\delta} - \sum_j b'_{1j} = \frac{1}{m}\sum_j \left(\frac{b'_{1j}}{b'_{1j}+\delta} - m\cdot b'_{1j}\right) \\
&\le \frac{1}{m}\sum_j \left(\frac{\sqrt{\delta/m}-\delta}{\sqrt{\delta/m}} - m\cdot(\sqrt{\delta/m}-\delta)\right) \\
&= \frac{1}{m}\sum_j \left(1 - 2\sqrt{m\cdot\delta} + m\cdot\delta\right) = \frac{1}{m}\sum_j \left(1-\sqrt{m\cdot\delta}\right)^2 \\
&= \frac{1}{m}\sum_j \left(\frac{1}{\sqrt{m}+1}\right)^2 = \delta = \mathbb{E}_{v_2}[u_1(\mathbf{b})].
\end{aligned}$$

The inequality comes from the fact that $\frac{b'_{1j}}{b'_{1j}+\delta} - m\cdot b'_{1j}$ is maximized for $b'_{1j} = \sqrt{\delta/m} - \delta$. So we conclude that $\mathbf{b}$ is a Bayesian equilibrium.

Finally we compute the PoA. The expected social welfare under $\mathbf{b}$ is $\mathbb{E}_{v_2}[\mathrm{SW}(\mathbf{b})] = \frac{\sqrt{\delta/m}-\delta}{\sqrt{\delta/m}-\delta+\delta} + \frac{1}{\sqrt{m}}\frac{\delta}{\sqrt{\delta/m}-\delta+\delta} = 1 - \sqrt{m\delta} + \sqrt{\delta} = \frac{2}{\sqrt{m}+1} < \frac{2}{\sqrt{m}}$. But the optimal social welfare is 1 by allocating to agent 1 all resources. So, PoA $\geq \frac{\sqrt{m}}{2}$. □

4 Subadditive Valuations

In this section, we focus on agents with subadditive valuations. We first show that the proportional allocation mechanism is at least 50 % efficient for coarse correlated equilibria and Bayesian Nash equilibria, i.e., PoA ≤ 2. Then we show that this bound is tight and cannot be improved by any simple mechanism.

Upper Bound. A common approach to prove PoA bounds is to find a deviation with proper utility bounds and then use the definition of Nash equilibrium to bound agents' utilities at equilibrium. The bidding strategy described in the following lemma is for this purpose.

Lemma 5. *Let* $\mathbf{v}$ *be any subadditive valuation profile and* $\mathbf{B}$ *be some randomized bidding profile. For any agent* i*, there exists a randomized bidding strategy* $a_i(\mathbf{v}, \mathbf{B}_{-i})$ *such that:* $\sum_i u_i(a_i(\mathbf{v}, \mathbf{B}_{-i}), \mathbf{B}_{-i}) \geq \frac{1}{2}\sum_i v_i(o_i^{\mathbf{v}}) - \sum_i \sum_j \mathbb{E}_{\mathbf{b}\sim\mathbf{B}}[b_{ij}]$.

Proof. Let p_{ij} be the sum of the bids of all agents except i on resource j, i.e., $p_{ij} = \sum_{k\neq i} b_{kj}$. Note that p_{ij} is a random variable that depends on $\mathbf{b}_{-i} \sim \mathbf{B}_{-i}$. Let P_i be the propability distribution of $p_i = (p_{ij})_j$. Inspired by [9], we consider the bidding strategy $a_i(\mathbf{v}, \mathbf{B}_{-i}) = (o_{ij}^{\mathbf{v}} \cdot b'_{ij})_j$, where $b'_i \sim P_i$. Then, $u_i(a_i(\mathbf{v}, \mathbf{B}_{-i}), \mathbf{B}_{-i})$ is

$$\mathbb{E}_{b'_i\sim P_i}\mathbb{E}_{p_i\sim P_i}\left[v_i\left(\left(\frac{o_{ij}^{\mathbf{v}} b'_{ij}}{o_{ij}^{\mathbf{v}} b'_{ij} + p_{ij}}\right)_j\right) - o_i^{\mathbf{v}} \cdot b'_i\right]$$

$$\geq \frac{1}{2} \cdot \mathbb{E}_{p_i\sim P_i}\mathbb{E}_{b'_i\sim P_i}\left[v_i\left(\left(\frac{o_{ij}^{\mathbf{v}} b'_{ij}}{o_{ij}^{\mathbf{v}} b'_{ij} + p_{ij}} + \frac{o_{ij}^{\mathbf{v}} p_{ij}}{o_{ij}^{\mathbf{v}} p_{ij} + b'_{ij}}\right)_j\right)\right] - \mathbb{E}_{p_i\sim P_i}[o_i^{\mathbf{v}} \cdot p_i]$$

$$\geq \frac{1}{2} \cdot \mathbb{E}_{p_i\sim P_i}\mathbb{E}_{b'_i\sim P_i}\left[v_i\left(\left(\frac{o_{ij}^{\mathbf{v}} (b'_{ij} + p_{ij})}{b'_{ij} + p_{ij}}\right)_j\right)\right] - \mathbb{E}_{p_i\sim P_i}[o_i^{\mathbf{v}} \cdot p_i]$$

$$= \frac{1}{2} \cdot v_i(o_i^{\mathbf{v}}) - \sum_j \sum_{k\neq i} \mathbb{E}_{\mathbf{b}\sim\mathbf{B}}[o_{ij}^{\mathbf{v}} \cdot b_{kj}]$$

The first inequality follows by swapping p_{ij} and b'_{ij} and using the subadditivity of v_i. The second inequality comes from the fact that $o_{ij}^{\mathbf{v}} \leq 1$. The lemma follows by summing up over all agents and the fact that $\sum_{i\in[n]} o_{ij}^{\mathbf{v}} = 1$. □

Theorem 6. *The coarse correlated PoA of the proportional allocation mechanism with subadditive agents is at most* 2.

Proof. Let $\mathbf{B}$ be any coarse correlated equilibrium (note that $\mathbf{v}$ is fixed). By Lemma 5 and the definition of the coarse correlated equilibrium, we have

$$\sum_i u_i(\mathbf{B}) \geq \sum_i u_i(a_i(\mathbf{v}, \mathbf{B}_{-i}), \mathbf{B}_{-i}) \geq \frac{1}{2}\sum_i v_i(o_i) - \sum_i \sum_j \mathbb{E}[b_{ij}]$$

By rearranging terms, $\mathrm{SW}(\mathbf{B}) = \sum_i u_i(\mathbf{B}) + \sum_i \sum_j \mathbb{E}[b_{ij}] \geq \frac{1}{2} \cdot \mathrm{SW}(\mathbf{o})$. □

Theorem 7. *The Bayesian PoA of the proportional allocation mechanism with subadditive agents is at most* 2.

Proof. Let $\mathbf{B}$ be any Bayesian Nash Equilibrium and let $v_i \sim D_i$ be the valuation of each agent i drawn independently from D_i. We denote by $\mathbf{C} = (C_1, C_2, \ldots, C_n)$ the bidding distribution in $\mathbf{B}$ which includes the randomness of both the bidding strategy $\mathbf{b}$ and of the valuations $\mathbf{v}$. The utility of agent i with valuation v_i can be expressed by $u_i(\mathbf{B}_i(v_i), \mathbf{C}_{-i})$. It should be noted that $\mathbf{C}_{-i}$ does *not* depend on some particular $\mathbf{v}_{-i}$, but merely on $\mathbf{D}_{-i}$ and $\mathbf{B}_{-i}$. For any agent i and any subadditive valuation $v_i \in V_i$, consider the deviation $a_i(v_i; \mathbf{w}_{-i}, \mathbf{C}_{-i})$ as defined in Lemma 5, where $\mathbf{w}_{-i} \sim \mathbf{D}_{-i}$. By the definition of the Bayesian Nash equilibrium, we obtain

$$\mathbb{E}_{\mathbf{v}_{-i}}[u_i^{v_i}(\mathbf{B}_i(v_i), \mathbf{B}_{-i}(\mathbf{v}_{-i}))] = u_i^{v_i}(\mathbf{B}_i(v_i), \mathbf{C}_{-i}) \geq \mathbb{E}_{\mathbf{w}_{-i}}[u_i^{v_i}(a_i(v_i; \mathbf{w}_{-i}, \mathbf{C}_{-i}), \mathbf{C}_{-i})].$$

By taking expectation over v_i and summing up over all agents,

$$\sum_i \mathbb{E}_{\mathbf{v}}[u_i(\mathbf{B}(\mathbf{v}))] \geq \sum_i \mathbb{E}_{v_i, \mathbf{w}_{-i}}[u_i^{v_i}(a_i(v_i; \mathbf{w}_{-i}, \mathbf{C}_{-i}), \mathbf{C}_{-i})]$$

$$= \mathbb{E}_{\mathbf{v}}\left[\sum_i u_i^{v_i}(a_i(\mathbf{v}, \mathbf{C}_{-i}), \mathbf{C}_{-i})\right] \geq \frac{1}{2} \cdot \sum_i \mathbb{E}_{\mathbf{v}}[v_i(o_i^{\mathbf{v}})] - \sum_i \sum_j \mathbb{E}[b_{ij}]$$

So, $\mathbb{E}_{\mathbf{v}}[\mathrm{SW}(\mathbf{B}(\mathbf{v}))] = \sum_i \mathbb{E}_{\mathbf{v}}[u_i(\mathbf{B}(\mathbf{v}))] + \sum_i \sum_j \mathbb{E}[b_{ij}] \geq \frac{1}{2} \cdot \mathbb{E}_{\mathbf{v}}[\mathrm{SW}(\mathbf{o}^{\mathbf{v}})]$. □

Lower Bound. Now, we show a lower bound that applies to all simple mechanisms, where the bidding space has size (at most) sub-doubly-exponential in m. More specifically, we apply the general framework of Roughgarden [19], for showing lower bounds on the price of anarchy for *all* simple mechanisms, via communication complexity reductions with respect to the underlying optimization problem. In our setting, the problem is to maximize the social welfare by allocating divisible resources to agents with subadditive valuations. We proceed by proving a communication lower bound for this problem in the following lemma.

Lemma 8. *For any constant $\varepsilon > 0$, any $(2-\varepsilon)$-approximation (non-deterministic) algorithm for maximizing social welfare in resource allocation problem with subadditive valuations, requires an exponential amount of communication.*

Proof. We prove this lemma by reducing the communication lower bound for combinatorial auctions with general valuations (Theorem 3 of [17]) to our setting (see also [7] for a reduction to combinatorial auctions with subadditive agents).

Nisan [17] used an instance with n players and m items, with $n < m^{1/2-\varepsilon}$. Each player i is associated with a set T_i, with $|T_i| = t$ for some $t > 0$. At every instance of this problem, the players' valuations are determined by sets I_i of bundles, where $I_i \subseteq T_i$ for every i. Given I_i, player i's valuation on some subset S of items is $v_i(S) = 1$, if there exists some $R \in I_i$ such that $R \subseteq S$, otherwise $v_i(S) = 0$. In [17], it was shown that distinguishing between instances with optimal social welfare of n and 1, requires t bits of communication. By choosing t exponential in m, their theorem follows.

We prove the lemma by associating any valuation v of the above combinatorial auction problem, to some appropriate subadditive valuation v' for our setting. For any player i and any fractional allocation $\mathbf{x} = (x_1, \ldots, x_m)$, let $A_{x_i} = \{j | x_{ij} > \frac{1}{2}\}$. We define $v'_i(x_i) = v_i(A_{x_i}) + 1$ if $x_i \neq \mathbf{0}$ and $v'_i(x_i) = 0$ otherwise. It is easy to verify that v'_i is subadditive. Notice that $v'_i(x) = 2$ only if there exists $R \in I_i$ such that player i is allocated a fraction higher than 1/2 for every resource in R. The value 1/2 is chosen such that no two players are assigned more than that fraction from the same resource. This corresponds to the constraint of an allocation in the combinatorial auction where no item is allocated to two players.

Therefore, in the divisible goods allocation problem, distinguishing between instances where the optimal social welfare is $2n$ and $n+1$ is equivalent to distinguishing between instances where the optimal social welfare is n and 1 in the corresponding combinatorial auction and hence requires exponential, in m, number of communication bits. □

The PoA lower bound follows the general reduction described in [19].

Theorem 9. *The PoA of ϵ-mixed Nash equilibria*[2] *of every simple mechanism, when agents have subadditive valuations, is at least* 2.

Remark. This result only holds for ϵ-mixed Nash equilibria. Considering exact Nash equilibria, we show a lower bound for all *scale-free* mechanisms including the proportional allocation mechanism in the full version.

5 Budget Constraints

In this section, we switch to scenarios where agents have budget constraints. We use as a benchmark the *effective welfare* similarly to [2,20]. We compare the effective welfare of the allocation at equilibrium with the optimal effective welfare. We prove an upper bound of $\phi + 1 \approx 2.618$ for coarse correlated equilibria, where $\phi = \frac{\sqrt{5}+1}{2}$ is the golden ratio. This improves the previously known 2.78 upper bound in [2] for a single resource and concave valuations.

To prove this upper bound, we use the fact that in the equilibrium there is no profitable unilateral deviation, and, in particular, the utility of agent i obtained by any pure deviating bid a_i should be bounded by her budget c_i, i.e., $\sum_{j \in [m]} a_{ij} \leq c_i$. We define v^c to be the valuation v suppressed by the budget c, i.e., $v^c(x) = \min\{v(x), c\}$. Note that v^c is also subadditive since v is subadditive. For a fixed pair $(\mathbf{v}, \mathbf{c})$, let $\mathbf{o} = (o_1, \ldots, o_n)$ be the allocation that maximizes the effective welfare. For a fixed agent i and a vector of bids $\mathbf{b}_{-i}$, we define the vector p_i as $p_i = \sum_{k \neq i} b_k$. We first show the existence of a proper deviation.

Lemma 10. *For any subadditive agent i, and any randomized bidding profile $\mathbf{B}$, there exists a randomized bid $a_i(\mathbf{B}_{-i})$, such that for any $\lambda \geq 1$, it is*

---

[2] A bidding profile $\mathbf{B} = \times_i B_i$ is called ϵ-mixed Nash equilibrium if, for every agent i and all bids b'_i, $\mathbb{E}_{\mathbf{b} \sim \mathbf{B}}[u_i(\mathbf{b})] \geq \mathbb{E}_{\mathbf{b} \sim \mathbf{B}}[u_i(b'_i, \mathbf{b}_{-i})] - \epsilon$.

$$u_i(a_i(\mathbf{B}_{-i}), \mathbf{B}_{-i}) \geq \frac{v_i^{c_i}(o_i)}{\lambda+1} - \frac{\sum_{j\in[m]}\sum_{k\in[n]} o_{ij}\mathbb{E}[b_{kj}]}{\lambda}.$$

Moreover, if $\hat{a}_i$ is any pure strategy in the support of $a_i(\mathbf{B}_{-i})$, then $\sum_j \hat{a}_{ij} \leq c_i$.

Proof. In order to find $a_i(B_{-i})$, we define the truncated bid vector $\tilde{\mathbf{b}}_{-i}$ as follows. For any set $S \subseteq [m]$ of resources, we denote by $\mathbf{1}_S$ the indicator vector w.r.t. S, such that $x_j = 1$ for $j \in S$ and $x_j = 0$ otherwise. For any vector p_i and any $\lambda > 0$, let $T := T(\lambda, p_i)$ be a *maximal* subset of resources such that, $v_i^{c_i}(\mathbf{1}_T) < \frac{1}{\lambda}\sum_{j\in T} o_{ij}p_{ij}$. For every $k \neq i$, if $j \in T$, then $\tilde{b}_{kj} = 0$, otherwise $\tilde{b}_{kj} = b_{kj}$. Similarly, $\tilde{p}_i = \sum_{k\neq i}\tilde{b}_k$. Moreover, if $\mathbf{b}_{-i} \sim \mathbf{B}_{-i}$, then p_i is an induced random variable with distribution denoted by P_i. We further define distributions $\tilde{\mathbf{B}}_{-i}$ and $\tilde{P}_i$, as $\tilde{\mathbf{B}}_{-i} = \{\tilde{\mathbf{b}}_{-i}|\mathbf{b}_{-i} \sim \mathbf{B}_{-i}\}$ and $\tilde{P}_i = \{\tilde{p}_i|p_i \sim P_i\}$.

Now consider the following bidding strategy $a_i(\mathbf{B}_{-i})$: sampling $b_i' \sim \tilde{P}_i$ and bidding $a_{ij} = \frac{1}{\lambda}o_{ij}b_{ij}'$ for each resource j. We first show $\sum_{j\in[m]} a_{ij} \leq c_i$. It is sufficient to show that $v_i^{c_i}(\mathbf{1}_{[m]\setminus T}) \geq \sum_{j\notin T} a_{ij}$ since $c_i \geq v_i^{c_i}(\mathbf{1}_{[m]\setminus T})$. For the sake of contradiction suppose $v_i^{c_i}(\mathbf{1}_{[m]\setminus T}) < \sum_{j\notin T} a_{ij}$. Then, by the definition of T and $\tilde{p}_i$, $v_i^{c_i}(\mathbf{1}_{[m]}) \leq v_i^{c_i}(\mathbf{1}_T) + v_i^{c_i}(\mathbf{1}_{[m]\setminus T}) < \sum_{j\in T} a_{ij} + \sum_{j\notin T} a_{ij} = \frac{1}{\lambda}\sum_{j\in[m]} o_{ij}p_{ij}$,which contradicts the maximality of T.

Next we show for any bid b_i and $\lambda > 0$,

$$v_i^{c_i}(x_i(b_i, \mathbf{B}_{-i})) + \frac{1}{\lambda}\sum_{j\in[m]} o_{ij}\mathbb{E}_{p_i\sim P_i}[p_{ij}] \geq v_i^{c_i}(x_i(b_i, \tilde{\mathbf{B}}_{-i})) + \frac{1}{\lambda}\sum_{j\in[m]} o_{ij}\mathbb{E}_{\tilde{p}_i\sim\tilde{P}_i}[\tilde{p}_{ij}] \tag{1}$$

Observe that $x_i(b_i, \tilde{\mathbf{b}}_{-i}) \leq x_i(b_i, \mathbf{b}_{-i}) + \mathbf{1}_T$. Therefore, and by the definitions of T and $\tilde{p}_i$, $v_i^{c_i}(x_i(b_i, \tilde{\mathbf{b}}_{-i})) \leq v_i^{c_i}(x_i(b_i, \mathbf{b}_{-i})) + v_i^{c_i}(\mathbf{1}_T) \leq v_i^{c_i}(x_i(b_i, \mathbf{b}_{-i})) + \frac{1}{\lambda}\sum_{j\in T} o_{ij}p_{ij} = v_i^{c_i}(x_i(b_i, \mathbf{b}_{-i})) + \frac{1}{\lambda}\sum_{j\in[m]} o_{ij}p_{ij} - \frac{1}{\lambda}\sum_{j\in[m]} o_{ij}\tilde{p}_{ij}$. The claim follows by rearranging terms and taking the expectation of $\mathbf{b}_{-i}$, $\tilde{\mathbf{b}}_{-i}$, p_i and $\tilde{p}_i$ over $\mathbf{B}_{-i}$, $\tilde{\mathbf{B}}_{-i}$, P_i and $\tilde{P}_i$, respectively.

$$\begin{aligned}
\mathbb{E}_{b_i'\sim\tilde{P}_i}\left[u_i\left(\frac{1}{\lambda}o_i b_i', \mathbf{B}_{-i}\right)\right] &= \mathbb{E}_{b_i'\sim\tilde{P}_i}\left[v_i\left(\frac{1}{\lambda}o_i b_i', \mathbf{B}_{-i}\right)\right] - \frac{1}{\lambda}\sum_{j\in[m]} o_{ij}\mathbb{E}_{b_i'\sim\tilde{P}_i}\left[b_{ij}'\right] \\
&\geq \mathbb{E}_{b_i'\sim\tilde{P}_i}\left[v_i^{c_i}\left(\frac{1}{\lambda}o_i b_i', \mathbf{B}_{-i}\right)\right] - \frac{1}{\lambda}\sum_{j\in[m]} o_{ij}\mathbb{E}_{\tilde{p}_i\sim\tilde{P}_i}\left[\tilde{p}_{ij}\right] \quad \text{(by definition of } v_i^{c_i}\text{)} \\
&\geq \mathbb{E}_{b_i'\sim\tilde{P}_i}\left[v_i^{c_i}\left(\frac{1}{\lambda}o_i b_i', \tilde{\mathbf{B}}_{-i}\right)\right] - \frac{1}{\lambda}\sum_{j\in[m]} o_{ij}\mathbb{E}_{p_i\sim P_i}\left[p_{ij}\right] \quad \text{(by Inequality (1))} \\
&\geq \frac{1}{2}\mathbb{E}_{b_i'\sim\tilde{P}_i}\mathbb{E}_{\tilde{p}_i\sim\tilde{P}_i}\left[v_i^{c_i}\left(\frac{o_i b_i'}{o_i b_i' + \lambda\tilde{p}_i} + \frac{o_i\tilde{p}_i}{o_i\tilde{p}_i + \lambda b_i'}\right)\right] - \frac{1}{\lambda}\sum_{j\in[m]} o_{ij}\sum_{k\neq i} B_{kj} \\
&\quad \text{(by swapping } b_i' \text{ with } \tilde{p}_i \text{ and the subadditivity of } v_i^{c_i}(\cdot)\text{)} \\
&\geq \frac{1}{2}\mathbb{E}_{b_i'\sim\tilde{P}_i}\mathbb{E}_{\tilde{p}_i\sim\tilde{P}_i}\left[v_i^{c_i}\left(o_i\left(\frac{b_i'}{b_i' + \lambda\tilde{p}_i} + \frac{\tilde{p}_i}{\tilde{p}_i + \lambda b_i'}\right)\right)\right] - \frac{1}{\lambda}\sum_{j\in[m]}\sum_{k\in[n]} o_{ij}\mathbb{E}[b_{kj}] \\
&\geq \frac{1}{2}v_i^{c_i}\left(\frac{2o_i}{\lambda+1}\right) - \frac{1}{\lambda}\sum_{j\in[m]} o_{ij}\sum_k B_{kj} \quad \text{(by monotonicity of } v_i^{c_i}\text{)} \\
&\geq \frac{1}{\lambda+1}v_i^{c_i}(o_i) - \frac{1}{\lambda}\sum_{j\in[m]} o_{ij}\sum_k B_{kj} \quad \text{(subadditivity of } v_i^{c_i}\text{; } \tfrac{2}{\lambda+1} \leq 1\text{)}
\end{aligned}$$

For the second inequality, notice that the second term doesn't depend on b'_i, so we apply Lemma 11 for every b'_i. For the forth and fifth inequalities, $o_i \leq 1$ and $\frac{b'_i}{b'_i+\lambda\tilde{p}_i} + \frac{\tilde{p}_i}{\tilde{p}_i+\lambda b'_i} \geq \frac{2}{\lambda+1}$ for every b'_i, $\tilde{p}_i$ and $\lambda \geq 1$. □

We are ready to show the PoA bound by using the above lemma.

Theorem 11. *The coarse correlated PoA for the proportional allocation mechanism when agents have budgets and subadditive valuations, is at most $\phi + 1 \approx$ 2.618.*

Proof. Suppose $\mathbf{B}$ is a coarse correlated equilibrium. Let A be the set of agents such that for every $i \in A$, $v_i(\mathbf{B}) \leq c_i$. For simplicity, we use $v_i^{c_i}(\mathbf{B})$ to denote $\min\{\mathbb{E}_{\mathbf{b}\sim\mathbf{B}}[v_i(x_i(\mathbf{b}))], c_i\}$. Then for all $i \notin A$, $v_i^{c_i}(\mathbf{B}) = c_i \geq v_i^{c_i}(o_i)$ and $v_i^{c_i}(\mathbf{B}) = c_i \geq \sum_{j\in[m]} \mathbb{E}[b_{ij}]$. The latter inequality comes from that agents do not bid higher than their budgets. Let $\lambda = \phi$. So $1 - 1/\lambda = 1/(1+\lambda)$. By taking the linear combination and summing up over all agents not in A, we get

$$\sum_{i\notin A} v_i^{c_i}(\mathbf{B}) \geq \frac{1}{\lambda+1}\sum_{i\notin A} v_i^{c_i}(o_i) + \frac{1}{\lambda}\sum_{i\notin A}\sum_{j\in[m]} \mathbb{E}[b_{ij}] \tag{2}$$

For every $i \in A$, we consider the deviating bidding strategy $a_i(\mathbf{B}_{-i})$ that is described in Lemma 10, then

$$\begin{aligned} v_i^{c_i}(\mathbf{B}) = v_i(x_i(\mathbf{B})) = u_i(x_i(\mathbf{B})) + \sum_{j\in[m]} \mathbb{E}[b_{ij}] &\geq u_i(a_i(\mathbf{B}_{-i}), \mathbf{B}_{-i}) + \frac{1}{\lambda}\sum_{j\in[m]} \mathbb{E}[b_{ij}] \\ &\geq \frac{1}{\lambda+1} v_i^{c_i}(o_i) - \frac{1}{\lambda}\sum_{j\in[m]}\sum_{k\in[n]} o_{ij}\mathbb{E}[b_{kj}] + \frac{1}{\lambda}\sum_{j\in[m]} \mathbb{E}[b_{ij}] \end{aligned}$$

By summing up over all $i \in A$ and by combining with inequality (2) we get

$$\begin{aligned} &\sum_{i\in[n]} \min\{v_i(x_i(\mathbf{B})), c_i\} \\ &\geq \frac{1}{\lambda+1}\sum_{i\in[n]} v_i^{c_i}(o_i) + \frac{1}{\lambda}\sum_{i\in[n]}\sum_{j\in[m]} \mathbb{E}[b_{ij}] - \frac{1}{\lambda}\sum_{i\in A}\sum_{j\in[m]}\sum_{k\in[n]} o_{ij}\mathbb{E}[b_{kj}] \\ &\geq \frac{1}{\lambda+1}\sum_{i\in[n]} v_i^{c_i}(o_i) \qquad (\text{since } \textstyle\sum_{i\in A} o_{ij} \leq 1) \end{aligned}$$

Therefore, the PoA with respect to the effective welfare is at most $\phi + 1$. □

By applying Jensen's inequality for concave functions, our upper bound also holds for the Bayesian case with single-resource and concave functions.

Theorem 12. *The Bayesian PoA of single-resource proportional allocation games is at most $\phi + 1 \approx 2.618$, when agents have budgets and concave valuations.*

Remark. Syrgkanis and Tardos [20], compared the social welfare in the equilibrium with the effective welfare in the optimum allocation. Caragiannis and Voudouris [2] also give an upper bound of 2 for this ratio in the single resource case. We can obtain the same upper bound by replacing λ with 1 in the proofs.

6 Polyhedral Environment

In this section, we study the efficiency of the proportional allocation mechanism in the polyhedral environment, that was previously studied by Nguyen and Tardos [16]. We show a *tight* price of anarchy bound of 2 for agents with subadditive valuations. Recall that, in this setting, the allocation to each agent i is now represented by a *single parameter* x_i, and not by a vector $(x_{i1}, \dots, x_{im})$. In addition, any feasible allocation vector $\mathbf{x} = (x_1, \dots, x_n)$ should satisfy a polyhedral constraint $A \cdot \mathbf{x} \leq \mathbf{1}$, where A is a non-negative $m \times n$ matrix and each row of A corresponds to a different resource, and $\mathbf{1}$ is a vector with all ones. Each agent aims to maximize her utility $u_i = v_i(x_i) - q_i$, where v_i is a subadditive function representing the agent's valuation. The proportional allocation mechanism determines the following allocation and payments for each agent:

$$x_i(\mathbf{b}) = \min_{j:a_{ij}>0} \left\{ \frac{b_{ij}}{a_{ij} \sum_{k \in [n]} b_{kj}} \right\}; \qquad q_i(\mathbf{b}) = \sum_{j \in [m]} b_{ij},$$

where a_{ij} is the (i,j)-th entry of matrix A. It is easy to verify that the above allocation satisfies the polyhedral constraints.

Theorem 13. *If agents have subadditive valuations, the pure PoA of the proportional allocation mechanism in the polyhedral environment is exactly* 2.

Proof. We first show that the PoA is at most 2. Let $\mathbf{o} = \{o_1, \dots, o_n\}$ be the optimal allocation, $\mathbf{b}$ be a pure Nash Equilibrium, and let $p_{ij} = \sum_{k \neq i} b_{ij}$. For each agent i, consider the deviating bid b'_i such that $b'_{ij} = o_i a_{ij} p_{ij}$ for all resources j. Since $\mathbf{b}$ is a Nash Equilibrium,

$$\begin{aligned} u_i(\mathbf{b}) \geq u_i(b'_i, b_{-i}) &= v_i \left(\min_{j:a_{ij}>0} \left\{ \frac{o_i a_{ij} p_{ij}}{a_{ij} \left(p_{ij} + o_i a_{ij} p_{ij}\right)} \right\} \right) - \sum_{j \in [m]} o_i a_{ij} p_{ij} \\ &\geq v_i \left(\min_{j:a_{ij}>0} \left\{ \frac{o_i}{1 + o_i a_{ij}} \right\} \right) - \sum_{j \in [m]} o_i a_{ij} p_{ij} \geq \frac{1}{2} v_i(o_i) - \sum_{j \in [m]} o_i a_{ij} p_{ij} \end{aligned}$$

The second inequality is true since $A \cdot \mathbf{x} \leq \mathbf{1}$, for every allocation $\mathbf{x}$, and therefore $o_i a_{ij} < 1$. The last inequality holds due to subadditivity of v_i. By summing up over all agents, we get $\sum_i u_i(\mathbf{b}) \geq \frac{1}{2} \sum_i v_i(o_i) - \sum_{j \in [m]} \sum_{i \in [n]} o_i a_{ij} p_{ij} \geq \frac{1}{2} \sum_i v_i(o_i) - \sum_{j \in [m]} \sum_{k \in [n]} b_{kj}$. The last inequality holds due to the fact that $p_{ij} \leq \sum_{k \in [n]} b_{kj}$ and $\sum_{i \in [n]} o_i a_{ij} \leq 1$. PoA ≤ 2 follows by rearranging the terms.

For the lower bound, consider a game with only two agents and a single resource where the polyhedral constraint is given by $x_1 + x_2 \leq 1$. The valuation of the first agent is $v_1(x) = 1 + \epsilon \cdot x$, for some $\epsilon < 1$ if $x < 1$ and $v_1(x) = 2$ if $x = 1$. The valuation of the second agent is $\epsilon \cdot x$. One can verify that these two functions are subadditive and the optimal social welfare is 2. Consider the bidding strategies $b_1 = b_2 = \frac{\epsilon}{4}$. The utility of agent 1, when she bids x and agent 2 bids $\frac{\epsilon}{4}$, is given by $1 + \epsilon \cdot \frac{x}{x + \epsilon/4} - x$ which is maximized for $x = \frac{\epsilon}{4}$.

The utility of agent 2, when she bids x and agent 1 bids $\frac{\epsilon}{4}$, is $\epsilon \cdot \frac{x}{x+\epsilon/4} - x$ which is also maximized when $x = \frac{\epsilon}{4}$. So (b_1, b_2) is a pure Nash Equilibrium with social welfare $1 + \epsilon$. Therefore, the PoA converges to 2 when ϵ goes to 0. □

References

1. Bhawalkar, K., Roughgarden, T.: Welfare guarantees for combinatorial auctions with item bidding. In: SODA 2011. SIAM (2011)
2. Caragiannis, I., Voudouris, A.A.: Welfare guarantees for proportional allocations. In: Lavi, R. (ed.) SAGT 2014. LNCS, vol. 8768, pp. 206–217. Springer, Heidelberg (2014)
3. Christodoulou, G., Kovács, A., Schapira, M.: Bayesian combinatorial auctions. In: Aceto, L., Damgård, I., Goldberg, L.A., Halldórsson, M.M., Ingólfsdóttir, A., Walukiewicz, I. (eds.) ICALP 2008, Part I. LNCS, vol. 5125, pp. 820–832. Springer, Heidelberg (2008)
4. Christodoulou, G., Kovács, A., Sgouritsa, A., Tang, B.: Tight bounds for the price of anarchy of simultaneous first price auctions. CoRR abs/1312.2371 (2013)
5. Chun, B.N., Culler, D.E.: Market-based proportional resource sharing for clusters. Technical report, University of California at Berkeley, Berkeley (2000)
6. Correa, J.R., Schulz, A.S., Stier-Moses, N.E.: The price of anarchy of the proportional allocation mechanism revisited. In: Chen, Y., Immorlica, N. (eds.) WINE 2013. LNCS, vol. 8289, pp. 109–120. Springer, Heidelberg (2013)
7. Dobzinski, S., Nisan, N., Schapira, M.: Approximation algorithms for combinatorial auctions with complement-free bidders. Math. Oper. Res. **35**(1), 1–13 (2010)
8. Evans, D.S., Heckman, J.J.: A test for subadditivity of the cost function with an application to the bell system. The American Economic Review pp. 615–623 (1984)
9. Feldman, M., Fu, H., Gravin, N., Lucier, B.: Simultaneous auctions are (almost) efficient. In: STOC 2013 (2012)
10. Feldman, M., Lai, K., Zhang, L.: A price-anticipating resource allocation mechanism for distributed shared clusters. In: EC 2005, pp. 127–136. ACM (2005)
11. Hassidim, A., Kaplan, H., Mansour, Y., Nisan, N.: Non-price equilibria in markets of discrete goods. In: EC 2011, pp. 295–296. ACM (2011)
12. Johari, R., Tsitsiklis, J.N.: Efficiency loss in a network resource allocation game. Math. Oper. Res. **29**(3), 407–435 (2004)
13. Johari, R., Tsitsiklis, J.N.: Efficiency of scalar-parameterized mechanisms. Oper. Res. **57**(4), 823–839 (2009)
14. Kelly, F.: Charging and rate control for elastic traffic. Eur. Trans. Telecomm. **8**(1), 33–37 (1997)
15. Koutsoupias, E., Papadimitriou, C.: Worst-case equilibria. In: Meinel, C., Tison, S. (eds.) STACS 1999. LNCS, vol. 1563, pp. 404–413. Springer, Heidelberg (1999)
16. Nguyen, T., Tardos, E.: Approximately maximizing efficiency and revenue in polyhedral environments. In: EC 2007. ACM (2007)
17. Nisan, N.: The communication complexity of approximate set packing and covering. In: Widmayer, P., Triguero, F., Morales, R., Hennessy, M., Eidenbenz, S., Conejo, R. (eds.) ICALP 2002. LNCS, vol. 2380, pp. 868–875. Springer, Heidelberg (2002)
18. Rosenbaum, R.A.: Sub-additive functions. Duke Math. J. **17**(3), 227–247 (1950)

19. Roughgarden, T.: Barriers to near-optimal equilibria. In: Proceedings of the 55th Annual IEEE Symposium on Foundations of Computer Science (FOCS) (2014)
20. Syrgkanis, V., Tardos, E.: Composable and Efficient Mechanisms. In: STOC 2013 (2013)
21. Zhang, L.: The efficiency and fairness of a fixed budget resource allocation game. In: Caires, L., Italiano, G.F., Monteiro, L., Palamidessi, C., Yung, M. (eds.) ICALP 2005. LNCS, vol. 3580, pp. 485–496. Springer, Heidelberg (2005)

On Existence and Properties of Approximate Pure Nash Equilibria in Bandwidth Allocation Games

Maximilian Drees(✉), Matthias Feldotto, Sören Riechers, and Alexander Skopalik

Heinz Nixdorf Institute & Department of Computer Science, University of Paderborn, Paderborn, Germany
maxdress@mail.upb.de

Abstract. In *bandwidth allocation games* (BAGs), the strategy of a player consists of various demands on different resources. The player's utility is at most the sum of these demands, provided they are fully satisfied. Every resource has a limited capacity and if it is exceeded by the total demand, it has to be split between the players. Since these games generally do not have pure Nash equilibria, we consider approximate pure Nash equilibria, in which no player can prove her utility by more than some fixed factor α through unilateral strategy changes. There is a threshold α_δ (where δ is a parameter that limits the demand of each player on a specific resource) such that α-approximate pure Nash equilibria always exist for $\alpha \geq \alpha_\delta$, but not for $\alpha < \alpha_\delta$. We give both upper and lower bounds on this threshold α_δ and show that the corresponding decision problem is NP-hard. We also show that the α-approximate price of anarchy for BAGs is $\alpha + 1$. For a restricted version of the game, where demands of players only differ slightly from each other (e.g. symmetric games), we show that approximate Nash equilibria can be reached (and thus also be computed) in polynomial time using the best-response dynamic. Finally, we show that a broader class of utility-maximization games (which includes BAGs) converges quickly towards states whose social welfare is close to the optimum.

1 Introduction

Nowadays, as cloud computing and other data intensive applications such as video streaming gain more and more importance, the amount of data processed in networks and compute centers is growing. Moore's law for data traffic [16] states that the overall data traffic doubles each year. This yields unique challenges for resource management, particularly bandwidth allocation. As technology cannot follow up with the data increase, bandwidth constraints are often a bottleneck of current systems.

This work was partially supported by the German Research Foundation (DFG) within the Collaborative Research Centre "On-The-Fly Computing" (SFB 901) and by the EU within FET project MULTIPLEX under contract no. 317532.

M. Hoefer (Ed.): SAGT 2015, LNCS 9347, pp. 178–189, 2015.
DOI: 10.1007/978-3-662-48433-3_14

In our paper, we cope with the problem that service providers often cannot satisfy the needs of all customers. That is, the overall size of connections between the provider and all customers exceeds the amount of data that the provider can process. By allowing different link sizes in network structures, connections between providers and customers with different capacities can be modeled. In case a provider cannot fulfill the requirements of all customers, the available bandwidth needs to be split. This results in customers not being supplied with their full capacity. In video streaming, for example, this may lead to a lower quality stream for certain customers. In our setting, we assume that each customer can choose the service providers she wants to use herself. While this aspect has recently been studied from the compute center's point of view [9], our work considers limited resources from the customers' point of view.

We study this scenario in a game theoretic setting called *bandwidth allocation games*. Here, we are interested in the effects of rational decision making by individuals. In our context, the customers act as the players. In contrast, we view the service providers as resources with a limited capacity. Each possible distribution of a player among the resources (which we view as network entrance points) is regarded as one of her strategies. Now, each player strives to maximize the overall amount of bandwidth that is supplied to her. Our main interest lies in states in which no customer wants to deviate from her current strategy, as this would yield no or only a marginal benefit under the given situation. These states are called (approximate) pure Nash equilibria. Instead of a global instance enforcing such stable states, they occur as the result of player-induced dynamics. At every point in time, exactly one player changes her strategy such that the amount of received bandwidth is maximized, assuming the strategies of the other players are fixed. We show that if we allow only changes which increase the received bandwidth by some constant factor, this indeed leads to stable states. We further analyze the quality of such states in regard to the total bandwidth received by all players and compare it to the state which maximizes this global payoff.

Related Work. Bandwidth allocation games can be considered to be a generalization of market sharing games [22], in which players choose a set of market in which they offer a service. Each market has a fixed cost and each player a budget. The set of markets a player can service is thus determined by a knapsack constraint. The utility of a player is the sum of utilities that she receives from each market that she services. Each market has a fixed total profit or utility that is evenly distributed among the players that service the market.

The utility functions of bandwidth allocation games are more general. In particular the influence of a player on the utility share others players receive is not uniform. Players with high demand have a much stronger influence on the bandwidth other players receive than player with small demands. This feature can also be found in demanded congestion games [28]. Players in a congestion game choose among subsets of resources while trying to minimize costs. The cost of a player is sum of the costs of the resources. In the undemanded version which was introduced by Rosenthal [30] the cost of each resource depends only on the number of players using that resource. In the demanded version each player has a demand and the cost of a resource is a function of the sum of demands of the players using the resource. In both model the cost caused by a resource is identical for each player that uses the resource. In the variant of player-specific congestion games, each player has her own set of cost functions [28] for each resource that map from the number of players using a resource to the cost incurred to that specific player Mavronicolas et al. [27] combined these two variations into demanded congestion games with player-specific constants, in which the cost functions are based on abelian group operations. Harks and Klimm [25] introduced a model in which each player not only picks a subset of resources, but also her single demand on them. A higher demand equals a higher utility for each player, but also increases the congestion at the chosen resources. The final payoff results from the difference between utility and congestion.

Both, market sharing games and congestion games always posses pure Nash eqilibria. Moreover they are potential games [29] which implies that every finite sequence of best response dynamics is guaranteed to converge to a pure Nash equilibrium. demanded congestion games are potential games only if the cost function are linear or exponential functions [24]. For demanded and player-specific games the existence of pure Nash equilibria this is guaranteed for the special case in which the strategy spaces of the players for the bases of a matroid [1].

Fabrikant et al. [20] showed that the problem of computing a pure Nash equilibrium is PLS-complete. This result implies that the improvement path could be exponentially long. In the case of demanded [19] or player-specific [2] congestion games it is NP-hard to decide if there exists a pure Nash equilibrium. These negative computational and existence results lead to the study of α-approximate pure Nash equilibria which are states in which no player can increases her utility (or decrease her cost) by a *factor* of more than α. Chien and Sinclair [13] showed that in symmetric undemanded congestions games and under a mild assumption on the cost functions every sequence of $(1+\varepsilon)$-improving steps convergence to $(1+\varepsilon)$-approximate equilibria in polynomial time in the number of players and ε^{-1}. This result cannot be generalized to asymmetric games as Skopalik and Vöcking [32] showed that the problem is still PLS-complete. However, for the case of linear or polynomial cost function Caragiannis et al. presented [10] an algorithm to compute approximate pure Nash equilibria in polynomial time which was slightly improved in [21].

For demanded congestion games it was shown that α-approximate pure equilibria with small values of α exist [23] and that they can be computed in polynomial time [11] albeit only for a larger values of α. Chen and Roughgarden [12] proved the existence of approximate equilibria in network design games with demanded players. The results have been used by Christodoulou et al. [15] to give tight bounds on the price of anarchy and price of stability of approximate pure Nash equilibria in undemanded congestion games.

To quantify the inefficiency of equilibrium outcomes the price of anarchy has been thoroughly analyzed for exact equilibria for undemanded [3,14,31] as well as for demanded congestion games [3,6,8,14]. Christodoulou et al. [15] also investigated the PoA for approximate pure Nash equilibria.

Recent work bounded the convergence time to states with a social welfare close to the optimum rather than equilibria. The concept of smoothness was first introduced by Roughgarden [31]. Several variants such as the concept of semi-smoothness [26] followed. Awerbuch et al. [7] proposed β-niceness which was reworked in [5]. It is the basis of the concept of nice games introduced in [4], which we use in our work.

Our Contribution. We introduce the notion of *δ-share bandwidth allocation games* (BAGs). The demand on a resource may not exceed that resource's capacity by a factor of more than δ. Building on a result from our previous paper [18], we show that no matter how small we choose δ, these games generally do not have pure Nash equilibria. We then turn to α-approximate pure Nash equilibria, in which no player can improve her utility by a factor of more than α through unilateral strategy changes. We are interested in the threshold α_δ (based on a given δ), such that for all $\alpha < \alpha_\delta$, there is a δ-share BAG without an α-approximate pure Nash equilibrium, and for all $\alpha \geq \alpha_\delta$, every δ-share BAG has an α-approximate pure Nash equilibrium. By using a potential function argument, we give both upper and lower bounds for α_δ. For a general δ-share BAG $\mathcal{B}$ and $\alpha < \alpha_\delta^l$, it is NP-complete to decide if $\mathcal{B}$ has an α-approximate pure Nash equilibria and NP-hard to compute it, if available. On the other hand, for $\alpha \geq \alpha_\delta^u$ and if the difference between the most-profitable strategies of the players can be bounded by some constant λ, then an $(\alpha+\varepsilon)$-approximate Nash equilibrium can be computed efficiently. We give an almost tight bound of $\alpha + 1$ for the α-approximate price of anarchy for BAGs and finally show that utility-maximization games with certain properties converge quickly towards states with a social welfare close to the optimum. We then adapt this general result to δ-share BAGs.

2 Model and Preliminaries

A *bandwidth allocation game* (BAG) $\mathcal{B}$ is a tuple $(\mathcal{N}, \mathcal{R}, (b_r)_{r\in\mathcal{R}}, (\mathcal{S}_i)_{i\in\mathcal{N}})$ where the set of players is denoted by $\mathcal{N} = \{1, \ldots, n\}$, the set of resources by $\mathcal{R} = \{r_1, \ldots, r_m\}$, the *capacity* of resource r by b_r and the strategy space of player i by $\mathcal{S}_i$. Each $s_i \in \mathcal{S}_i$ has the form $(s_i(r_1), \ldots, s_i(r_m)) \in \mathbb{R}^m_{\geq 0}$, with $s_i(r_j) \in \mathbb{R}_{\geq 0}$ being the *demand* of s_i on the resource r_j. We say that a strategy s_i *uses* a resource r_j if $s_i(r_j) > 0$. $\mathcal{S} = \mathcal{S}_1 \times \ldots \times \mathcal{S}_n$ is the set of strategy profiles and $u_i : \mathcal{S} \to \mathbb{R}_{\geq 0}$ denotes the private utility function player i strives to maximize. For a strategy profile $\mathsf{s} = (s_1, \ldots, s_n)$, let $u_{i,r}(\mathsf{s}) \in \mathbb{R}_{\geq 0}$ denote the utility of player i from resource r, which is defined as

$$u_{i,r}(\mathsf{s}) := \min\left(s_i(r), \frac{b_r \cdot s_i(r)}{\sum_{j\in\mathcal{N}} s_j(r)}\right).$$

The total utility of i is then defined as $u_i(\mathsf{s}) := \sum_{r\in\mathcal{R}} u_{i,r}(\mathsf{s})$.

Let $\delta > 0$. We call a bandwidth allocation game a *δ-share bandwidth allocation game* if for every strategy s_i and every resource r, the restriction $s_i(r) \leq \delta b_r$ holds.

Let s be an arbitrary strategy profile and $i \in \mathcal{N}$. We denote with $\mathsf{s}_{-i} := (s_1, \ldots, s_{i-1}, s_{i+1}, \ldots, s_n)$ the strategy vector of all players except i. For any $s_i \in \mathcal{S}_i$, we can extend this to the strategy profile $(\mathsf{s}_{-i}, s_i) := (s_1, \ldots, s_{i-1}, s_i, s_{i+1}, \ldots, s_n)$. We denote with $s_i^b \in \mathcal{S}_i$ the *best response* of i to s_{-i} if $u_i(\mathsf{s}_{-i}, s_i^b) \geq u_i(\mathsf{s}_{-i}, s_i)$ for all $s_i \in \mathcal{S}_i$.

Let $\alpha \geq 1$ and s_i a strategy of player i. If there is a strategy $s_i' \in \mathcal{S}_i$ with $\alpha \cdot u_i(\mathsf{s}_{-i}, s_i) < u_i(\mathsf{s}_{-i}, s_i')$, then we call the switch from s_i to s_i' an *α-move*. For $\alpha = 1$, we simply use the term *move*. A strategy profile s is called an α-approximate pure Nash equilibrium (α-NE) if $\alpha \cdot u_i(\mathsf{s}) \geq u_i(\mathsf{s}_{-i}, s_i')$ for every $i \in \mathcal{N}$ and $s_i' \in \mathcal{S}_i$. For $\alpha = 1$, s is simply called a pure Nash equilibrium (NE). If a bandwidth allocation game eventually reaches an (α-approximate) pure Nash equilibrium after a finite number of (α-)moves from any initial strategy profile s, we say that the game has the *finite improvement property*.

The *social welfare* of a strategy profile s is defined as $u(\mathsf{s}) = \sum_{i \in \mathcal{N}} u_i(\mathsf{s})$. Let *opt* be the strategy profile with $u(opt) \geq u(\mathsf{s})$ for all $\mathsf{s} \in \mathcal{S}$. If $\mathcal{S}^\alpha \subseteq \mathcal{S}$ is the set of all α-approximate pure Nash equilibria in a bandwidth allocation game $\mathcal{B}$, then $\mathcal{B}$'s *α-approximate price of anarchy* (α-PoA) is the ratio $\max_{\mathsf{s} \in \mathcal{S}^\alpha} \frac{u(opt)}{u(\mathsf{s})}$. Again, we simply use the term *price of anarchy* (PoA) for $\alpha = 1$.

Throughout the paper, we are going to use a potential function $\phi : \mathcal{S} \to \mathbb{R}$ to analyze the properties of bandwidth allocation games. Let $T_r(\mathsf{s}) := \sum_{i \in \mathcal{N}} s_i(r)$ be the total demand on resource r under strategy profile s. We define $\phi(\mathsf{s}) := \sum_{r \in \mathcal{R}} \phi_r(\mathsf{s})$ with

$$\phi_r(\mathsf{s}) := \begin{cases} T_r(\mathsf{s}) & \text{if } T_r(\mathsf{s}) \leq b_r \\ b_r + \int_{b_r}^{T_r(\mathsf{s})} \frac{b_r}{x}\, \mathrm{d}x & \text{else} \end{cases}$$

3 Pure Nash Equilibria

The δ-share BAGs in this paper resemble the standard budget games from our previous work [18] in which δ was unbounded. This allowed arbitrarily large demands for the strategies. In particular, the demand of a strategy on a resource r could exceed the capacity b_r. In δ-share BAGs, that demand is restricted to the interval $[0, \delta b_r]$ for a fixed $\delta > 0$. We now show that our previous result concerning the existence of NE still holds for any restriction on the demands.

Definition 1. *Let $\delta > 0$ be arbitrary, but fixed. Choose $\gamma, \sigma > 0$ and $n \in \mathbb{N}_0$ s.t. $\gamma < \delta$, $\sigma \leq \delta$ and $n \cdot \sigma + \delta = 1$. Let $\mathcal{B}_0$ be a δ-share bandwidth allocation game with $|\mathcal{N}_0| = n+2$, $\mathcal{R}_0 = \{r_1, r_2, r_3, r_4\}$ resources with capacity 1 and the strategy spaces $\mathcal{S}_1 = \{s_1^1 = (\gamma, \delta, 0, 0), s_1^2 = (0, 0, \delta, \gamma)\}$, $\mathcal{S}_2 = \{s_2^1 = (\delta, 0, \gamma, 0), s_2^2 = (0, \gamma, 0, \delta)\}$ and $\mathcal{S}_i = \{s_i = (\sigma, \sigma, \sigma, \sigma)\}$ for $i \in \{3, \ldots, n+2\}$.*

The players $i \in \{3, \ldots, n+2\}$ serve as auxiliary players to reduce the available capacity of the resources. Each can only play strategy s_i, so we focus on the two remaining players 1 and 2, which we regard as the main players of the game.

In every strategy profile, one of them has a utility of $u := \frac{\gamma}{\delta+\gamma+n\cdot\sigma} + \delta$ while the other one has a utility of $u' := \frac{\delta}{\delta+\gamma+n\cdot\sigma} + \gamma$. Assume $\delta \leq 1$. Since $\delta+\gamma+n\cdot\sigma > 1$, we obtain $\frac{\delta-\gamma}{\delta+\gamma+n\cdot\sigma} < \delta - \gamma$ and therefore $u' < u$. Since the player with utility u' can always change strategy to swap the two utilities, $\mathcal{B}_0$ does not have a pure Nash equilibrium. For $\delta > 1$, we choose $n = 0$ and $\gamma > 1$. In this case, $u = \frac{\gamma}{\delta+\gamma} + 1$ and $u' = \frac{\delta}{\delta+\gamma} + 1$ with $u < u'$. Again, the player with the lower utility u can always improve her utility. A visualisation of the game $\mathcal{B}_0$ for $\delta > 1$ can be found in the full version of the paper [17]. We conclude the following result.

Corollary 1. *For every $\delta > 0$, there is a δ-share bandwidth allocation game which does not yield a pure Nash equilibrium.*

4 Approximate Pure Nash Equilibria

The previous section has shown that we cannot expect any δ-share BAG to have a pure Nash equilibrium. Therefore, we turn our attention to α-approximate pure Nash equilibria. If α is chosen large enough, any strategy profile becomes an α-NE, whereas we know that there may not be an α-NE for $\alpha = 1$. Hence, there has to be a threshold α_δ for a guaranteed existence of these equilibria in dependency of δ. In this section, we give both upper and lower bounds on α_δ. We start with the upper bound α_δ^u, which we define as follows.

Definition 2. *Let $\delta > 0$. We define the upper bound α_δ^u on α_δ as*

$$\alpha_\delta^u := w \cdot \frac{\ln(w) - w + \delta + 1}{\delta} \quad \text{with } w = \left(-\frac{1}{2}W_{-1}\left(-2e^{(-\delta)-2}\right)\right).$$

Here, W_{-1} is the lower branch of the Lambert W function. Table 1 shows a selection of values of α_δ^u.

Theorem 1. *Let $\delta > 0$ and $\mathcal{B}$ be a δ-share bandwidth allocation game. For $\alpha \geq \alpha_\delta^u$, $\mathcal{B}$ reaches an α-approximate pure Nash equilibrium after a finite number of α-moves.*

Proof. For this proof, we use the potential function ϕ introduced in Sect. 2. We also need some additional concepts. For a resource r, let $\phi_r(\mathsf{s}_{-i})$ be the potential of r omitting the demand of player i. Now, $\phi_{i,r}(\mathsf{s}) := \phi_r(\mathsf{s}) - \phi_r(\mathsf{s}_{-i})$ is the part of r's potential due to strategy s_i if s_i is the last strategy to be considered when evaluating ϕ_r (cf. Fig. 1). Note that we always have $u_{i,r}(\mathsf{s}) \leq \phi_{i,r}(\mathsf{s})$. We are going to show that any strategy change of a player i improving her personal utility by a factor of more than α also results in an increase of ϕ if α is chosen accordingly. This implies that the game does not possess any cycles and thus always reaches an α-NE after finitely many steps (finite improvement property), as the total number of strategy profiles is finite.

For now, let $\alpha \geq \max_{i,r}\left(\frac{\phi_{i,r}(\mathsf{s})}{u_{i,r}(\mathsf{s})}\right)$ which trivially implies $\phi_{i,r}(\mathsf{s}) \leq \alpha u_{i,r}(\mathsf{s})$ $\forall\, i,r$. Assume that under the strategy profile s, player i changes her strategy from s_i to s_i', increasing her overall utility by a factor of more than α in the process. We denote the resulting strategy profile by s'. It follows

$$\Delta\phi = \phi(\mathsf{s}') - \phi(\mathsf{s}) = \sum_{r\in\mathcal{R}} \phi_{i,r}(\mathsf{s}') - \sum_{r\in\mathcal{R}} \phi_{i,r}(\mathsf{s}) \geq \sum_{r\in\mathcal{R}} u_{i,r}(\mathsf{s}') - \alpha \cdot \sum_{r\in\mathcal{R}} u_{i,r}(\mathsf{s})$$
$$= u_i(\mathsf{s}') - \alpha \cdot u_i(\mathsf{s}) > \alpha \cdot u_i(\mathsf{s}) - \alpha \cdot u_i(\mathsf{s}) = 0$$

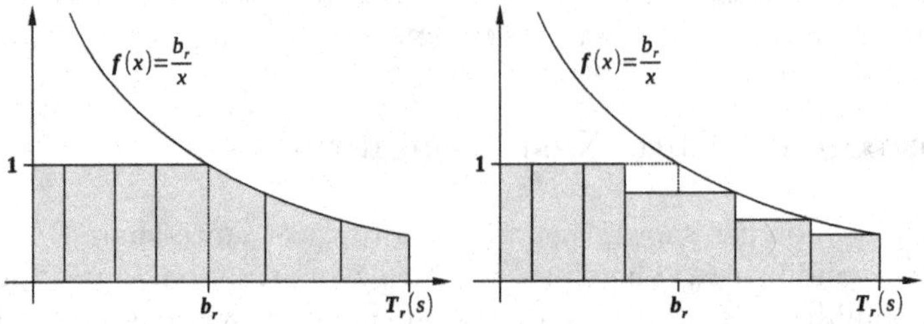

Fig. 1. The left side shows the potential of resource r, divided over the players. Each block represents a player currently using r. The order of the players does not affect the potential as a whole, but the amount caused by each individual player. The right side shows how much utility each player initially receives if they arrive at r according to their order.Therefore, the utility of the last player in the graph is her actual utility from r under the strategy profile s.

Therefore, the potential ϕ of $\mathcal{B}$ indeed grows with every α-move. It remains to be shown that $\alpha_\delta^u \geq \max_{i,r}\left(\frac{\phi_{i,r}(\mathsf{s})}{u_{i,r}(\mathsf{s})}\right)$. For a resource r, define $T_{-i,r}(\mathsf{s}) := T_r(\mathsf{s}) - s_i(r)$ as the total demand on r excluding player i. When the situation is clear from the context, we also write t_{-i} instead of $T_{-i,r}(\mathsf{s})$. We make a case distinction based on the size of t_{-i} and look at the two cases $t_{-i} < b_r$ and $t_{-i} \geq b_r$. We start with the first one. Note that we can assume $t_{-i} + s_i > b_r$, because otherwise the ratio between potential and utility of i at r would be 1. The ratio looks as follows:

$$\frac{\phi_{i,r}(\mathsf{s})}{u_{i,r}(\mathsf{s})} = \frac{b_r - t_{-i} + \int_{b_r}^{t_{-i}+s_i} \frac{b_r}{x}\,\mathrm{d}x}{\frac{b_r \cdot s_i}{t_{-i}+s_i}} = (t_{-i} + s_i) \cdot \frac{b_r - t_{-i} + b_r \cdot \ln(\frac{t_{-i}+s_i}{b_r})}{b_r \cdot s_i}$$

One can show that this ratio is at most α_δ^u as defined in Definition 2. For the second case, $t_{-i} \geq b_r$, this ratio only becomes smaller. We refer to the extended proof in the full version of the paper [17] for details. □

Now that we have an upper bound on α_δ, we give a lower bound α_δ^l, as well.

Definition 3. *Let $\delta > 0$. We define the lower bound α_δ^l on α_δ as*

$$\alpha_\delta^l := \frac{2\sqrt{\delta^2(\delta+2)} + \delta - 1}{4\delta - 1}$$

Again, we list some values for α_δ^l in Table 1.

Theorem 2. *Let $\delta > 0$ and $\alpha < \alpha_\delta^l$. There is a δ-share bandwidth allocation game without an α-approximate pure Nash equilibrium.*

Proof. We refer to the δ-share BAG from Definition 1. If we fix δ, the ratio between u and u' becomes a function f in γ.

$$f(\gamma) := \frac{\delta + \frac{\gamma}{\delta+\gamma+n\cdot\sigma}}{\gamma + \frac{\delta}{\delta+\gamma+n\cdot\sigma}} = \frac{\gamma + \delta(\delta+\gamma+n\cdot\sigma)}{\delta + \gamma(\delta+\gamma+n\cdot\sigma)} = \frac{\gamma+\delta(\gamma+1)}{\delta+\gamma(\gamma+1)}$$

Deriving f with respect to γ yields

$$f'(\gamma) = \frac{\delta^2 - \delta\gamma(\gamma+2) - \gamma^2}{(\delta+\gamma^2+\gamma)^2} = 0 \text{ for } \gamma_0 = \frac{\sqrt{\delta^3+2\delta^2}-\delta}{\delta+1}$$

One can check that this is indeed the only local maximum of f for $\gamma > 0$.

$$f(\gamma_0) = \frac{2\sqrt{\delta^2(\delta+2)}+\delta-1}{4\delta-1} = \alpha_\delta^l$$ □

The smaller δ is chosen, the better our result, i.e. the gap between α_δ^u and α_δ^l becomes smaller and α_δ^u decreases. A value of $\delta = 1$ is a realistic assumption as it states that the demand on a resource may not exceed its capacity, but it also means that one player is able to fully occupy any resource. However, if we think back to our motivation, it usually takes several thousand clients to exhaust the capacity of a provider. In this context, α_δ^u-approximate Nash equilibria are close to the definition of (regular) Nash equilibria. A visualisation of α_δ^u depending on $\delta \in]0,1]$ can be found in the full version of the paper [17]. Theorem 2 states that for α below α_δ^l, an α-NE cannot be guaranteed in general. The following result shows that below this lower bound, it is computationally hard to both check for a given δ-share BAG whether it has such an equilibrium and to compute it. The corresponding proof can be found in the full version of the paper [17].

Theorem 3. *Let $\delta > 0$ and $\alpha < \alpha_\delta^l$. Computing an α-approximate Nash equilibrium for any δ-share bandwidth allocation game is* NP*-hard.*

The proof also shows that the decision version of this problem is NP-complete. However, for $\alpha \geq \alpha_\delta^u$ and if the utilities $u_i^{\text{opt}} := \max_{s_i \in \mathcal{S}_i} \sum_{r \in \mathcal{R}} \min(s_i(r), b_r)$ of the most-profitable strategies of the players do not differ too much from each other, approximate Nash equilibria can be computed efficiently. For example, symmetric games always have this property. We do not impose any restriction on how much the demands of a single player may deviate from another between

Table 1. Upper and lower bounds for α_δ derived from δ.

δ	0.1	0.2	0.3	0.4	0.5	0.6	0.7	0.8	0.9	1
α_δ^u	1.0485	1.0946	1.1388	1.1816	1.2232	1.2637	1.3033	1.3422	1.3804	1.4181
α_δ^l	1.0170	1.0335	1.0497	1.0656	1.0811	1.0964	1.1114	1.1261	1.1405	1.1547

her different strategies. However, we can assume that u_i^{opt} and the potential utility of any other strategy differ by a factor of at most $n\delta$. Otherwise, that strategy would never be chosen.

Lemma 1. *Let $\mathcal{B}$ be a δ-share BAG. Then $u_i(\mathsf{s}) \geq \frac{u_i^{opt}}{(n\delta)^2}$ for all players $i \in \mathcal{N}$ and strategy profile s.*

Proof. Let s_i^{opt} be the strategy of i associated with u_i^{opt}. First, we show that $u_{i,r}(\mathsf{s}_{-i}, s_i^{\mathrm{opt}}) \geq \frac{s_i^{\mathrm{opt}}(r)}{n\delta}$ for all s_{-i} and r. If $T_r(\mathsf{s}_{-i}, s_i^{\mathrm{opt}}) \leq b_r$, then the claim holds, as $u_{i,r}(\mathsf{s}_{-i}, s_i^{\mathrm{opt}}) = s_i^{\mathrm{opt}}(r)$. For $T_r(\mathsf{s}_{-i}, s_i^{\mathrm{opt}}) > b_r$, $u_{i,r}(\mathsf{s}_{-i}, s_i^{\mathrm{opt}}) = \frac{s_i^{\mathrm{opt}}(r)\cdot b_r}{s_i^{\mathrm{opt}}+T_r(\mathsf{s}_{-i})} \geq \frac{s_i^{\mathrm{opt}}(r)\cdot b_r}{n\delta b_r} = \frac{s_i^{\mathrm{opt}}(r)}{n\delta}$. By summing up over all resources, we obtain $u_i(\mathsf{s}_{-i}, s_i^{\mathrm{opt}}) = \sum_{r\in\mathcal{R}} \frac{s_i^{\mathrm{opt}}(r)}{n\delta} \geq \frac{u_i^{\mathrm{opt}}}{n\delta}$. So we can assume wlog that for all strategies $s_i \in \mathcal{S}_i$, $\sum_{r\in\mathcal{R}} \min(s_i(r), b_r) \geq \frac{u_i^{\mathrm{opt}}}{n\delta}$. Otherwise, the strategy s_i^{opt} would yield a higher utility in all situations. By the same arguments made above, this implies $u_i(\mathsf{s}_{-i}, s_i) \geq \frac{u_i^{\mathrm{opt}}}{(n\delta)^2}$. □

We further need an additional lemma to bound the potential of a BAG in respect to its social welfare.

Lemma 2. *For any δ-share BAG and any strategy profile s, $(1+\log(n\delta))\cdot u(\mathsf{s}) \geq \phi(\mathsf{s})$.*

Theorem 4. *Let $\mathcal{B}$ be a δ-share BAG for $\delta \leq 1$, $\varepsilon > 0$ and $\lambda \in \,]0,1]$ such that for all players i, j $u_i^{opt} \geq \lambda u_j^{opt}$. Then $\mathcal{B}$ reaches an $(\alpha_\delta^u + \varepsilon)$-approximate NE in $\mathcal{O}\left(\log(n) \cdot n^5 \cdot (\varepsilon\lambda)^{-1}\right)$ $(\alpha_\delta^u + \varepsilon)$-moves.*

Proof. Let i be the player performing an $(\alpha_\delta^u+\varepsilon)$-move under the strategy profile s, leading to the strategy profile s'. We can bound the increase in the potential:

$$\Phi(\mathsf{s}') - \Phi(\mathsf{s}) \geq \varepsilon u_i(\mathsf{s}) \overset{(1)}{\geq} \frac{\varepsilon}{(n\delta)^2} u_i^{\mathrm{opt}} \overset{(2)}{\geq} \frac{\varepsilon\lambda}{n(n\delta)^2} u(\mathsf{s}) \overset{(3)}{\geq} \frac{\varepsilon \cdot \lambda}{n(1+\log(n))(n\delta)^2} \Phi(\mathsf{s})$$

Inequalities (1) and (3) follow by Lemmas 1 and 2 respectively while (2) holds due to $u(\mathsf{s}) = \sum_{j\in\mathcal{N}} u_j(\mathsf{s}) \leq \sum_{j\in\mathcal{N}} u_j^{\mathrm{opt}} \leq \frac{n}{\lambda} u_i^{\mathrm{opt}}$. For convenience, we define $\beta := \frac{\varepsilon\cdot\lambda}{n(1+\log(n))(n\delta)^2}$. Assume that we need t steps to increase the potential from $\Phi(\mathsf{s})$ to $2\Phi(\mathsf{s})$. Then $\Phi(\mathsf{s}) = 2\Phi(\mathsf{s}) - \Phi(\mathsf{s}) \geq \beta \cdot t \cdot \Phi(\mathsf{s}) \Leftrightarrow t \leq \beta^{-1}$. So in order to double the current potential of $\mathcal{B}$, we need at most β^{-1} improving moves. Therefore, the game has to reach a corresponding equilibrium after at most $\log\left(\frac{\Phi_{\max}}{\Phi_{\min}}\right) \cdot \beta^{-1}$ improving moves, with $\Phi_{\max}$ and $\Phi_{\min}$ denoting the maximum and minimum potential of $\mathcal{B}$, respectively. Since $\Phi_{\max} \leq \sum_{i\in\mathcal{N}} u_i^{\mathrm{opt}}$ due to $\delta \leq 1$ and $\Phi_{\min} \geq \sum_{i\in\mathcal{N}} \frac{u_i^{\mathrm{opt}}}{(n\delta)^2}$, we can bound $\log\left(\frac{\Phi_{\max}}{\Phi_{\min}}\right) \leq (n\delta)^2$. □

To conclude this section, we turn towards the quality of α-approximate Nash equilibria. Although no player has an incentive to change her strategy, the social welfare, which is the total utility of all players combined, may not be optimal. To express how well Nash equilibria perform in comparison to a globally determined optimal solution, the price of anarchy has been introduced.

Theorem 5. *The α-approximate price of anarchy of any δ-share bandwidth allocation game is at most $\alpha + 1$. For every $\varepsilon > 0$, there is a δ-share bandwidth allocation game with an α-approximate Price of Anarchy of $\alpha + 1 - \varepsilon$.*

Due to space constraints, the proof can be found in the full version of the paper [17]. Note that our result even holds for $\alpha < \alpha_\delta^l$, provided the BAG has an α-NE. This result matches our previous work [18], where we have shown that the price of anarchy for pure Nash equilibria ($\alpha = 1$) is 2.

5 Approximating the Optimal Social Welfare

In this final section, we look at how fast certain utility-maximization games converge towards socially *good* states, i.e. strategy profiles with a social welfare close to $u(opt)$ if the players keep performing α-moves. We then apply this result to bandwidth allocation games. For this, we use the concept of nice games introduced in [4]. A utility-maximization game is (λ, μ)*-nice* if for every strategy profile s, there is a strategy profile s' with $\sum_{i \in \mathcal{N}} u_i(\mathsf{s}_{-i}, s'_i) \geq \lambda \cdot u(opt) - \mu \cdot u(\mathsf{s})$ for constants λ, μ. The proof of the following theorem is also derived from [4].

Theorem 6. *Let $\mathcal{B}$ be a utility-maximization game with a potential function $\phi(\mathsf{s})$ such that for some $A, B, C \geq 1$, we have that $A \cdot \phi(\mathsf{s}) \geq u(\mathsf{s}) \geq \frac{1}{B} \cdot \phi(\mathsf{s})$, $\phi(\mathsf{s}_{-i}, s_i^b) - \phi(\mathsf{s}) \geq u_i(\mathsf{s}_{-i}, s_i^b) - C \cdot u_i(\mathsf{s})$ and which is (λ, μ)-nice. Let $\rho = \frac{\lambda}{C+\mu}$. Then, for any $\varepsilon > 0$ and any initial strategy profile s^0, the best-response dynamic reaches a state s^t with $u(\mathsf{s}^t) \geq \frac{\rho(1-\varepsilon)}{AB} u(opt)$ in at most $\mathcal{O}\left(\frac{n}{A(C+\mu)} \log \frac{1}{\varepsilon}\right)$ steps. All future states reached via best-response dynamics will satisfy this approximation factor as well.*

When adapting this result for bandwidth allocation games, note that the players have to perform α-moves when following the best-response dynamic. Otherwise, we cannot guarantee that ϕ is strictly monotone.

Corollary 2. *Let $\mathcal{B}$ be a δ-share BAG and $\alpha \geq \alpha_\delta^u$. For any $\varepsilon > 0$ and any initial strategy profile s^0, the best-response dynamic using only α-moves reaches a state s^t with $u(\mathsf{s}^t) \geq \frac{1-\varepsilon}{(\alpha^2+1)(\ln(n\delta)+1)} u(opt)$ in at most $\mathcal{O}\left(\frac{n}{\alpha+\alpha^{-1}} \log \frac{1}{\varepsilon}\right)$ steps. All future states reached via best-response dynamics will satisfy this approximation factor as well.*

Proof. First we show that any δ-share BAG is $(\alpha^{-1}, \alpha^{-1})$-nice. Let s be an arbitrary strategy profile. We show that $\sum_{i \in \mathcal{N}} u_i(\mathsf{s}_{-i}, s_i^b) \geq \alpha^{-1} \cdot u(opt) - \alpha^{-1} \cdot u(\mathsf{s})$. Note that $u_i(\mathsf{s}_{-i}, s_i^b) \geq u_i(\mathsf{s}_{-i}, opt_i)$ by definition of s_i^b. This implies $u_i(\mathsf{s}_{-i}, s_i^b) \geq \alpha^{-1} \cdot u_i(\mathsf{s}_{-i}, opt_i)$ and we can therefore copy the proof of Theorem 5 to show that $\sum_{i \in \mathcal{N}} u_i(\mathsf{s}_{-i}, s_i^b) \geq \alpha^{-1} \cdot u(opt) - \alpha^{-1} \cdot u(\mathsf{s})$.

We now use our potential function $\phi(\mathsf{s})$, for which we already know that $\phi(\mathsf{s}) \geq u(\mathsf{s}) \geq \frac{1}{1+\ln(n\delta)} \phi(\mathsf{s})$ (see Lemma 2) and $\phi(\mathsf{s}_{-i}, s_i^b) - \phi(\mathsf{s}) \geq u_i(\mathsf{s}_{-i}, s_i^b) - \alpha \cdot u_i(\mathsf{s})$. So we obtain $A = 1$, $B = 1 + \ln(n\delta)$ and $C = \alpha$. Using these values together with Theorem 6 directly leads to our result. □

Computing *opt* is NP-hard and prior to this result, an approximation algorithm was only known for games in which the strategy spaces consist of the bases of a matroid over the resources [18]. Following the best-response dynamic, we can now approximate the optimal solution for arbitrary strategy space structures. While reaching an actual α-approximate NE by this method may take exponentially long, we obtain an $\mathcal{O}(\alpha^2 \log(n))$-approximation of the worst-case equilibrium after a linear number of strategy changes.

References

1. Ackermann, H., Röglin, H., Vöcking, B.: Pure Nash equilibria in player-specific and weighted congestion games. Theoret. Comput. Sci. **410**(17), 1552–1563 (2009)
2. Ackermann, H., Skopalik, A.: Complexity of pure Nash equilibria in player-specific network congestion games. Internet Math. **5**(4), 323–342 (2008)
3. Aland, S., Dumrauf, D., Gairing, M., Monien, B., Schoppmann, F.: Exact price of anarchy for polynomial congestion games. SIAM J. Comput. **40**(5), 1211–1233 (2011)
4. Anshelevich, E., Postl, J., Wexler, T.: Assignment games with conflicts: price of total anarchy and convergence results via semi-smoothness. In: CoRR abs/1304.5149 (2013)
5. Augustine, J., Chen, N., Elkind, E., Fanelli, A., Gravin, N., Shiryaev, D.: Dynamics of profit-sharing games. In: Proceedings of IJCAI, pp. 37–42. IJCAI/AAAI (2011)
6. Awerbuch, B., Azar, Y., Epstein, A.: The price of routing unsplittable flow. SIAM J. Comput. **42**(1), 160–177 (2013)
7. Awerbuch, B., Azar, Y., Epstein, A., Mirrokni, V.S., Skopalik, A.: Fast convergence to nearly optimal solutions in potential games. In: Proceedings of EC, pp. 264–273. ACM (2008)
8. Bhawalkar, K., Gairing, M., Roughgarden, T.: Weighted congestion games: the price of anarchy, universal worst-case examples, and tightness. ACM TEAC **2**(4), 14 (2014)
9. Brinkmann, A., Kling, P., Meyer auf der Heide, F., Nagel, L., Riechers, S., Süß, T.: Scheduling shared continuous resources on many-cores. In: Proceedings of 26th ACM SPAA, pp. 128–137. ACM, New York (2014)
10. Caragiannis, I., Fanelli, A., Gravin, N., Skopalik, A.: Efficient computation of approximate pure Nash equilibria in congestion games. In: 52nd FOCS, pp. 532–541. IEEE Computer Society (2011)
11. Caragiannis, I., Fanelli, A., Gravin, N., Skopalik, A.: Approximate pure Nash equilibria in weighted congestion games: existence, efficient computation, and structure. In: Proceedings of 13th EC, pp. 284–301. ACM (2012)
12. Chen, H., Roughgarden, T.: Network Design with Weighted Players. Theory Comput. Syst. **45**(2), 302–324 (2009)
13. Chien, S., Sinclair, A.: Convergence to approximate Nash equilibria in congestion games. Games Econ. Behav. **71**(2), 315–327 (2011)
14. Christodoulou, G., Koutsoupias, E.: The price of anarchy of finite congestion games. In: Proceedings of 37th STOC, pp. 67–73. ACM (2005)
15. Christodoulou, G., Koutsoupias, E., Spirakis, P.G.: On the performance of approximate equilibria in congestion games. Algorithmica **61**(1), 116–140 (2011)

16. Coffman, K., Odlyzko, A.: Internet growth: is there a moores law for data traffic? In: Panos, J.A., Mauricio, M.P., Resende, G.C. (eds.) Handbook of Massive Data Sets, Massive Computing, vol. 4, pp. 47–93. Springer, New York (2002)
17. Drees, M., Feldotto, M., Riechers, S., Skopalik, A.: On existence and properties of approximate pure nash equilibria in bandwidth allocation games. In: CoRR abs/1507.02908 (2015)
18. Drees, M., Riechers, S., Skopalik, A.: Budget-restricted utility games with ordered strategic decisions. In: Lavi, R. (ed.) SAGT 2014. LNCS, vol. 8768, pp. 110–121. Springer, Heidelberg (2014)
19. Dunkel, J., Schulz, A.S.: On the complexity of pure-strategy nash equilibria in congestion and local-effect games. Math. Oper. Res. **33**(4), 851–868 (2008)
20. Fabrikant, A., Papadimitriou, C.H., Talwar, K.: The complexity of pure Nash equilibria. In: Proceedings of 36th STOC, pp. 604–612. ACM (2004)
21. Feldotto, M., Gairing, M., Skopalik, A.: Bounding the potential function in congestion games and approximate pure Nash equilibria. In: Liu, T.-Y., Qi, Q., Ye, Y. (eds.) WINE 2014. LNCS, vol. 8877, pp. 30–43. Springer, Heidelberg (2014)
22. Goemans, M.X., Li, L., Mirrokni, V.S., Thottan, M.: Market sharing games applied to content distribution in Ad-Hoc networks. IEEE J. Sel. Areas Commun. **24**(5), 1020–1033 (2006)
23. Hansknecht, C., Klimm, M., Skopalik, A.: Approximate pure Nash equilibria in weighted congestion games. In: APPROX/RANDOM, pp. 242–257 (2014)
24. Harks, T., Klimm, M.: On the existence of pure nash equilibria in weighted congestion games. In: Abramsky, S., Gavoille, C., Kirchner, C., Meyer auf der Heide, F., Spirakis, P.G. (eds.) ICALP 2010. LNCS, vol. 6198, pp. 79–89. Springer, Heidelberg (2010)
25. Harks, T., Klimm, M.: Congestion games with variable demands. In: Proceedings of 13th TARK, pp. 111–120. ACM (2011)
26. Lucier, B., Leme, R.P.: GSP auctions with correlated types. In: Proceedings of 12th EC, pp. 71–80. ACM (2011)
27. Mavronicolas, M., Milchtaich, I., Monien, B., Tiemann, K.: Congestion games with player-specific constants. In: Kučera, L., Kučera, A. (eds.) MFCS 2007. LNCS, vol. 4708, pp. 633–644. Springer, Heidelberg (2007)
28. Milchtaich, I.: Congestion games with player-specific payoff functions. Games Econ. Behav. **13**(1), 111–124 (1996)
29. Monderer, D., Shapley, L.S.: Potential games. Games Econ. Behav. **14**(1), 124–143 (1996)
30. Rosenthal, R.W.: A class of games possessing pure-strategy Nash equilibria. Int. J. Game Theory **2**(1), 65–67 (1973)
31. Roughgarden, T.: Intrinsic robustness of the price of anarchy. In: Proceedings of 41st STOC, pp. 513–522. ACM (2009)
32. Skopalik, A., Vöcking, B.: Inapproximability of pure nash equilibria. In: Proceedings of 40th STOC, pp. 355–364. ACM (2008)

Can Bandwidth Sharing Be Truthful?

Yukun Cheng[1], Xiaotie Deng[2], Yifan Pi[3], and Xiang Yan[2](✉)

[1] School of Mathematics and Statistics, Zhejiang University of Finance and Economics, Hangzhou, China
ykcheng@amss.ac.cn

[2] Department of Computer Science, Shanghai Jiaotong University, Shanghai, China
deng-xt@cs.sjtu.edu.cn, xyansjtu@163.com

[3] IIIS, Tsinghua University, Beijing, China
piyifan@gmail.com

Abstract. Peer to peer (P2P) and Crowd Sourcing systems have built their success on resource sharing protocols such as BitTorrent and Open Garden. While previous studies addressed the issue of fairness, we discuss prevention of manipulative actions a player may lie to take the advantage of the protocols. We prove that, under a proportional response protocol, a player deviating from the protocol by reporting false broken links will not make any gain in its utility eventually. This result establishes the strategic stability of a popular resource sharing P2P system in the Internet.

Keywords: Bandwidth sharing · Peer to peer system · Strategic behavior

1 Introduction

We consider a decentralized and distributed network architecture where vertices (called "peers") act as both suppliers and consumers of resources. Commonly referred to as peer to peer (P2P) networks, or crowd sourcing systems, participants use plum-for-peach (or tit-for-tat) mechanisms to make a portion of their resources directly available to other network participants [7]. While peers gain convenience and benefit created with the system, fairness or manipulativeness become possible issues of the protocol design.

Wu and Zhang [8], motivated by P2P systems such as BitTorrent, have pioneered a model of proportional response (PR) for the consideration of fairness among the participating agents. Under the PR protocol, each agent responds to bandwidth provided by its neighbors by allocating its bandwidth in proportion to its received shares. In recent years, in the context of crowd-sourcing mobile net such as Opengarden and Karma and so on, the issue of incentive schemes for resource sharing protocols has recently been put forth as a key challenge in the

This research was partially supported by the National Nature Science Foundation of China (No. 11301475, 61173011).

M. Hoefer (Ed.): SAGT 2015, LNCS 9347, pp. 190–202, 2015.
DOI: 10.1007/978-3-662-48433-3_15

field [4–6]. Further, Godfrey, et al., considered a problem of congestion control in TCP system queueing [3] and proved that efficiency and truthfulness cannot hold at the same time.

In this work, we take up the challenge with respect to the PR protocol. A natural strategic move to misreport utility is to remove an edge or more by an agent, faked as happened naturally where a connection is broken, or shut down. We confirm strategy stability of the PR protocol under such manipulative moves by any agent: No agent can increase its bandwidth obtained from its neighbors by unilaterally cutting off some of its links under the PR protocol.

To the best of our knowledge, this is the first concrete result for bandwidth sharing protocols to be proven truthful under a natural strategic behavior.

Our work is based on an elegant bottleneck decomposition structure, and the equivalence of fairness (with the proportional response protocol) and competitiveness (market equilibrium) originally proven in [8]. By considering the alternation on the structure caused by an agent's manipulative act, the relationship of the original one and the altered one are shown to have interesting properties. Then participants' incentives are taken as the next line of analysis. Together, we derive a set of localized properties where an agent may deviate from the standard PR protocol by cutting of some edges it connects to the network.

Despite of the equivalence discussed above, those results under the limited manipulative behavior space are in a sharp contrast to the negative results for the linear market equilibrium where there are ample possibilities for an agent to cheat [1,2]. One may observe that the P2P setting provides a model with a unique cheating behavior. By cutting of an edge, a cheater will change its own utility function as well as its neighbors on the other ends of its deleted edges. That is, the deviation behavior in the PR protocol is no longer equivalent to that in the market equilibrium model.

In the next section, we introduce some useful notations and some known results pertinent to the subsequent studies. In Sect. 3, we first present some structure properties of bottleneck decomposition which are crucial for our study on the strategy stability of PR protocol. Next, we discuss our main result and conclude to make remarks on future directions in Sect. 4.

2 Definitions and Terminologies

In this section, we present the basic model of [8] and discuss its basic properties.

P2P Bandwidth Sharing Problem: Consider a network, represented by an undirected graph $G = (V, E; w)$, where $w_v : V \rightarrow R_+$ is the upload bandwidth capacity of player v to be shared with its neighbors. Each edge represents a communication link. Let $\Gamma(v)$ denote the set of vertices adjacent to v in graph G and x_{vu} denote the fraction of bandwidth v allocated for use by its neighbor u. Therefore, the upload bandwidth a vertex v provides to a neighboring vertex u is $w_v \cdot x_{vu}$. $X = (x_{vu})_{(u,v)\in E}$ is called an *allocation*. We use vertex agents to refer to the players.

Specially, for any subset $S \subseteq V$ we define $w(S) = \sum_{u \in S} w_u$, and let $\Gamma(S) = \cup_{v \in S} \Gamma(v)$ where $\Gamma(v)$ is the neighborhood of v in G. It is obvious that $S \cap \Gamma(S) \neq \emptyset$ if S is not independent. Define $\alpha(S) = w(\Gamma(S))/w(S)$, referred to as the inclusive expansion ratio of S, or the α-ratio of S for short.

Proportional Response Protocol: For each vertex u, the allocation $(x_{uv} : v \in \Gamma(u))$ of its bandwidth w_u is proportional to what it receives from its neighbors $(w_v \cdot x_{vu} : \Gamma(u))$.

Alternatively, we may consider an economy model where each player v sells its own bandwidth to its neighbors, and at the same time each player is also a buyer of bandwidth from its neighbors for its own communication needs. We assume that each player's utility is the sum of bandwidth it acquires from its neighbors, i.e., the utility of v for an allocation X is $U_v(X) = \sum_{u \in \Gamma(v)} x_{uv} w_u$. One supplies its bandwidth only for the communication needs of its neighbors.

Market Equilibrium: Let p_v be the price for bandwidth of player $v \in V$. The price vector $p = (p_v)_{v \in V}$, together with the allocation $X = (x_{vu})$ is called a *market equilibrium* if for any agent $v \in V$ the following holds.

- Market clearance: $(\sum_{u \in \Gamma(v)} x_{vu} - 1) \cdot p_v = 0$. All bandwidth of v are allocated or priced null. The latter may occur in the case of an isolated vertex in the P2P bandwidth sharing problem.
- Budget constraint: $\sum_{u \in \Gamma(v)} x_{uv} p_u \leq p_v$. The money player v should pay out must not exceed its budget
- Individual optimality: The solution $X = (x_{vu})$ maximizes utility $\sum_{u \in \Gamma(v)} x_{uv} w_u$, subject to $\sum_{u \in \Gamma(v)} x_{uv} p_u \leq p_v$ and $x_{uv} \geq 0$ for each vertex v. That is, each player is optimally happy for its allocation at the current price.

In [8], an elegant bottleneck decomposition structure from which we can obtain a market equilibrium is proposed.

Maximal Bottleneck: A vertex subset $B \subseteq V$ is called a *bottleneck* of G if $\alpha(B) = \min_{S \subseteq V} \alpha(S)$. B is a *maximal bottleneck* if $\forall \widetilde{B}$, $B \subset \widetilde{B} \subseteq V$ implies $\alpha(\widetilde{B}) > \alpha(B)$. We also call $(B, \Gamma(B))$ the maximal bottleneck pair.

Bottleneck Decomposition: Start with $V_1 = V$, $G_1 = G$ and $i = 1$. Find the maximal bottleneck B_i of G_i and let G_{i+1} be the induced subgraph on the vertex set $V_{i+1} = V_i - (B_i \cup C_i)$, where $C_i = \Gamma(B_i) \cap V_i$, the neighbor set of B_i in the subgraph G_i. Repeat if $G_{i+1} \neq \emptyset$ and set $k = i$ if $G_{i+1} = \emptyset$. Then we call $\mathcal{B} = \{(B_1, C_1), \cdots, (B_k, C_k)\}$ the *bottleneck decomposition* of G, α_i the i-th *α-ratio* and $(\alpha_i = \frac{w(C_i)}{w(B_i)} : i = 1, 2, \cdots, k)$ the *α-ratio vector.*

The following properties of the bottleneck decomposition will be useful in our discussion.

Proposition 1 *[8]. Given graph G, the bottleneck decomposition is unique and*

1. $0 \leq \alpha_1 < \alpha_2 < \cdots < \alpha_k \leq 1$;
2. *if there is a pair (B, C) in G_i with $w(C)/w(B) = \alpha_i$, then $B \subseteq B_i$ and $C \subseteq C_i$;*

3. if $\alpha_i = 1$*, then* $i = k$ *and* $B_i = C_i$*; otherwise* B_i *is independent and* $B_i \cap C_i = \emptyset$*;*
4. $\cup_{i:\alpha_i<1} B_i$ *is an independent set.*

Proposition 2 *[8]. For the bottleneck decomposition* $\mathcal{B} = \{(B_1, C_1), \cdots, (B_k, C_k)\}$,

1. *there is no edge between* B_i *and* B_j*,* $i \neq j = 1, 2, \cdots, k$*;*
2. *there is no edge between* B_i *and* C_j*, where* $j > i$*.*

With those properties, a connection was established in [8] from the bottleneck decomposition to the market equilibrium. First a price vector is established as: for any $u \in B_i$, let $p_u = \alpha_i w_u$; for any $u \in C_i$, let $p_u = w_u$; and if $\alpha_k = 1$, then for any $u \in B_k = C_k$ let $p_u = w_u$. The corresponding allocations of bandwidth of the vertices are determined as follows in three cases.

- $\alpha_i < 1$: consider the bipartite graph $\widehat{G}_i = (B_i, C_i; E_i)$ where $E_i = (B_i \times C_i) \cap E$. Let $\widehat{x}_{uv}$ be the amount of bandwidth that vertex $u \in B_i$ uploads to $v \in C_i$ along edge $(u, v) \in E_i$. By the max-flow min-cut theorem, there exist $\widehat{x}_{uv} \geq 0$ for $u \in B_i$ and $v \in C_i$ such that $\sum_{v \in \Gamma(u) \cap C_i} \widehat{x}_{uv} = w_u$ and $\sum_{u \in \Gamma(v) \cap B_i} \widehat{x}_{uv} = w_v / \alpha_i$. Let $\widehat{x}_{vu} = \alpha_i \widehat{x}_{uv}$ which means that $\sum_{u \in \Gamma(v) \cap B_i} \widehat{x}_{vu} = w_v$.
- $B_k = C_k$ with $\alpha_k = 1$: Construct a bipartite graph $\widehat{G} = (B_k, B'_k; E'_k)$ such that B'_k is a copy of B_k. There is an edge $(u, v') \in E'_k$ iff $u, v \in B_k$ and $(u, v) \in E[B_k]$. Then, by Hall's theorem, for any edge $(u, v) \in E[B_k]$, there exist $\widehat{x}_{uv'}$ such that $\sum_{v' \in \Gamma(u) \cap B'_k} \widehat{x}_{uv'} = w_u$.
- For any other edge, $(u, v) \notin B_i \times C_i$, $i = 1, 2, \cdots, k$, define $\widehat{x}_{uv} = 0$

The allocation defined above assigns all the bandwidth of each vertex to its neighbors, called a feasible allocation. For this particular feasible allocation, all available bandwidth are uploaded along edges in $B_i \times C_i$, $i = 1, 2, \cdots, k$. In addition, if we define $x_{uv} = \widehat{x}_{uv} / w_u$ and $X = (x_{uv})$, then

Proposition 3 *[8].* (p, X) *is a market equilibrium.* $\forall u \in B_i, v \in C_i$*,* $i = 1, 2, \cdots, k$*, the utilities of* u *and* v *are* $U_u = w(u) \cdot \alpha_i$ *and* $U_v = w(v)/\alpha_i$*, respectively.*

Furthermore, Wu and Zhang [8] show that the bandwidth allocation under the market equilibrium derived from the bottleneck decomposition is a PR protocol. So in the subsequent discussion, we shall speak of the strategy stability under the PR protocol in terms of the market equilibrium considered above.

3 Incentive Compatibility in the Bandwidth Sharing

In Sect. 2, a relationship between the market equilibrium and the bottleneck decomposition is presented. Such a bandwidth allocation approach is derived from the market equilibrium as a resource allocation rule for P2P network bandwidth resources. A possibility arises that, under the distributed network protocol, a player may or may not follow it in its execution. Can players make strategic moves by misrepresenting their utility functions for gains in their true utilities?

Specific to the bandwidth allocation problem, we consider a way to lie that one player may do by removing edges adjacent to itself. Such a deceit way would change the bottleneck decomposition which leads to the alteration of the market equilibrium. And then the agent may obtain a different utility.

For simplicity of the presentation, we consider the case one edge is removed. We observe that this would not reduce the generality of the result as we can remove edges one by one at each of which steps the true utility of the cheater would be proven to be non-increasing.

To prepare for the discussion, we first present several useful structure properties important to our analysis in Subsect. 3.1, including the Basic lemma, the Key Lemma and the Main Lemma. Next, we will study the incentive properties of an agent who would like to cut an incident edge to increase its utility in the bottleneck decompositions of G and $G' = G - (u, v)$ in the original and the resulting network before or after an agent may change the network by deleting an edge of its own.

3.1 Structure Properties of Bottleneck Decomposition $\mathcal{B}$ and $\mathcal{B}'$

Definition 1 (Bottleneck Decomposition). *Let $\mathcal{B} = \{(B_1, C_1), \cdots, (B_k, C_k)\}$ be the bottleneck decomposition of the original graph G and let the α-ratio of (B_i, C_i) be $\alpha_i = w(C_i)/w(B_i)$, $i = 1, 2, \cdots, k$. For pair (B_i, C_i) with $\alpha_i < 1$, each vertex in B_i is called a B-class vertex, and each vertex in C_i is called a C-class vertex. For the special case $B_k = C_k$, i.e., $\alpha_k = 1$, all vertices in B_k are categorized as both B-class and C-class. Define $V_1 = V$, $V_{i+1} = V_i - (B_i \cup C_i)$ for $i = 1, 2, \cdots, k-1$ and G_i for the induced subgraph on V_i, $i = 1, 2, \cdots, k$.*

Similarly, for the new graph G' after removing one edge (u, v) from G, its bottleneck decomposition may be different from that of G. Let it be $\mathcal{B}' = \{(B'_1, C'_1), \cdots, (B'_{k'}, C'_{k'})\}$ with α-ratio α'_j, $j = 1, 2, \cdots, k'$. Similarly, $V'_1 = V$, $V'_{j+1} = V'_j - (B'_j \cup C'_j)$ for $j = 1, 2, \cdots, k' - 1$ and $G'_j = G'[V'_j]$, $j = 1, 2, \cdots, k'$. Call vertices in B'_j, $1 \le j \le k'$, B'-class, and those of C'_j's, C'-class.

Therefore, a vertex in V_k with $\alpha_k = 1$ (or $\alpha'_{k'} = 1$) could simultaneously be B-class and C-class (or B'-class and C'-class), in the case $B_k = C_k$ (or $B'_{k'} = C'_{k'}$).

By Proposition 1, B-class vertices, except those in B_k with $\alpha_k = 1$, form an independent subset of vertices. Edges are either between vertices from B_i to C_j ($i \ge j$), or between vertices from C_i to C_j. Thus at least one of u and v must be a C-class vertex in G. The following is an immediate result of the definition of the maximal bottleneck.

Lemma 1. *Given pair (B_i, C_i), if $B \subseteq V_i$ and $C = \Gamma(B) \cap V_i$, then $w(C)/w(B) \ge \alpha_i$. Specially, if $B \subseteq B_i$ and $C = \Gamma(B) \cap C_i$, then $w(C)/w(B) \ge \alpha_i$. Further, if let $B^c = B_i - B$ and $C^c = C_i - C$, then $w(C^c)/w(B^c) \le \alpha_i$.*

Definition 2 (Edge (u, v) and Indices i_u, i_v, j_u, j_v). *Let $(u, v) \in E$ be cut to obtain $G' = G - (u, v)$. If u and v are in different classes, w.l.o.g, we assume v*

is in C-class. Let vertex u (and v, respectively) appear in pair (B_l, C_l) *at step* $l = i_u$ *(and* $l = i_v$*, respectively) of the bottleneck decomposition of G. Specifically, if both u and v are both in C-class, let* $i_v \le i_u$*. Similarly, let vertex u (and v, respectively) appear in pair* (B'_l, C'_l) *at step* $l = j_u$ *(and* $l = j_v$*, respectively) of the bottleneck decomposition of G'. Define* $j_* = \min\{j_u, j_v\}$*. In addition, let* $\Gamma'(x)$ *be the neighborhood of x in G' and define* $\Gamma'(S) = \cup_{x \in S}\Gamma'(x)$ *for set S.*

Based on Definitions 1 and 2, the following results can be derived directly.

Lemma 2 (Basic Lemma). *For the bottleneck decompositions* $\mathcal{B}$ *and* $\mathcal{B}'$,

1. *for any* $1 \le t < j_*$, $(B'_t, C'_t) = (B_t, C_t)$.
2. *for any* $1 \le t \le j_*$, $V'_t = V_t$.
3. *if* $V'_t = V_t$*, then* $\alpha'_t \le \alpha_t$.
4. *for any* $1 \le t < j_*$, $B'_t \cap (\cup_{i=1}^k C_i) = \emptyset$.
5. *for any* $1 \le t < j_*$, $B_t \cap (\cup_{i=1}^k C'_i) = \emptyset$.
6. $j_* \le i_v \le i_u$.

Proof. We first show the correctness of Item 1. Suppose to the contrary that $j < j_*$ is the smallest index such that $(B'_j, C'_j) \ne (B_j, C_j)$. So $(B'_1, C'_1) = (B_1, C_1), \cdots, (B'_{j-1}, C'_{j-1}) = (B_{j-1}, C_{j-1})$ and $V'_1 = V_1, \cdots, V'_j = V_j$. Furthermore, the assumption that $j < j_* = \min\{j_u, j_v\}$ implies that u and v cannot be in B'_j and C'_j.

On the one hand, the fact that $u, v \notin B'_j \cup C'_j$ implies that $B'_j \subseteq V_j$ and $C'_j = \Gamma(B'_j) \cap V_j$. So (B'_j, C'_j) is a pair in G_j which guarantees $w(C'_j)/w(B'_j) \ge w(C_j)/w(B_j) = \alpha_j$ by Lemma 1. On the other hand since $V'_j = V_j$, $\Gamma'(B_j) \cap V'_j = C_j$ or $C_j - \{v\}$. The second case happens when $u \in B_j$ and any other vertex in B_j is not incident to v in G'_j. So (B_j, C_j) or $(B_j, C_j - \{v\})$ is a pair in G'_j and $w(C'_j)/w(B'_j) = \alpha'_j \le w(C_j)/w(B_j)$. Combining above two inequalities, we know $w(C'_j)/w(B'_j) = w(C_j)/w(B_j)$. So Proposition 1-(2) tells us that $B'_j \subseteq B_j$, $C'_j \subseteq C_j$ in G_j and $B_j \subseteq B'_j$, , $C_j \subseteq C'_j$ in G'_j. Thus $(B'_j, C'_j) = (B_j, C_j)$ which contradicts the assumption.

Item 2 to Item 5 follow from Item 1. For Item 6, if $i_v < j_*$, then $(B'_{i_v}, C'_{i_v}) = (B_{i_v}, C_{i_v})$ by Item 1. But the fact that v appears in (B_{i_v}, C_{i_v}) ensures that $j_* \le i_v$ which is a contradiction. Further, Definition 2 promises the correctness of the second inequality of Item 6. □

Next we introduce a technique, *dense kernel removal*, for subsequent discussions. The goal is to derive a contradiction to the minimality of the α-ratio of pair (B_i, C_i) in the bottleneck decomposition if we suppose the desired result does not hold.

Definition 3 (Dense kernel removal). *Given* $B \subseteq B_i$ *and* $C \subseteq C_i$ *for a pair* (B_i, C_i) *in the bottleneck decomposition, let* $C_i^c = C_i - C$ *and* $B_i^c = B_i - B$*. If* $\Gamma(B_i^c) \cap V_i \subseteq C_i^c$ *and* $w(C)/w(B) > \alpha_i$*, then removing the pair* (B, C) *from* (B_i, C_i) *would render a pair* (B_i^c, C_i^c) *such that* $w(C_i^c)/w(B_i^c) < \alpha_i$*, a contradiction to the minimum inclusive expansion ratio. We denote this technique by* $(B_i^c, C_i^c) = DKR(B, C; B_i, C_i)$ *for simplicity.*

Lemma 3. *If $\emptyset \neq C \subseteq C_i$ and $B = \Gamma(C) \cap B_i$, then $w(B)/w(C) \geq 1/\alpha_i$.*

The following are two important results.

Lemma 4 (Key Lemma). *Consider the bottleneck decompositions $\mathcal{B}$ and $\mathcal{B}'$. Each of the following conditions implies that for any $1 \leq t \leq k'$ with $\alpha'_t < 1$, $B'_t \cap (\cup_{i=1}^k C_i) = \emptyset$:*

Case 1. $(u, v) \in B_k \times C_k$ *with* $\alpha_k = 1$ *and* u, v *are both in* C'*-class;*
Case 2. $(u, v) \in B_i \times C_i$ *with* $\alpha_i < 1$, $i = 1, \cdots, k$, *and* v *is in* C'*-class;*
Case 3. $(u, v) \notin B_i \times C_i$, $i = 1, \cdots, k$.

Proof. Let t be the smallest index such that B'_t with $\alpha'_t < 1$ contains a C-class vertex x and suppose $x \in B'_t \cap C_l \subseteq V'_t$ for some $l \in \{1, 2, \cdots, k\}$. We derive a contradiction to the minimality of t in the choice of B'_t.

First we prove that $\Gamma(x) \cap B_l \subseteq C'_t$. The condition that v and u are in C'-class in Case 1 shows that $x \neq u, v$ because x is B'-class. For Case 2, condition that u is a B-class vertex and v is in C'-class guarantees that $x \neq u, v$. Therefore, $\Gamma(x) = \Gamma'(x)$ in Case 1 and Case 2. As $x \in C_l$, if $x \in \{u, v\}$, $B_l \cap \Gamma(x) \cap \{u, v\} = \emptyset$ in Case 3. We have $\Gamma(x) \cap B_l = \Gamma'(x) \cap B_l$ in all three cases.

If $\Gamma(x) \cap B_l \nsubseteq C'_t$, then at least one vertex $y \in \Gamma(x) \cap B_l = \Gamma'(x) \cap B_l$ is not in C'_t. As $x \in B'_t$, such a vertex y must be deleted in one of the first $t-1$ steps in the construction of $\mathcal{B}'$ and it should be in either B'_h or C'_h, $h \in \{1, 2, \cdots, t-1\}$.

If y is in B'_h, then $x \in \Gamma'(y)$ must be in $C'_{h'}$ and should have been deleted at step $h' \leq h < t$. So x could not be in V'_t and this is a contradiction. If y is in C'_h, then there must be another vertex z such that $z \in B'_h$ and $(z, y) \in E$. The fact $y \in \Gamma(x) \cap B_l$ which means that y is a B-class vertex indicates that z is a C-class vertex (B-class vertices being independent) or both $l = k$ and $B_l = C_l$ (the only other possibility). In both cases, we have found another B'_h with index $h < t$ which contains a C-class vertex z, a contradiction implying that $\Gamma(x) \cap B_l \subseteq C'_t$ for any vertex $x \in B'_t \cap C_l$, $l \in \{1, 2, , \cdots, k\}$.

Second, we partition B'_t (and C'_t, respectively) into two disjoint subsets $B'_{t1} \cup B'_{t2}$ (and $C'_{t1} \cup C'_{t2}$) as follows.

$$B'_{t1} = \bigcup_{l=1}^{k} \bigcup_{x \in B'_t \cap C_l} \{x\}; \quad B'_{t2} = B'_t - B'_{t1};$$

$$C'_{t1} = \bigcup_{l=1}^{k} \bigcup_{x \in B'_t \cap C_l} (\Gamma(x) \cap B_l); \quad C'_{t2} = C'_t - C'_{t1}.$$

Fixing l: $1 \leq l \leq k$, let $C_{sub} = \bigcup_{x \in B'_t \cap C_l} \{x\} \subseteq C_l$ and $B_{sub} = \bigcup_{x \in B'_t \cap C_l} (\Gamma(x) \cap B_l) = \Gamma(C_{sub}) \cap B_l$. Thus we have $w(B_{sub})/w(C_{sub}) \geq 1/\alpha_l \geq 1$ by Lemma 3. As $\alpha'_t < 1$, it follows that $w(C'_{t1})/w(B'_{t1}) \geq 1 > \alpha'_t$. Therefore, $(B'_{t2}, C'_{t2}) = DKR(B'_{t1}, C'_{t1}; B'_t, C'_t)$ which in turn implies $w(C'_{t2})/w(B'_{t2}) < \alpha'_t$.

Further, by definition all C-class vertices of B'_t are contained in B'_{t1}. Therefore, B'_{t2} only contains B-class vertices and its neighbors are all C-class.

As all vertices in C'_{t1} are B-class, $(\Gamma(B'_{t2}) \cap V'_t) \cap C'_{t1} = \emptyset$ and $(\Gamma(B'_{t2}) \cap V'_t) \subseteq C'_{t2}$. Then,

$$\frac{w(\Gamma(B'_{t2}) \cap V'_t)}{w(B'_{t2})} \leq \frac{w(C'_{t2})}{w(B'_{t2})} < \alpha'_t.$$

Here another pair $(B'_{t2}, \Gamma(B'_{t2}) \cap V'_t)$ with $B'_{t2} \subseteq V'_t$ is found such that the α-ratio is strictly less than α'_t. Such a contradiction concludes the Key Lemma. □

Key Lemma illustrates that if the conditions in Lemma 4 are satisfied, then there is no C-class vertex in B'-class with $\alpha' < 1$. Symmetrically, we introduce the following Main Lemma.

Lemma 5 (Main Lemma). *Consider the bottleneck decompositions $\mathcal{B}$ and $\mathcal{B}'$. If the case that $(u, v) \in B_i \times C_i$, and u, v are both in B'-class does not happen, then $B_t \cap \left(\cup_{\alpha'_l<1} C'_l\right) = \emptyset$, $t = 1, \cdots, k$.*

Proof. If the claim does not hold, we choose the minimum t such that the above result fails, i.e. $B_t \cap (\cup_{\alpha'_l<1} C'_l) \neq \emptyset$. Thus there is an index l with $\alpha'_l < 1$, $1 \leq l \leq k'$, such that $x \in B_t \cap C'_l$.

We claim that $\Gamma'(x) \cap B'_l \subseteq V_t$. Otherwise, let $y \in \Gamma'(x) \cap B'_l - V_t$ be deleted in the first $t-1$ steps. Therefore, $y \in C_h$ with $h < t$; otherwise if $y \in B_h$, then x should be in C_h and be deleted before step t. Hence there exists $z \in B_h$ such that $(z, y) \in B_h \times C_h$. If $(z, y) \in E(G')$, then the fact that $y \in B'_l$ and $\alpha'_l < 1$ implies that z is in C'-class, a contradiction to the choice of t and x. Otherwise if $(z, y) \notin E(G')$, then $z = u$ and $y = v$ by the assumption that v is C-class. So the condition that $(u, v) \in B_i \times C_i$, and u, v are not both in B'-class ensures that z must be in C'-class because $y = v$ is in $\Gamma'(x) \cap B'_l \subseteq B'_l$. So $z \in B_h \cap (\cup_{\alpha'_l<1} C'_l)$ with $h < t$, which is a contradiction to the choice of t and x.

Therefore, for any $x \in B_t \cap C'_l$ with $\alpha'_l < 1$, we have $\Gamma'(x) \cap B'_l \subseteq V_t$. It follows that $\bigcup_{x \in B_t \cap C'_l, \alpha'_l<1} \Gamma'(x) \cap B'_l \subseteq C_t$. Let us partition B_t (and C_t, respectively) into two disjoint subsets $B_{t1} \cup B_{t2}$ (and $C_{t1} \cup C_{t2}$) as follows.

$$B_{t1} = \bigcup_{l=1}^{k'} \bigcup_{x \in B_t \cap C'_l, \alpha'_l<1} \{x\}; \quad B_{t2} = B_t - B_{t1};$$

$$C_{t1} = \bigcup_{l=1}^{k'} \bigcup_{x \in B_t \cap C'_l, \alpha'_l<1} (\Gamma'(x) \cap B'_l); \quad C_{t2} = C_t - C_{t1}.$$

Fixing l: we have $C'_{sub} = \cup_{x \in B_t \cap C'_l, \alpha'_l<1} \{x\} \subseteq C'_l$ and $B'_{sub} = \cup_{x \in B_t \cap C'_l, \alpha'_l<1} (\Gamma'(x) \cap B'_l) = \Gamma'(C'_{sub}) \cap B'_l$. Then we obtain $w(B'_{sub})/w(C'_{sub}) \geq 1/\alpha'_l > 1$ by Lemma 3. It follows that $w(C_{t1})/w(B_{t1}) > 1 \geq \alpha_t$. Therefore, $(B_{t2}, C_{t2}) = DKR(B_{t1}, C_{t2}; B_t, C_t)$ which in turn implies $w(C_{t2})/w(B_{t2}) < \alpha_t$.

In addition, all B'-class vertices of B_t are in B_{t2} and all its neighbors are C'-class. Specially if B_{t2} contains vertex x from $B'_{k'}$, $\alpha'_{k'} = 1$, then x does not has neighbors in B'_l, $\alpha'_l < 1$. We know that all vertices have the same neighborhoods in G and G' except for u and v. Further condition in lemma excludes the case that $u \in B_{t2}$, $v \in C_{t1}$ which leads to $\Gamma(u) \nsubseteq C_{t2}$. So $(\Gamma(B_{t2}) \cap V_t) \cap C_{t1} = \emptyset$ and $(\Gamma(B_{t2}) \cap V_t) \subseteq C_{t2}$. Then,

$$\frac{w\left(\Gamma(B_{t2}) \cap V_t\right)}{w(B_{t2})} \leq \frac{w(C_{t2})}{w(B_{t2})} < \alpha_t.$$

Here another pair $(B_{t2}, \Gamma(B_{t2}) \cap V_t)$ with $B_{t2} \subseteq V_t$ is found such that its α-ratio is strictly less than α_t. Such a contradiction concludes Lemma 5. □

From Main Lemma, the following corollary can be obtained directly.

Corollary 1. *If $u \in B_{i_u}$, then u cannot be in C'-class with $\alpha' < 1$.*

The following theorem shows that the decomposition of G' is equal to that of G if $(u, v) \notin B_i \times C_i$, $i = 1, \cdots, k$.

Theorem 1. *If $(u, v) \notin B_i \times C_i$, $i = 1, \cdots, k$, then $\mathcal{B} = \mathcal{B}'$.*

The proof shall be shown in the full version. We should emphasize that $(u, v) \notin B_i \times C_i$ includes two cases: Case $u \in B_{i_u}$, $v \in C_{i_v}$ with $i_v < i_u$ and Case $u \in C_{i_u}$, $v \in C_{i_v}$ with $i_v \leq i_u$.

3.2 Incentive Properties

In this subsection we propose the strategy-proof properties of agent u and v by analyzing different cases. It is easy to compute that there are totally 16 cases because agent u can be in B_{i_u}, C_{i_u}, B'_{j_u} or C'_{j_u} and v can be in B_{i_v}, C_{i_v}, B'_{j_v} or C'_{j_v}. But according to the assumption that u and v are in B-class and C-class respectively if edge (u, v) is between different classes, it is not necessary to consider such cases in which $u \in C_{i_u}$ and $v \in B_{i_v}$. In addition, for the cases in which $u \in B_{i_u}$ and $v \in B_{i_v}$, it is only possible that $(u, v) \in B_k \times C_k$ with $\alpha_k = 1$. Thus u and v have the same statuses in G and Case $u \in B_{i_u}$, $v \in B_{i_v}$, $u \in B'_{j_u}$, $v \in C'_{j_v}$ is equivalent to Case $u \in B_{i_u}$, $v \in B_{i_v}$, $u \in C'_{j_u}$, $v \in B'_{j_v}$. It is enough to discuss one of them. Here we should point out that when such cases in which $u \in C_{i_u}$ or $v \in C_{i_v}$ ($u \in C'_{j_u}$ or $v \in C'_{j_v}$ respectively) are discussed, we mean its α-ratio (α'-ratio) less than 1. Because if its α-ratio (α'-ratio) equals to 1, such cases would come down to Case $u \in B_k$ or $v \in B_k$ with $\alpha_k = 1$ (Case $u \in B'_{k'}$ or $v \in B'_{k'}$ with $\alpha' = 1$, respectively). In the following we will try our best to prove that u and v has no incentive to cheat by removing edge (u, v) in the rest cases.

First we know that if $u \in C_{i_u}$, $v \in C_{i_v}$ and $u \in B_{i_u}$, $v \in C_{i_v}$ with $i_v < i_u$, then $(u, v) \notin B_i \times C_i$ and $\mathcal{B} = \mathcal{B}'$ by Theorem 1. So $U'_u = U_u$ and $U'_v = U_v$ for such cases in which that $u \in C_{i_u}$, $v \in C_{i_v}$ and $u \in B_{i_u}$, $v \in C_{i_v}$ with $i_v < i_u$.

We also get that any agent has its utility larger than or equal to its bandwidth if it is in C-class (or in C'-class) and has its utility smaller than or equal to its bandwidth if it is in B-class (or in B'-class) by Proposition 3. Furthermore, if u and v are both in B-class, it is only possible that $(u, v) \in B_k \times C_k$ with $\alpha_k = 1$

which implies that $U_u = w_u$ and $U_v = w_v$. So $U'_u \leq U_u$ and $U'_v \leq U_v$ for Case $u \in B_k$, $v \in B_k$ with $\alpha_k = 1$ and $u \in B'_{j_u}$, $v \in B'_{j_v}$.

Further Corollary 1 ensures the following cases cannot exist because u is in B-class and $u \in C'_{j_u}$ with $\alpha' < 1$.

- Case $u \in B_k$, $v \in B_k$ with $\alpha_k = 1$, $u \in C'_{j_u}$, $v \in C'_{j_v}$;
- Case $u \in B_k$, $v \in B_k$ with $\alpha_k = 1$, $u \in C'_{j_u}$, $v \in B'_{j_v}$;
- Case $u \in B_i$, $v \in C_i$, $u \in C'_{j_u}$, $v \in C'_{j_v}$;
- Case $u \in B_i$, $v \in C_i$, $u \in C'_{j_u}$, $v \in B'_{j_v}$.

Next, we shall prove our main result for the rest of 2 cases in the following lemmas.

Lemma 6. *If $u \in B_i$, $v \in C_i$, $u \in B'_{j_u}$ and $v \in B'_{j_v}$, then u and v both cannot gain more utilities by cutting one of adjacent edges.*

Proof. Because $U'_v = \alpha'_{j_v} w_v \leq w_v/\alpha_{i_v} = U_v$, agent v has no incentive to cheat. For agent u if $j_u \leq j_v$, then $j_u = j_* \leq i$ by Basic Lemma-(6). Because Basic Lemma-(3) shows that $\alpha'_{j_*} \leq \alpha_{j_*}$ and $j_* \leq i$ which implies $\alpha_{j_*} \leq \alpha_i$, we have $\alpha'_{j_*} \leq \alpha_i$ and $U'_u = \alpha'_{j_*} w_u \leq \alpha_i w_u = U_u$. Thus agent u cannot get benefit by cheating in this case.

It is more complicated to discuss the utility of agent u in G' if $j_v < j_u$. Our main idea is to construct a pair containing u in V'_{j_u} whose α-ratio is smaller than α_i. Thus we can get that $\alpha'_{j_u} < \alpha_i$ and $U'_u < U_u$. In order to obtain this result, the following results are necessary and their proofs are shown in the full version.

Fact 1. *For each $1 \leq l < i$, $C_l \cap B'_j = \emptyset$ and $B_l \cap C'_j = \emptyset$, $j = j_v, \cdots, j_u$, if $(u, v) \in B_i \times C_i$ with $\alpha_i < 1$ and $j_v < j_u$.*

Fact 1 shows that no vertex in C_l belongs to B'_j and no vertex in B_l belongs to C'_j, for $1 \leq l \leq i-1$ and $j_v \leq j \leq j_u$. Therefore, we can define

$$B_l^h = B_l - \bigcup_{j=j_v}^{h} B'_j,\ C_l^h = C_l - \bigcup_{j=j_v}^{h} C'_j,\ h = j_v, \cdots, j_u - 1.$$

For each $h = j_v, \cdots, j_u - 1$, if let $B = \bigcup_{j=j_v}^{h} (B'_j \cap B_l)$ and $C = \bigcup_{j=j_v}^{h} (C'_j \cup C_l)$, then $\Gamma(B) \cap C_l = C$. So if let $B_l^h = B^c = B_l - B$ and $C_l^h = C^c = C_l - C$, then Lemma 1 tells us, for $l = j_v, \cdots, i-1$, $h = j_v, \cdots, j_u - 1$,

$$\frac{w(C_l^h)}{w(B_l^h)} = \frac{w(C^c)}{w(B^c)} \leq \alpha_l < \alpha_i. \tag{1}$$

But if $l = i$, there is no such a nice result in Fact 1. Thus for $h = j_v, \cdots, j_u - 1$, we define

$$B_i^h = B_i - \bigcup_{j=j_v}^{h} (B'_j \cup C'_j),\quad C_i^h = C_i - \bigcup_{j=j_v}^{h} (B'_j \cup C'_j),$$

and have

Fact 2. *If* $(u,v) \in B_i \times C_i$ *with* $\alpha_i < 1$ *and* $j_v < j_u$*, then* $w(C_i^h)/w(B_i^h) < \alpha_i$*,* $h = j_v, \cdots, j_u - 1$.

Now let us focus on V'_{j_u}. Since $B_l^{j_u-1} \subseteq V'_{j_u}$ and $C_l^{j_u-1} \subseteq V'_{j_u}$, $l = j_v, \cdots, i$, we define $\widehat{B} = \bigcup_{l=j_v}^{i} B_l^{j_u-1}$ and $\widehat{C} = \bigcup_{l=j_v}^{i} C_l^{j_u-1}$. Then (1) and Fact 2 tells

$$\frac{w(\widehat{C})}{w(\widehat{B})} = \frac{\sum_{l=j_v}^{i} w(C_l^{j_u-1})}{\sum_{l=j_v}^{i} w(B_l^{j_u-1})} < \alpha_i.$$

Further, all vertices in $B_l^{j_u-1}$ only can be adjacent to vertices in $C_h^{j_u-1}$, $h = j_v, \cdots, l$, in V'_{j_u}. Thus $\Gamma'(B_l^{j_u-1}) \cap V'_{j_u} \subseteq \widehat{C}$ for each $l = j_v, \cdots, i$. So $\Gamma'(\widehat{B}) \cap V'_{j_u} = C$ and $(\widehat{B}, \widehat{C})$ is a pair in V'_{j_u} whose α-ratio is less than α_i. Thus $\alpha'_{j_u} \leq w(\widehat{C})/w(\widehat{B}) < \alpha_i$. □

Lemma 7. *If* $u \in B_i$*,* $v \in C_i$*,* $u \in B'_{j_u}$ *and* $v \in C'_{j_v}$*, then u and v both cannot improve their utilities by cutting one of adjacent edges.*

Proof. First we shall claim that if $j_v \leq j_u$, then it must be that $j_u = j_v = i$ and $(B_i, C_i) = (B'_i, C'_i)$. On the one hand, Basic Lemma-(3) ensures that $\alpha'_{j_v} \leq \alpha_{j_v}$ because $V'_{j_v} = V_{j_v}$. On the other hand by Basic Lemma-(6), it is true that $j_v \leq i$ if $j_v \leq j_u$. So at step j_v of the decomposition in G, if $u \notin B'_{j_v}$, then all vertices in B'_{j_v} have the same neighborhoods in G and G' which means that $\Gamma(B'_{j_v}) \cap V_{j_v} = C'_{j_v}$. If $u \in B'_{j_v}$, then we still have $\Gamma(u) \cap V_{j_v} = (\Gamma'(u) \cap V_{j_v}) \cup \{v\} \subseteq C'_{j_v}$ as $v \in C'_{j_v}$ and $v \notin \Gamma'(u)$. Thus (B'_{j_v}, C'_{j_v}) is a pair in V_{j_v}. It follows that $\alpha'_{j_v} \geq \alpha_{j_v}$. Therefore $\alpha'_{j_v} = \alpha_{j_v}$ and $B'_{j_v} \subseteq B_{j_v}$, $C'_{j_v} \subseteq C_{j_v}$. Further the condition $v \in C'_{j_v} \subseteq C_{j_v}$ promises that $v \in C_{j_v}$ which means that $j_v = i$. So at step $i = j_v$ of the decomposition in G', (B_i, C_i) is a pair in V'_i. The result $\alpha_i = \alpha'_i$ shows that $B_i \subseteq B'_i$ and $C_i \subseteq C'_i$. Combining the previous result, we have $(B_i, C_i) = (B'_i, C'_i)$ and $j_v = j_u = i$. Under this case, $U'_u = U_u$ and $U'_v = U_v$.

Next let us discuss the case that $j_u < j_v$. Thus $U'_u \leq U_u$ since $j_u \leq i$ and $\alpha'_{j_u} \leq \alpha_i$. But it's much more difficult to analyze the characterization of agent v. In order to get the ideal result, the following fact is necessary.

Fact 3. *If* $(u,v) \in B_i \times C_i$ *and* $j_v > j_u$*,* $v \in C'_{j_v}$*, then there is an index* $j_u < j \leq j_v$ *satisfying*

1. $B_h \cap V'_j = \emptyset$ *and* $C_h \cap V'_j = \emptyset$*,* $h = 1, 2, \cdots, i-1$*;*
2. $B_h \subseteq V'_j$ *and* $C_h \subseteq V'_j$*,* $h = i+1, \cdots, k$*;*
3. $\alpha'_j \geq \alpha_i$*;*
4. $v \in V'_j$.

The proof of Fact 3 is presented in the full version. Fact 3 indicates that V'_j does not have any vertex from $\bigcup_{h=1}^{i-1}(B_h \cup C_h)$, all vertices in $\bigcup_{h=i+1}^{k}(B_h \cup C_h)$ are contained in V'_j and some vertices including v in $B_i \cup C_i$ belong to V'_j. Based on the result of Fact 3-(3), we know that $\alpha'_{j_v} \geq \alpha'_j \geq \alpha_i$ since $j_v \geq j$. Thus $U'_v = w_v/\alpha'_{j_v} \leq w_v/\alpha_i = U_v$. □

Based on the analysis above, the main result can be induced directly.

Theorem 2 (Main Result). *Given the bandwidth allocation mechanism obtained by the market equilibrium from the bottleneck decomposition, no agent has incentive to cheat by cutting any of its incident edges.*

4 Conclusion

In this article, we discuss the issue of possible cheating strategies of agent with respect to the proportional response protocol for the application of bandwidth sharing. We show that, if every agent follow this strategy, no player could gain by removing a connection arc from its neighbors. In other words, no player could gain eventually, if it removes one or more of its edge from the network environment. In order to obtain this main result, we need to analyze 11 cases one by one and list them in the following table.

	u in G	v in G	u in G'	v in G'	Reason for impossibility/treatment
1	B_{i_u}	B_{i_v}	B'_{j_u}	B'_{j_v}	$U'_u \leq U_u, U'_v \leq U_v$
2	B_{i_u}	B_{i_v}	C'_{j_u}	B'_{j_v}	Cannot exist by Corollary 1
3	B_{i_u}	B_{i_v}	C'_{j_u}	C'_{j_v}	Cannot exist by Corollary 1
4	C_{i_u}	C_{i_v}	B'_{j_u}	B'_{j_v}	Cannot exist by Theorem 1
5	C_{i_u}	C_{i_v}	C'_{j_u}	B'_{j_v}	Cannot exist by Theorem 1
6	C_{i_u}	C_{i_v}	B'_{j_u}	C'_{j_v}	Cannot exist by Theorem 1
7	C_{i_u}	C_{i_v}	C'_{j_u}	C'_{j_v}	$(B_{i_u}, C_{i_u}) = (B'_{i_u}, C'_{i_u})$, $(B_{i_v}, C_{i_v}) = (B'_{i_v}, C'_{i_v})$ by Theorem 1
8	B_{i_u}	C_{i_v}	C'_{j_u}	C'_{j_v}	Cannot exist by Corollary 1
9	B_{i_u}	C_{i_v}	C'_{j_u}	B'_{j_v}	Cannot exist by Corollary 1
10	B_{i_u}	C_{i_v}	B'_{j_u}	B'_{j_v}	Cannot exist by Theorem 1 if $i_u > i_v$
					Please check Lemma 6 if $i_u = i_v$
11	B_{i_u}	C_{i_v}	B'_{j_u}	C'_{j_v}	$(B_{i_u}, C_{i_u}) = (B'_{i_u}, C'_{i_u}), (B_{i_v}, C_{i_v}) = (B'_{i_v}, C'_{i_v})$ by Theorem 1 if $i_u > i_v$
					Please check Lemma 7 if $i_u = i_v$

The result, building on discrete mathematical techniques, resolves a long time unsolved problem whether the proportional response protocol is truthful. It is the first, to the best of our knowledge, non-trivial result on truthful mechanism for a practical network bandwidth resource sharing scheme.

References

1. Adsul, B., Babu, C.S., Garg, J., Mehta, R., Sohoni, M.: Nash equilibria in fisher market. In: Kontogiannis, S., Koutsoupias, E., Spirakis, P.G. (eds.) SAGT 2010. LNCS, vol. 6386, pp. 30–41. Springer, Heidelberg (2010)

2. Chen, N., Deng, X., Zhang, H., Zhang, J.: Incentive ratios of fisher markets. In: Czumaj, A., Mehlhorn, K., Pitts, A., Wattenhofer, R. (eds.) ICALP 2012, Part II. LNCS, vol. 7392, pp. 464–475. Springer, Heidelberg (2012)
3. Godfrey, P.B., Schapira, M., Zohar, A., Shenker, S.: Incentive compatibility and dynamics of congestion control. ACM SIGMETRICS Perform. Eval. Rev. **38**(1), 95–106 (2010)
4. Iosifidis, G., Gao, L., Huang, J.W., Tassiulas, L.: Enabling crowd-sourced mobile internet access. In: 2014 Proceedings IEEE INFOCOM, pp. 451–459 (2014)
5. Iosifidis, G., Gao, L., Huang, J.W., Tassiulas, L.: Incentive mechanisms for user-provided networks. IEEE Commun. Mag. **52**(9), 20–27 (2014)
6. Levin, D., LaCurt, K., Spring, N., Bhattacharjee, B.: BitTorrent is an auction: analyzing and improving BitTorrent incentives. ACM SIGCOMM Comput. Commun. Rev. **38**(4), 243–254 (2008)
7. Schollmeier, R.: A definition of peer-to-peer networking for the classification of peer-to-peer architectures and applications. In: Proceedings of the First International Conference on Peer-to-Peer Computing, IEEE Computer Society (2001)
8. Wu, F., Zhang, L.: Proportional response dynamics leads to market equilibrium. In: Proceedings of the Thirty-ninth Annual ACM Symposium on Theory of Computing, STOC 2007, pp. 354–363. ACM, New York, NY, USA (2007)

The Web Graph as an Equilibrium

Georgios Kouroupas[1], Evangelos Markakis[1](✉), Christos Papadimitriou[2], Vasileios Rigas[1], and Martha Sideri[1]

[1] Department of Informatics, Athens University of Economics and Business, Athens, Greece
markakis@gmail.com
[2] Department of EECS, University of California - Berkeley, Berkeley, USA

Abstract. We present a game-theoretic model for the creation of content networks such as the worldwide web. The action space of a node in our model consists of choosing a set of outgoing links as well as click probabilities on these links. A node's utility is then the product of the *traffic* through this node, captured by its PageRank in the Markov chain created by the strategy profile, times the *quality of the node,* a surrogate for the website's utility per visit, such as repute or monetization potential. The latter depends on the intrinsic quality of the node's content, as modified by the chosen outgoing links and probabilities. We only require that the quality be a concave function of the node's strategy (the distribution over outgoing links), and we suggest a natural example of such a function. We prove that the resulting game always has a pure Nash equilibrium. Experiments suggest that these equilibria are not hard to compute, avoid the reciprocal equilibria of other such models, have characteristics broadly consistent with what we know about the worldwide web, and seem to have favorable price of anarchy.

1 Introduction

During the past quarter of a century, a great variety of networks of towering magnitude, and paramount economic, social, and scientific importance, have emerged. Understanding, ex post, these networks — their origins, properties, operation, and destiny — has been a most important and central theme of scientific inquiry. Many simple *generative models* have been proposed as an important tool of such inquiry, sometimes successfully predicting many of the observed properties of networks, see for example [5] and the related work section. Several of these models ascribe certain incentives, or behaviors, to individual nodes, and the network emerges as the equilibrium, or otherwise as the sum total of these behaviors. For example, several early network creation games, motivated by the

Research supported by the European Union (European Social Fund- ESF) and Greek national funds through the Operational Program "Education and Lifelong Learning" of the National Strategic Reference Framework (NSRF) - Research Funding Program: THALES, investing in knowledge society through the European Social Fund. Also supported by NSF grants CCF0964033 and CCF1408635.

M. Hoefer (Ed.): SAGT 2015, LNCS 9347, pp. 203–215, 2015.
DOI: 10.1007/978-3-662-48433-3_16

network infrastructure of the Internet, assumed that nodes optimize a combination of connection cost and topology advantage [2,11,12,29]. Following a different perspective, certain web-creation models have postulated a *link-copying* behavior by nodes, which leads to "the rich get richer" phenomena of web growth [5,21]. Often, an important goal of network models has been to predict the power-law degree distributions and other quantitative features observed in real networks [7,24,26]. In a different domain, for *social* networks, there are by now several competing generative models, attempting to capture important aspects of those complex systems [15,25].

In this paper we propose a *game-theoretic* model for the creation of *content networks*, such as the worldwide web. The basic decision made by each node is to determine *the set of outgoing links, as well as their weights.* We assume, in other words, that a node can decide which other nodes to link to, but can also influence the precise percentage of outgoing traffic on each outgoing link. Mathematically, the action set of a node/player consists of all distributions over the remaining nodes. A node in a content network chooses outgoing links so as to accomplish two goals: outgoing links affect the volume of *traffic*, i.e., the number of users who frequent your site. But also well-chosen links can improve your content's quality, variety, informativeness, and prestige, or even its direct profitability through paid outgoing traffic, and therefore enhance the overall *utility per unit of traffic.* These two objectives, traffic and quality, will be of course conflicting in general. Ideally our model should postulate that each node maximizes the *product* of the two, i.e., traffic times utility per traffic, since this is the true payoff received (there have been models proposed in the literature where one of the two objectives reigned, see the subsection on related work). But what hope is there for such a complicated utility function to result in a game between the nodes that has good properties — for example, always possess a pure Nash equilibrium? As we shall see in Sect. 5, even best response computation is highly nontrivial under our model.

Our main result is that, quite surprisingly, this multi-player game, as described above, *does always possess a pure Nash equilibrium* (Theorem 1). The proof entails looking at the expression for the player's utility, as a function of the actions of everybody — that is to say, of the resulting Markov chain — and observing that, rather unexpectedly, this complicated expression turns out to be quasi-concave in the actions of the player. Existence then follows from the Nash-Debreu-Fan theorem [9,13,27]. The result is true as long as the quality of a node is a continuous concave function of the node's action — of the distribution of outgoing flows. We view this as a fundamental positive result in the economic analysis of networks, whose importance goes beyond the particular model proposed here.

Our theorem raises certain intriguing questions: Is there a polynomial-time algorithm for computing these equilibria? Or is the problem PPAD-hard? We have no result, or even intelligent guess, to offer in this regard, even though we did succeed in computing the Nash equilibria of quite large games of this sort. And is the price of anarchy of these equilibria favorable? Again, we have no definitive results here, even though some quite encouraging conclusions can be drawn from experimentation, summarized below.

In Sect. 5, we summarize the results of extensive experimentation with our model. In the absence of a provably efficient algorithm, we find Nash equilibria through simulations of best-response dynamics; with very modest computational resources, we are able to solve networks with a number of nodes in the thousands (the algorithm is described in Sect. 4). In these experiments we do observe in-degree distributions that are quite power law-like, as well as a giant strongly connected component, existence of dense bipartite subgraphs, and several other features commonly observed in the web graph. We also experiment with a gradient ascent heuristic for estimating the social optimum, and notice that the local optima of the social welfare function found are usally no more than 5 % better than the worst Nash equilibrium computed by our algorithm.

Related Work. Several well known studies [7,23,24] have analyzed big parts of the worldwide web, providing important insights about macroscopic properties of the web graph, such as degree distribution, clustering, connectivity and many more. Generative models were then developed to explain and predict these properties; such models are typically *probabilistic*, and only occasionally *game-theoretic*. From the first category, the *preferential attachment* model [4], is the most popular one. A key feature is that it creates a power law degree distribution, a known property of the web graph [5]. Several variants have been proposed, elaborating on the initial model of [4], see e.g., among others, [1,8,21].

Several network formation games with strategic nodes have been studied, starting with [12] (see also the survey [29]), but these works are meant to model the Internet, and are much less suitable for modeling the web. Among the first game-theoretic models tailored for content graphs is the work by marketing researchers Katona and Sarvary [18]. They model what they call "the commercial web"; sites can sell or buy advertising links, at prices per click that are determined by the seller's "quality". The utility function is the traffic minus advertising costs plus advertising revenue. Subsequent work [10] considers networks of content sites, with very limited exchange of user traffic. A generalization of [18] is given in [19], where the effects of sticky content (content that induces return traffic) are taken into consideration.

Another game-theoretic model, a bit closer to our spirit, is by Hopcroft and Sheldon [17], see also their earlier work [16]. They are interested in modeling *reputation networks*, where the utility function of a site is just traffic. They consider two objectives: maximizing PageRank [6], or minimizing hitting time; both utility functions result in rich sets of Nash equilibria, but the latter is manipulation-resistant, as it does not depend on a website's outgoing links. However, hitting time offers poor predictions; for example, *every* graph is a Nash equilibrium. PageRank as utility yields a more refined class of equilibrium graphs, but these equilibria possess certain symmetries and reciprocities that are not observed in practice — for example, at equilibrium non-source sites will link only to sites pointing to them. Related to these models, [3] studied the Nash equilibria of PageRank-related games on undirected graphs modeling social networks. Finally, an earlier model of the web in the same category is [20], in which economic incentives create a web-like network of documents, whose properties are then analyzed.

2 Our Model

There are n players, alternatively referred to as *nodes* or *websites*; the set of all players is denoted by $[n]$. The set of *actions* available to player i is the creation of an arbitrary set of hyperlinks, where each hyperlink connects i to some $j \neq i$. Player i also has to choose a weight x_{ij} for each hyperlink, equal to the "click probability" of this hyperlink, that is, the probability that a user visiting website i will end up following this (i, j) link and land in website j next. We assume that all possible distributions on $[n] - \{i\}$ are available actions of player i; that is, through manipulation of the precise positions of the hyperlinks, of the anchor texts, and other characteristics, the player can achieve any action of the form $\mathbf{x}_i = (x_{i1}, \ldots x_{in})$, where $x_{ij} \geq 0$ for all j, $x_{ii} = 0$, and $\sum_{j \neq i} x_{ij} = 1$. Hence, the set of actions (or pure strategies), A_i, available to player i, is the $(n-2)$-dimensional simplex Δ_{n-1}. Note that a strategy profile $\mathbf{x} = (\mathbf{x}_1, \ldots, \mathbf{x}_n) \in \prod_{i=1}^{n} A_i$ can be seen as a Markov chain on the n nodes without self loops.

We define the *payoff* of a player to be the product of two terms: the *quality* of its content multiplied by the *traffic:*

$$P_i(\mathbf{x}) = Q_i(\mathbf{x}_i) \cdot T_i(\mathbf{x}) \tag{1}$$

Note that the quality Q_i only depends on the action of i — its weighted outgoing links — while the traffic T_i depends, quite naturally, on the whole Markov chain created by the actions of all players. We next define these two factors.

The quality of a player, $Q_i(\mathbf{x}_i)$, is defined as:

$$Q_i(\mathbf{x}_i) = q_i + \sum_j d_{ij} x_{ij} - c_i \cdot \sum_j x_{ij}^2, \tag{2}$$

The first term, q_i, is the *inherent quality* of i's content. The second term denotes contributions made by the outgoing links, where it is assumed that a unit of traffic from i to j affects the quality of i by the parameter d_{ij} (which may be negative). Intuitively, d_{ij} captures the relevance and affinity between the two websites, or any kind of benefit — such as advertising payment — to i from a user following the link to j. It is assumed for now that the parameters q_i and d_{ij} are known and fixed; in the description of our experiments in Sect. 5, we shall explain how these parameters can be obtained from more primitive data, namely the parameters of a *topic model* for the contents of these websites. Finally, the third term in (2) captures the fact that a website benefits from the *diversity* in its outgoing links. The parameter $c_i \geq 0$ is called the *branching factor* of node i, and is a measure of the extent to which the quality of website i benefits from such diversity. The parameter c_i can be zero — in other words, the third term is included for generality and modeling realism, and is not needed for the validity of our theorem.

Coming now to the traffic factor, $T_i(\mathbf{x})$, the *traffic* of node i could be the *stationary probability* of node i in the Markov chain $\mathbf{x}$. In fact, we use a slight variant known as PageRank [6], in which there is *teleportation:* at any step the random walk may, with a small probability, proceed not to a neighbor, but to a node selected uniformly at random (our main result is unaffected if we adopt the stationary probability as our measure of traffic). More precisely, let us denote as t_{ji} the expected number of hops that a user needs to travel from website j to website i in the random surfer model. For $j \neq i$:

$$t_{ji} = (1-\alpha)(1 + \textstyle\sum_{k \neq i} x_{jk} t_{ki}) + \alpha(1 + \frac{1}{n} \sum_{k \neq i} t_{ki})$$

Here α is the teleportation factor, i.e., the random restart probability. Similarly, we can also define t_{ii}, the expected number of steps that a user needs to travel from one website back to it:

$$t_{ii} = (1-\alpha)(1 + \sum_{j \neq i} x_{ij} t_{ji}) + \alpha(1 + \frac{1}{n} \sum_{j \neq i} t_{ji}) = 1 + \sum_{j \neq i} ((1-\alpha)x_{ij} + \frac{\alpha}{n}) t_{ji} \quad (3)$$

Finally, the traffic $T_i(\mathbf{x})$ (or PageRank of i) of a website is defined as:

$$T_i(\mathbf{x}) = \frac{1}{t_{ii}} = \frac{1}{1 + \sum_{j \neq i} ((1-\alpha)x_{ij} + \frac{\alpha}{n}) t_{ji}(\mathbf{x})} \quad (4)$$

Note that we write $t_{ji}(\mathbf{x})$ in T_i, since these quantities are parameters of the Markov chain $\mathbf{x}$; Hence, overall, the payoff of a player i under a strategy profile $\mathbf{x}$ is:

$$P_i(\mathbf{x}) = Q_i(\mathbf{x}_i) \cdot T_i(\mathbf{x}) = \frac{q_i + \sum_j d_{ij} x_{ij} - c_i \cdot \sum_j x_{ij}^2}{1 + \sum_{j \neq i} ((1-\alpha)x_{ij} + \frac{\alpha}{n}) t_{ji}(\mathbf{x})} \quad (5)$$

3 The Main Theorem

Let $\mathcal{G}$ denote a game on n players, as defined in the previous section. Note that it is not a finite game, as the set of pure strategies is uncountably infinite. Also, the utility function of each player at a strategy profile $\mathbf{x}$ is not (multi)-linear in the x_{ij} terms, which would have allowed us to use the theorem of Nash [27] (recall that pure strategies here are probability distributions). But still, rather remarkably, we can prove the following:

Theorem 1. *Every game $\mathcal{G}$, as defined above, always has a pure Nash equilbrium.*

Proof. We rely on an old theorem, usually referred to as *Debreu's Theorem* but based on ideas due to Nash, Debreu, and Fan [9,13,27] from the early 1950s. We refer the reader to [14] for a better exposition. The theorem states that in any game with finitely many players, in which the action sets are all non-empty, convex and compact, and in which, for each player i, the utility $u_i(\mathbf{x})$

is continuous and quasi-concave, *when considered as a function of* $\mathbf{x}_i$, *with the actions in* $\mathbf{x}_{-i}$ *considered as fixed*, then the game is guaranteed to have a pure Nash equilibrium.

Clearly, the action sets in our case satisfy the above requirements. To complete the proof, we need to check that the function

$$\frac{q_i + \sum_j d_{ij}x_{ij} - c_i \cdot \sum_j x_{ij}^2}{1 + \sum_{j \neq i}((1-\alpha)x_{ij} + \frac{\alpha}{n})t_{ji}(\mathbf{x})}$$

is quasi-concave. This is established by the following two observations:

- The numerator is a linear function of $\mathbf{x}_i$ minus a sum of squares from $\mathbf{x}_i$, and is therefore concave.
- The denominator is a *linear function of* $\mathbf{x}_i$ (and not quadratic, as it seems superficially, due to the t_{ji} terms). To see this, recall that the coefficients $t_{ji}(\mathbf{x})$ stand for the expected number of steps it will take the random walk $\mathbf{x}$ to return, for the first time, from node j to node i. *However*, these quantities are all independent of $\mathbf{x}_i$: $t_{ji}(\mathbf{x}) = t_{ji}(\mathbf{x}_{-i})$. This is because the first-time return of the walk from node j to node i *does not use* any edge coming out of node i, and hence it does not depend on $\mathbf{x}_i$. It follows that the denominator of the payoff function for player i is indeed linear in $\mathbf{x}_i$..

Therefore, the payoff function is the ratio of a concave function of $\mathbf{x}_i$ divided by a linear function of $\mathbf{x}_i$. It follows that it is a quasi-concave function of $\mathbf{x}_i$, completing the proof. □

4 Computing Best Responses

We know of no polynomial-time algorithm for computing a Nash equilibrium guaranteed by our main result, and so we resort to a best-response heuristic for approximating Nash equilibria. As it turns out, even implementing a best response dynamics heuristic has its own challenges, as explained next.

Our best response dynamics proceed in rounds. In every round, each player is given the chance to update his strategy, given the current strategies of all players. Each round consists of n steps so that all players get to update their strategy (either in a fixed or random order). Every update is a best response to the strategy profile at the present step (i.e., taking into account all updates that have happened in previous steps by other players).

Consider a strategy profile $\mathbf{x}$, and suppose we want to run one step of the best response dynamics. From now on, let us fix a player i, for whom, we want to compute his best response. Let $\mathbf{x}_{-i} = (\mathbf{x}_1, ..., \mathbf{x}_{i-1}, \mathbf{x}_{i+1}, ..., \mathbf{x}_n)$ denote the strategy profile of the other players. Given $\mathbf{x}_{-i}$, the problem that we need to solve is to find a strategy $\mathbf{x}_i$ that maximizes the function $P_i(\mathbf{x}) = P_i(\mathbf{x}_i, \mathbf{x}_{-i})$. Note that P_i also depends on the traffic terms t_{ji} for $j \neq i$. However, these terms have been computed in the previous step, they are dependent on $\mathbf{x}_{-i}$, and are not affected by any changes we make to the x_{ij} variables at the current step.

The way we optimize $P_i(\mathbf{x})$ in an iterative fashion is as follows: if the best response to a given $\mathbf{x}_{-i}$ for player i, is a strategy $\mathbf{x}_i = (x_{i1}, ..., x_{in})$, then this means that there is some value K for which it holds that $P_i(x_{i1}, x_{i2}, .., x_{in}, \mathbf{x}_{-i}) = K$, or

$$\frac{q_i + \sum_j x_{ij} d_{ij} - c_i \sum_j x_{ij}^2}{1 + \sum_{j \neq i}((1-\alpha)x_{ij} + \frac{a}{n})t_{ji}} = K$$

By expanding the above equation we get:

$$q_i + \sum_j x_{ij} d_{ij} - c_i \sum_j x_{ij}^2 - K \sum_{j \neq i}((1-a)x_{ij} + \frac{a}{n})t_{ji} = K \tag{6}$$

Let us denote the left hand side by $H_i(\mathbf{x}_i)$. This means that when we maximize $H_i(\cdot)$, the optimal value is equal to K. To compute the best response, we compute K through binary search. Starting from some initial value for K we maximize $H_i(\cdot)$. If $K < \max H_i$, we increase K and repeat (respectively, we decrease it in the other case). We continue like this until the optimal value for H_i is equal to K (or ϵ-close to K).

So far, we have reduced the problem of computing the best response at a given step of our heuristic, to the problem of maximizing the function $H_i(\cdot)$. This is a nonlinear optimization problem, hence, we will make use of the KKT conditions. Note also that after ignoring constant terms, the function we want to optimize is in the following form:

$$\sum_j \beta_{ij} x_{ij} - c_i \sum_j x_{ij}^2, \text{ where } \beta_{ij} = d_{ij} - K(1-\alpha)t_{ji} \tag{7}$$

We can now plug in our functions in the KKT conditions and obtain that in order to maximize $H_i(\cdot)$, we have to meet the following conditions for player i:

$$\frac{\partial - H_i(x_{i1}, x_{i2}, .., x_{in})}{\partial x_{ij}} + \lambda - \mu_{ij} = 0 \Leftrightarrow -\beta_{ij} + 2c_i x_{ij} + \lambda - \mu_{ij} = 0, j \in [n] \tag{8}$$

$$\sum_j x_{ij} = 1 \tag{9}$$

$$\mu_{ij} = 0 \text{ or } x_{ij} = 0, j \in [n] \tag{10}$$

$$x_{ij} \geq 0, \mu_{ij} \geq 0, j \in [n] \tag{11}$$

Here, the λ_i's and μ_j's are multipliers, arising from the KKT conditions. Condition (10) is the complementarity condition of the system, implying that for every j, either $x_{ij} = 0$ or $\mu_{ij} = 0$. Overall, there are 2^n possible cases for deciding the *support* of $\mathbf{x}_i$, i.e., the set of non-zero values in the probability distribution $\mathbf{x}_i$. If we can identify the support, then we can find the value of each x_{ij} using Eq. (8). Fortunately, we are able to avoid this exponential search by using the complementarity condition and structural properties of an optimal solution, as we describe now.

Finding the Support of $\mathbf{x}_i$. Suppose for the moment that we knew the support of an optimal solution $\mathbf{x}_i$. Without loss of generality, let us reorder the variables so that all the positive x_{ij} variables are in the beginning of the sequence, say from $j = 1$ to $j = M$ for some M. Thus, $x_{ij} > 0, j = 1, .., M$, and $x_{ij} = 0, j = M + 1, .., n$.

By the complementarity condition, this implies that $\mu_{ij} = 0$, for $j \in \{1, ..., M\}$. Using (8), we get that:

$$x_{ij} = \frac{1}{2c_i}(\beta_{ij} - \lambda), \; j = 1, .., M \tag{12}$$

Since, $\sum_j x_{ij} = 1$ and $x_{ij} = 0$ for $j \geq M + 1$, by summing up (12), we have the following property:

$$\sum_{j=1}^{M} \beta_{ij} - M\lambda = 2c_i \tag{13}$$

The analysis above reveals that if we know the support of $\mathbf{x}_i$, we can then first use (13) to compute λ and use (12) to compute the strategy $\mathbf{x}_i$. To proceed, we first show that the support of an optimal strategy $\mathbf{x}_i$ has to respect the ordering of the β_{ij} values. In particular, let π be a permutation such that $\beta_{i,\pi(1)} \geq \beta_{i,\pi(2)} \geq \dots \geq \beta_{i,\pi(n)}$.

Claim. There is always an optimal solution where the support of $\mathbf{x}_i$ is a prefix of the sequence $\{\pi(1), \pi(2), ..., \pi(n)\}$, i.e., we give a non-zero value to the x_{ij}'s that correspond to the highest values of the β_{ij} terms.

Algorithm 1. Best Response Dynamics

Input : The parameters q_i and d_{ij} for each player i
Output: A graph that represents a NE

Initialization: Start with a random or with a uniform distribution $\mathbf{x}_i$ for each $i \in N$;
while *not in an ϵ-Nash equilibrium* **do**
  for *each player i* **do**
    Compute the t_{ji}'s;
    Initialize K;
    while $K! = \max H_i$ **do**
      change K appropriately;
      find the new β_{ij}'s, and the new M, and λ as described above ;
      compute the new $\mathbf{x}_i$ and the maximum of $H_i(\cdot)$;

The above claim is proved by a simple exchange argument. This implies that we only need to search n different possible supports for $\mathbf{x}_i$ and use the KKT conditions. Hence, we can try in polynomial time one by one the n possible values of M, compute λ using (13) and $\mathbf{x}_i$ by (12) and check where our function is maximized.

The description of the best response dynamics is summarized below. In our experiments, we terminate the process when we reach an ϵ-Nash Equilibrium, where ϵ is set to 10^{-9} (in fact our algorithm behaves even better, with ϵ coming even closer to 0).

5 Experiments

We have conducted simulations, constructing many hundreds of instances, with n ranging from 20 to 1,000 nodes. For each instance, we derived networks that are (approximate) Nash equilibria, through the best response dynamics discussed in Sect. 4.

Generating the Parameters q_i and d_{ij}. One of the challenges is to generate in a principled manner the q_i's and the d_{ij}'s needed for the definition of the game $\mathcal{G}$. To this end, we employ a *topics model.*

We start by assuming that we have a set of *topics* $[k] = \{1, ..., k\}$. These correspond to possible subjects which could be present in a website's content, e.g., sports, entertainment, news, etc. We associate to each website a vector characterizing its relevance/expertise to each one of these topics. For each $i \in [n]$, let $\mathbf{v}_i = (v_i(1), v_i(2), ..., v_i(k))$ be the vector that corresponds to the values of player i on each of the k topics.

Recall now that, when a player i creates a hyper-link to a player $j \neq i$, player i derives a value of d_{ij} from the link $i \to j$. We define this value according to the topic where i has his highest strength as follows:

$$d_{ij} = v_i(m_i) \cdot v_j(m_i)$$

where $m_i = \arg\max_\ell \, v_i(\ell)$. Notice that d_{ij} is not necessarily equal to d_{ji}.

Finally, we define q_i, the inherent value of website i, to be $q_i = v_i(m_i)$.

Experimental Setup. We implemented all our experiments in Java under Windows environment. We conducted several experiments in total, which vary in the input parameters. The values we used for each parameter are the following:

- **Number of nodes:** $n \in \{50, 100, 300, 500, 1000\}$,
- **Number of topics:** $k \in \{10, 20, 50\}$,
- **Branching factor:** $c \in \{1, 2, 5\}$ (for simplicity we used $c_i = c \; \forall i$),
- **Update Method:** Random permutation or serial update according to a fixed order,
- **Teleportation factor:** $\alpha \in \{0.10, 0.15\}$ (as is done in the related literature too).

For each combination of these parameters, we ran 20 experiments with different initializations for the topic vector $\mathbf{v}_i$ of each player. In a similar way we used either a uniform or a random assignment for the initial strategy profile $\mathbf{x}_i$ of player i. For every instance, we applied our best response dynamics algorithm, that was described earlier.

Table 1. Average number of iterations till convergence.

Branching factor (c)	Number of nodes (n)				
	50	100	300	500	1000
$c = 1$	48	20	26	40	42
$c = 2$	40	32	35	37	50
$c = 5$	6	7	10	15	19

Results. We briefly comment below on our findings. Further discussion and more supporting material will be made available in the full version of our work.

1. **Convergence and number of iterations.** The first positive news is that our algorithm always converges and in many cases it does so with a quite small number of iterations. As shown in Table 1, the number of iterations for various settings is mostly in the range [30, 40], and does not worsen significantly as the number of nodes gets bigger. The maximum number observed in our experiments was 108 and the minimum was 5.
2. **Convergence to approximate Nash equilibria and uniqueness.** Interestingly, the strategy profiles obtained by our algorithm always formed an ϵ-approximate equilibrium, where in our worst case $\epsilon = 10^{-9}$ (in most cases, we even had convergence with $\epsilon = 10^{-14}$). We also tried to find out how often we have convergence to a unique equilibrium. For this, we ran the same instance with different initializations. Our main conclusion is that in the vast majority of our instances the best response heuristic identified a unique approximate equilibrium. The cases where we did not have uniqueness were very rare, approximately 1 in every 200 instances.
3. **Number of strongly connected components (SCC).** There is typically a small number of strongly connected components in our graphs which is in accordance to what is observed in the web graph. The number of components in our instances is often 1 or a very small number for $c = 2$, almost always 1 when $c = 5$, and in the range [5,20], when $c = 1$.
4. **Degree distribution.** The graphs we produce are weighted by the probabilities, hence we focused on measuring the weighted in-degree of each node (the weighted out-degree always equals 1). We grouped together nodes that had close enough weighted in-degrees, Our main observation is that there is a strong correlation between power law behavior and the ratio $\frac{\#nodes}{\#topics} = \frac{n}{k}$. We obtained power law distributions in instances where this ratio was at least 50 or higher (say $n = 1000, k = 10$). Such ratios of n/k are closer to reality, as usually the number of topics is small relatively to the number of nodes in the web. In the log-log plots of Fig. 1, we present two instances with $n = 1000, k = 10, n/k = 100$. Data fitting with a straight line (using Matlab methods) yields an exponent that is approximately $\gamma \approx 2.91$ in the first instance and $\gamma \approx 2.7$ in the second instance.
5. **Existence of communities.** It has been argued in [23] (see also [22]), that complete, or generally dense bipartite subgraphs in the worldwide web, such

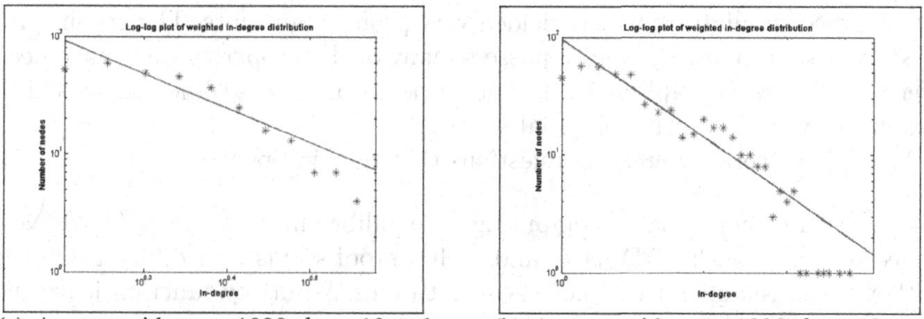

(a) A game with $n = 1000$, $k = 10$ and $c = 2$. Data fitting yields a slope of $\gamma \approx 2.91$.

(b) A game with $n = 1000$, $k = 10$ and $c = 5$. Data fitting yields a slope of $\gamma \approx 2.7$.

Fig. 1. Log-Log plots of weighted in degree distributions.

as $K_{3,3}$, is evidence of "cyber-communities." Furthermore, it was noted that such subgraphs occur far more frequently on the web than one would expect on a random directed graph with the same number of edges. In our experiments, indeed, for large values of n, the number of $K_{3,3}$'s in our graphs is always bigger than expectation by a factor ranging from 10^3 to 10^5, showing an abundance of such communities.

6. **Price of Anarchy.** For this, we first need an algorithm to estimate the optimal welfare. Given that the sum of quasi-concave functions is not necessarily a quasi-concave function, we cannot hope to apply ideas similar to the analysis presented in Sect. 4, where we optimized a single payoff function. Instead, we ran a simple gradient ascent heuristic to identify local maxima of the social welfare, and thus estimate bounds on the true maximum. Because of the difficulty of finding local maxima, we have experimented only with relatively small graphs, with up to 100 nodes. However, our findings were more than encouraging. On average, the welfare at a local maximum was no more than 2-8 % better than the welfare of an equilibrium found by our algorithm.

6 Conclusions

Strategic behavior is quintessential in modeling and understanding the explosive emergence of networks of all kinds over the past twenty-five years, and in particular of content networks such as the worldwide web. Here we have proposed a game-theoretic model for web-like networks that is novel in several ways. It is based on a concept of quality, which captures not only the intrinsic content quality of the website, but also the contributions to quality coming from the relevance, informativeness, and even per-click advertising income, of the outgoing links. It also takes into account the traffic through the website, captured by its pagerank We believe that game-theoretic approaches to the web need to account for both quantities, namely traffic and quality. On the experimental side, we performed simulations which yielded very promising results. The graphs produced by best response dynamics possess many of the properties that have been observed in the web graph and other networks in previous studies, and seem to be doing very well in terms of social welfare.

There are several interesting questions that remain open:

- What is the complexity of computing an equilibrium in our model? We have tried to prove it is PPAD-hard, and such a proof seems very difficult. On the other hand, the generically nonlinear nature of the utility function leaves no handles for the development of clever algorithms.
- Is there a nice upper bound on the price of anarchy of this game, perhaps based on smooth analysis [28]. We strongly suspect that this is the case.
- In order to extend our experiments to truly large graphs, an algorithm is needed for the best response problem that scales better with n.
- Finally, let us articulate an important challenge that we see further afield: Social networks constitute an even more challenging class from the game-theoretic modeling point of view. Can game-theoretic modeling help here? Is there a realistic model of a node's utility, as a function of its neighborhood, which predicts these intriguing phenomena?

References

1. Adamic, L., Huberman, B.: Power law distribution of the world wide web. Science **287**(5461), 2115–2115 (2000)
2. Albers, S., Eilts, S., Even-Dar, E., Mansour, Y., Roditty, L.: On Nash equilibria for a network creation game. ACM Trans. Econ. Comput. **2**(1), 1–27 (2014)
3. Avis, D., Iwama, K., Paku, D.: Verifying Nash equilibria in PageRank games on undirected web graphs. In: Asano, T., Nakano, S., Okamoto, Y., Watanabe, O. (eds.) ISAAC 2011. LNCS, vol. 7074, pp. 415–424. Springer, Heidelberg (2011)
4. Barabasi, A.L., Albert, R.: Emergence of scaling in random networks. Science **286**(5439), 509–512 (1999)
5. Bollobás, B., Riordan, O., Spencer, J., Tusnády, G.E.: The degree sequence of a scale-free random graph process. Random Struct. Algorithms **18**(3), 279–290 (2001)
6. Brin, S., Page, L.: The anatomy of a large-scale hypertextual web search engine. Comput. Netw. ISDN Syst. **30**(1–7), 107–117 (1998)
7. Broder, A.Z., Kumar, R., Maghoul, F., Raghavan, P., Rajagopalan, S., Stata, R., Tomkins, A., Wiener, J.L.: Graph structure in the web. Comput. Netw. **33**(1–6), 309–320 (2000)
8. Cooper, C., Frieze, A.M.: A general model of web graphs. Random Struct. Algorithms **22**(3), 311–335 (2003)
9. Debreu, G.: A social equilibrium existence theorem. Proc. Natl. Acad. Sci. (PNAS) **38**(10), 886–893 (1952)
10. Dellarocas, C., Katona, Z., Rand, W.: Media, aggregators, and the link economy: strategic hyperlink formation in content networks. Manage. Sci. **59**(10), 2360–2379 (2013)

11. Fabrikant, A., Koutsoupias, E., Papadimitriou, C.: Heuristically optimized trade-offs: a new paradigm for power laws in the internet. In: Widmayer, P., Triguero, F., Morales, R., Hennessy, M., Eidenbenz, S., Conejo, R. (eds.) ICALP 2002. LNCS, vol. 2380, p. 110. Springer, Heidelberg (2002)
12. Fabrikant, A., Luthra, A., Maneva, E., Papadimitriou, C., Shenker, S.: On a network creation game. In: Proceedings of the 22nd Symposium on Principles of Distributed Computing (PODC 2003), pp. 347–351 (2003)
13. Fan, K.: Fixed-point and minimax theorems in locally convex topological linear spaces. Proc. Natl. Acad. Sci. (PNAS) **38**, 121–126 (1952)
14. Fan, K.: Applications of a theorem concerning sets with convex sections. Math. Ann. **163**, 189–203 (1966)
15. Foudalis, I., Jain, K., Papadimitriou, C., Sideri, M.: Modeling social networks through user background and behavior. In: Frieze, A., Horn, P., Prałat, P. (eds.) WAW 2011. LNCS, vol. 6732, pp. 85–102. Springer, Heidelberg (2011)
16. Hopcroft, J., Sheldon, D.: Manipulation-resistant reputations using hitting time. In: Bonato, A., Chung, F.R.K. (eds.) WAW 2007. LNCS, vol. 4863, pp. 68–81. Springer, Heidelberg (2007)
17. Hopcroft, J., Sheldon, D.: Network reputation games. Techical report, Cornell University (2008)
18. Katona, Z., Sarvary, M.: Network formation and the structure of the commercial world wide web. Mark. Sci. **27**(5), 764–778 (2008)
19. Kominers, S.D.: Sticky content and the structure of the commercial web. Technical report, Harvard University (2009)
20. Kouroupas, G., Koutsoupias, E., Papadimitriou, C., Sideri, M.: Experiments with an economic model of the worldwide web. In: Deng, X., Ye, Y. (eds.) WINE 2005. LNCS, vol. 3828, pp. 46–54. Springer, Heidelberg (2005)
21. Kumar, R., Raghavan, P., Rajagopalan, S., Sivakumar, D., Tomkins, A., Upfal, E.: Stochastic models for the web graph. In: Proceedings of the 41st Annual Symposium on Foundations of Computer Science (FOCS 2000), pp. 57–65 (2000)
22. Kumar, R., Raghavan, P., Rajagopalan, S., Sivakumar, D., Tomkins, A., Upfal, E.: The web as a graph. In: Proceedings of the 19th ACM SIGMOD-SIGACT-SIGART Symposium on Principles of Database Systems (PODS 2000), pp. 1–10 (2000)
23. Kumar, R., Raghavan, P., Rajagopalan, S., Tomkins, A.: Trawling the web for emerging cyber-communities. In: Proceedings of the 8th International Conference on the World Wide Web (WWW 1999), pp. 1481–1493 (1999)
24. Laura, L., Leonardi, S., Millozzi, S., Meyer, U., Sibeyn, J.F.: Algorithms and experiments for the webgraph. In: Di Battista, G., Zwick, U. (eds.) ESA 2003. LNCS, vol. 2832, pp. 703–714. Springer, Heidelberg (2003)
25. Leskovec, J., Kleinberg, J., Faloutsos, C.: Graphs over time: densification laws, shrinking diameters and possible explanations. In: Proceedings of the 11th ACM SIGKDD International Conference on Knowledge Discovery and Data Mining (KDD 2005), pp. 177–187 (2005)
26. Mitzenmacher, M.: A brief history of generative models for power law and lognormal distributions. Internet Math. **1**(2), 226–251 (2004)
27. Nash, J.F.: Non-cooperative games. Ann. Math. **54**, 286–295 (1951)
28. Roughgarden, T.: Intrinsic robustness of the price of anarchy. Commun. ACM **55**(7), 116–123 (2012)
29. Tardos, E., Wexler, T.: Network formation games and the potential function method. In: Nisan, N., Roughgarden, T., Tardos, E., Vazirani, V. (eds.) Algorithmic Game Theory, pp. 487–516. Cambridge University Press, Cambridge (2007). (Chap. 19)

Routing and Fairness

Excluding Braess's Paradox in Nonatomic Selfish Routing

Xujin Chen(✉), Zhuo Diao, and Xiaodong Hu

Institute of Applied Mathematics, AMSS, Chinese Academy of Sciences,
Beijing 100190, China
{xchen,diaozhuo,xdhu}@amss.ac.cn

Abstract. Braess's paradox exposes a counterintuitive phenomenon that when travelers selfishly choose their routes in a network, removing links can improve overall network performance. Under the model of nonatomic selfish routing, we characterize the topologies of k-commodity undirected and directed networks in which Braess's paradox never occurs. Our results generalize Milchtaich's series-parallel characterization for the single-commodity undirected case.

Keywords: Nonatomic selfish routing · Braess's paradox · Single-commodity network · Multicommodity network · Series-parallel graph

1 Introduction

A basic task of network management is routing traffic to achieve the best possible network performance, e.g., to minimize the maximum latency. However, it is usually difficult or even impossible to implement centralized optimal routing in many large systems, as modeled by *selfish routing* games [12]. In these games, a number of players (network users) selfishly choose routes in the network for traveling from their origins to their destinations, aiming to minimize their own latencies. The equilibria of the selfish choices might not be socially optimal. A dazzling example is Braess's paradox [2], which exposes the seemingly counterintuitive phenomenon that less route options for the players lead to shorter travel time at the equilibrium – subnetworks have better performance under the selfish behaviors. The natural question arises as to which topologies of networks are immune to the inefficiency due to the occurrence of Braess's paradox. The characterization of network topologies, which model relatively fixed infrastructures, is independent of the relatively changeable latency functions and traffic demands. Once such a paradox-free network is established, no matter how the latency functions and traffic demands change, the entire network remains the best venue for all selfish players. The goal of this paper is to characterize

Research supported in part by NNSF of China under Grant No. 11222109, 11021161 and 10928102, by 973 Project of China under Grant No. 2011CB80800, and by CAS Program for Cross & Cooperative Team of Science & Technology Innovation.

M. Hoefer (Ed.): SAGT 2015, LNCS 9347, pp. 219–230, 2015.
DOI: 10.1007/978-3-662-48433-3_17

paradox-free network topologies for nonatomic players each routing a negligible portion of the overall traffic, where the nonatomic routing can be viewed as a mathematical idealization of a very large population of individuals.

1.1 Nonatomic Selfish Routing

We concern with both undirected and directed networks, and model them by multigraph or multidigraph $G = (V, E)$ with *vertex* set V and *link* set E. Loops are not allowed, while more than one link can join the same pair of vertices. Each link $e \in E$ is associated with a *nonnegative, continuous, nondecreasing* latency function $\ell_e(\cdot)$ specifying the time needed to traverse e as a function of the link congestion on e. We call G a *graph* (resp. *digraph*) if it is undirected (resp. directed). Undirected links are called *edges* while directed ones are called *arcs*. Throughout the paper, by a path (or a cycle) in a digraph we mean a directed one. Let $u, v \in V$, a path in G from u to v is called a u-v path. For convenience, graphs and digraphs are collectively referred to as (di)graphs.

Let $k \geq 1$ be a positive integer. Given k distinct origin-destination pairs of vertices (s_i, t_i) with $s_i \neq t_i$, $i = 1, \ldots, k$, in G, we call G a k-*commodity network* if G contains at least an s_i-t_i path for each $i \in [k] \coloneqq \{1, \ldots, k\}$. Given a nonnegative *traffic demand* (vector) $\mathbf{r} = (r_i)_{i=1}^k$, the traffic in G comprises k flows, where, for each $i \in [k]$, *the flow of commodity* i has an amount of r_i, and is formed by an infinite number of players traveling from s_i to t_i. Each player selects a single path from his origin to his destination, given the congestion imposed by the rest of players. Assuming a continuum of players, the choice of each individual player has a negligible impact on the experiences of others.

Formally, let the triple $(G, \mathbf{r}, \ell)$ denote a k-commodity selfish routing instance, where latency functions $\ell_e(\cdot)$, $e \in E$, are collectively represented by ℓ. For each $i \in [k]$, let $\mathcal{P}_i$ be the set of s_i-t_i paths in G; a *flow of commodity* i is a nonnegative vector $\mathbf{f}_i = (f_i(P))_{P \in \mathcal{P}_i}$ with $\sum_{P \in \mathcal{P}_i} f_i(P) = r_i$. The combination of $\mathbf{f}_1, \ldots, \mathbf{f}_k$ gives rise to a k-*commodity flow* $\mathbf{f} = (\mathbf{f}_i)_{i=1}^k$ for $(G, \mathbf{r})$. Under $\mathbf{f}$, each link e that is contained by some path in $\mathcal{P} = \cup_{i=1}^k \mathcal{P}_i$ experiences a *congestion* $f(e) = \sum_{i=1}^k \sum_{P \in \mathcal{P}_i : e \in P} f_i(P)$, and thus a *link latency* $\ell_e(f(e))$. Accordingly, each path P contained by $\cup_{Q \in \mathcal{P}} Q$ and any player traveling through P suffer from a *path latency* $\ell_P(\mathbf{f}) = \sum_{e \in P} \ell_e(f(e))$. In this nonatomic routing game, Nash equilibrium is characterized by Wardrop's principle in a way that all players travel only on the minimum latency paths from their own origins to their own destinations.

Definition 1. *We call* $\mathbf{f}$ *a* Nash equilibrium *(NE) of* $(G, \mathbf{r}, \ell)$ *if for each* $i \in [k]$ *and each* $P \in \mathcal{P}_i$ *with* $f_i(P) > 0$, *it holds that* $\ell_P(\mathbf{f}) = \min_{Q \in \mathcal{P}_i} \ell_Q(\mathbf{f})$.

By the classical result of Beckmann et al. [1] (see also [7,12]), the NE of $(G, \mathbf{r}, \ell)$ exist, and are essentially unique in the sense that the link latencies are invariant under any NE of $(G, \mathbf{r}, \ell)$. Thus, for each $i \in [k]$, the common latency experienced by all players traveling from s_i to t_i in any NE of $(G, \mathbf{r}, \ell)$ is also an invariant, which we denote by $\ell_i(G, \mathbf{r})$, and refer to as *the equilibrium latency of the commodity* i for $(G, \mathbf{r}, \ell)$.

1.2 Braess's Paradox

The formal definition of Braess's paradox involves specific meaning of subnetworks. Let $(G, \mathbf{r}, \ell)$ be a k-commodity selfish routing instance with origin-destination pairs $(s_i, t_i)_{i=1}^k$. For any index set $\Pi \subseteq [k]$, a sub(di)graph H of G is referred to as a *subnetwork* of (G, Π) if H is a $|\Pi|$-commodity network with origin-destination pairs $(s_i, t_i)_{i \in \Pi}$, Following [7], a subnetwork of $(G, [k])$ is simply called a *subnetwork* of G.

Let $\Pi(\mathbf{r}) = \{i \in [k] : r_i > 0\}$, H be a subnetwork of $(G, \Pi(\mathbf{r}))$, and ℓ' be the restriction of ℓ to the link set of H. In a mild abuse of notation, we use $(H, \mathbf{r}, \ell)$ to denote the selfish routing instance $(H, (r_i)_{i \in \Pi(\mathbf{r})}, \ell')$, and, for each $i \in \Pi(\mathbf{r})$, use $\ell_i(H, \mathbf{r})$ to denote the equilibrium latency of commodity i for $(H, \mathbf{r}, \ell)$. For simplicity, in case of single-commodity, i.e., $k = 1$, we often write $\mathbf{r}$ as r, and drop the subscript 1 in notions $s_1, t_1, \mathcal{P}_1$, and $\ell_1(\cdot, r)$ in our discussions. In case of $\mathbf{r}$ being an all-one vector we often write it as $\mathbf{1}$.

Definition 2. *We say that* Braess paradox occurs in $(G, \mathbf{r}, \ell)$ *if there exists subnetwork H of $(G, \Pi(\mathbf{r}))$ and $h \in \Pi(\mathbf{r})$ such that $\ell_h(H, \mathbf{r}) < \ell_h(G, \mathbf{r})$ and $\ell_i(H, \mathbf{r}) \le \ell_i(G, \mathbf{r})$ for all $i \in \Pi(\mathbf{r}) - \{h\}$. A k-commodity network G is said to be* Braess's paradox free, *or simply* paradox-free, *if for any traffic demand vector* $\mathbf{r}$ *and any nonnegative, continuous, nondecreasing latency functions ℓ, Braess's paradox does not occur in $(G, \mathbf{r}, \ell)$. A k-commodity network that is not Braess's paradox free is called* Braess's paradox ridden, *or simply* paradox-ridden.[1]

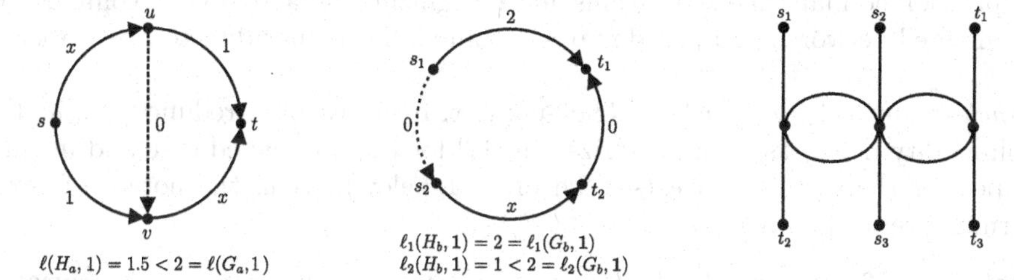

(a) Single-commodity network G_a that is paradox-ridden

(b) 2-commodity network G_b that is paradox-ridden

(c) 3-commodity network G_c that is paradox-free

Fig. 1. Braess's paradox in nonatomic selfish routing.

For example, Braess's paradox occurs in the routing instances (a) and (b) depicted in Fig. 1(a) and (b), respectively, where one unit flow is to be routed from each origin to its corresponding destination. The latency function $\ell_e(x)$ on arc e of network G_a or network G_b is either x or a constant 0, 1 or 2 as represented by the symbol beside the arc in the figure. In either instance, the subnetwork H_a

[1] We remark that the definition of paradox-ridden network here is a substantial relaxation the ones given by Roughgarden [11] and Fotakis et al. [4], which admit instances suffering from *the most severe* performance loss in terms of Braess's paradox.

(resp. H_b) is obtained from G_a (resp. G_b) by deleting the dotted arc. It is easy to see that, in the single-commodity case (a), at the NE of $(G_a, 1, \ell)$ all players go through the path *suvt*, while at the NE of $(H_a, 1, \ell)$ half of players go through path *sut* and the other half go through path *svt*; this gives $\ell(H_a, 1) = 1.5 < 2 = \ell(G_a, 1)$. Similarly, in the 2-commodity case (b), at the NE of $(G_b, \mathbf{1}, \ell)$ no player uses arc (s_1, t_1), while at the NE of $(H_b, \mathbf{1}, \ell)$ the flows of commodities 1 and 2 use disjoint paths s_1t_1 and s_2t_2, respectively; it follows that $\ell_1(H_b, \mathbf{1}) = 2 = \ell_1(G_b, \mathbf{1})$ and $\ell_2(H_b, \mathbf{1}) = 1 < 2 = \ell_2(G_b, \mathbf{1})$. Hence, both networks G_a and G_b are paradox-ridden. In contrast, the 3-commodity network G_c depicted in Fig. 1(c) is paradox-free, as our main result (Theorem 2) below guarantees.

1.3 Paradox-Free Networks

Considering a k-commodity network G with origin-destination pairs $(s_i, t_i)_{i=1}^k$, it would be convenient to think of G being *irredundant* in the sense that each link and each vertex of G are contained in at least an s_i-t_i path for some $i \in [k]$.

Milchtaich [10] established a series-parallel characterization for excluding Braess's paradox in irredundant single-commodity undirected networks (see Theorem 1), which partially solved the open question of characterizing paradox-free networks, proposed by Roughgarden [11]. In the paper, we almost complete the solution by characterizing *all k-commodity undirected networks* and *all irredundant k-commodity directed networks* with the series-parallel and coincident conditions (see Theorem 2). In particular, our results on multicommodity networks answer the open question raised by Milchtaich [10]. The theoretical results imply polynomial time algorithms for recognizing paradox-free k-commodity undirected networks and paradox-free k-commodity planar directed networks.

Single-commodity networks. The paradox-freeness of irredundant single-commodity networks is characterized in [10] for the undirected case and in this paper for the directed case (see Theorem 1 below), using the notion of two-terminal series-parallel.

Definition 3. *An irredundant single-commodity network G with origin-destination pair (s, t) is said to be* two-terminal series-parallel, *or s-t* series-parallel *to be more specific, if one of the following conditions holds.*

(i) G is undirected, and there do not exist two s-t paths in G that pass a common edge in opposite directions.
(ii) G is directed, and its underlying graph is s-t series-parallel.

Theorem 1. *An irredundant single-commodity network is paradox-free if and only if it is two-terminal series-parallel.*

Given the result by Milchtaich [10] for undirected networks, intuitively, one might expect a graphical characterization less restrictive than the two-terminal series-parallel one for directed networks, where players lose the flexibility to traverse a link in either direction, and Braess's paradox might have fewer chances to occur, and one might expect a graphical characterization less restrictive than the two-terminal series-parallel one. However, the intuition is disproved by our result on directed networks.

Despite the similarity of the necessary and sufficient condition for graphs and that for digraphs in Theorem 1, the necessity proof for digraphs constitutes our first technical contribution. Milchtaich's model [10] allows a class of latency functions wider than ours; two (different) functions are defined for each edge $e = uv$, one specifying the latency of passing e from u to v and the other specifying the latency of passing e from v to u. This class of latency functions brings a kind of directionality into Milchtaich's model; the topological efficiency result essentially means that, regardless of how the edges in the two-terminal series-parallel network are directed and their latencies are defined, Braess's paradox does not occur. This is useful for proving the sufficiency of a characterization for single-commodity directed networks to be paradox-free, although our short sufficiency proof (see Sect. 2) does not take advantage of it. On the other hand, Milchtaich's model does preclude predetermined directionality, and the wider class of latency functions does not contribute to the necessity proof for digraphs. Neither the results nor their proofs in [10] can imply the necessity in the directed case.

In the half-page necessity proof for graphs [10], Milchtaich derived a Braess's paradox from an undirected network that is not s-t series-parallel by finding a special pair of vertices whose existence relies on the property that the graph has two s-t paths which goes through an edge in opposite directions. Such a property is lost when considering digraphs. This is the main hurdle to extending Milchtaich's proof to digraphs. New ideas and approaches are required to overcome the difficulty due to more complicate structures of digraphs.

- We use the existence of cycles in digraphs to obtain these special pairs of vertices (see Lemma 1), which in turn lead us to Braess's paradoxes (see Lemmas 2 and 3).
- We translate edge traverses in opposite directions in graphs into the existences of s-t paths through reversed arcs in digraphs. Instead of relying on special pairs of vertices as Milchtaich did, we derive Braess's paradox directly in acyclic digraphs that are not s-t series-parallel using an inductive argument that carefully exploits the properties of digraphs (see Theorem 3).

Multicommodity Network. To characterize paradox-free multicommodity networks, we need the concept of a block [13]. Let H be a (di)graph. We say that H is 2-connected if it is connected and has no cut-vertices. A *block* of H is a maximal 2-connected sub(di)graph of H. If H is two-terminal series-parallel, then each block B of H is also two-terminal series-parallel, where the terminals of B are uniquely determined by the structure of H, and are referred to as *the terminals of B in H*.

For each $i \in [k]$, let $G_i = (V_i, E_i)$ be the maximum (in terms of the number of links) irredundant subnetwork of $(G, \{i\})$. Then G_i, consisting of all s_i-t_i paths in G, is an irredundant single-commodity network with origin-destination pair (s_i, t_i). We call s_i and t_i the *terminals* of G_i.

Definition 4. *If G_i and G_j ($i \neq j$) are two-terminal series-parallel, B is a block of both G_i and G_j, and the set of terminals of B in G_i and that in G_j are the same, then B is called a coincident block of G_i and G_j.*

It is worth noting that a common block B of G_i and G_j is coincident if G is directed, but it is not necessarily coincident if G is undirected, as the terminal sets of B in undirected G_i and G_j might be different. In addition, the order of a coincident block's terminals ("who is the origin and who is the destination") makes a difference between directed and undirected networks. Given a coincident block B of G_i and G_j, in the directed case, the order of B's terminals is unique; while in the undirected case, it is possible that the origin (resp. destination) of B in G_i is the destination (resp. origin) of B in G_j. This case of different order can be seen from G_1 and G_2 for $G = G_c$ in Fig. 1(c).

The next theorem summarizes the main results of this paper, which concerns both the directed and undirected case.

Theorem 2 (Main result). *An irredundant k-commodity network G is paradox-free if and only if G satisfies the following conditions:*

(i) Series-parallel Condition*: for each $i \in [k]$, G_i is an s_i-t_i series-parallel network; and*
(ii) Coincident Condition*: for any distinct $i, j \in [k]$, either $E_i \cap E_j = \emptyset$ or the (di)graph induced by $E_i \cap E_j$ consists of all coincident blocks of G_i and G_j.*

The theorem clearly generalizes Theorem 1. This generalization is by no means straightforward. Indeed, the coincident condition specifies the interactions of players with different origins and destinations, capturing in the context of paradox-freeness a key property of asymmetric nonatomic selfish routing, which, to the best of our knowledge, was not studied previously.

The configuration depicted in Fig. 1(b) and its undirected underlying configuration give visualizations of how Braess's paradox occurs in multicommodity networks. Moreover, we show that the they are essentially all the forbidden structures for the paradox-freeness in multicommodity networks. This complements the result that Wheatstone network (G_a in Fig. 1(a) and its underlying graph) is the only forbidden configuration for paradox-free single-commodity undirected networks.

The proofs for the necessity of the coincident condition constitute our second technical contribution.

- In the undirected case, coping with the nonidentical sets of terminals for a common block of G_i and G_j turns out to be the key for obtaining the paradox configurations. The careful path selections to avoid unnecessary intersections, which utilize techniques from graph connectivity theory, discover the essence of the problem.
- In the directed case, the difficulty lies on the fact that pure graph theory cannot enforce two intersecting blocks to be identical as it does for undirected networks. More elaborate inductive method is applied for reducing the proof

to smaller networks (obtained by arc deletions and contractions). As arc contractions might create Braess's paradox which is not possessed by the original network, the challenging task is to guarantee that the selected arc contraction does not destroy the paradox-freeness.

Inour proofs, the constructions of Braess's paradoxes only use linear latency functions of form $\ell_e(x) = a_e x + b_e$ for constants $a_e, b_e \geq 0$. An immediate corollary is that our results all hold even if the paradox-freeness is defined with respect to this kind of linear latency functions.

Our characterizations for paradox-free (paradox-ridden) networks show that the multicommodity cases are natural and nontrivial extensions from their single-commodity counterparts, which stands in contrast to the dichotomy between single- and 2-commodity networks in terms of severity of Braess's paradox [7,11]. It is known that Braess's paradox can be dramatically more severe for multi-commondity networks than single-commodity networks.

1.4 Related Work

Most literature on characterizing network topologies for various properties of NE in selfish routing is restricted to the single-commodity case. The most related work is the aforementioned two-terminal series-parallel characterization by Milchtaich [10] for paradox-free undirected networks in nonatomic selfish routing. In the same paper [10], the author proved that the undirected networks which guarantee all NE to be weakly Pareto efficient are exactly those with *linearly independent routes*, meaning that every *s*-*t* path has at least an edge that does not belong to any other *s*-*t* path. Networks with *heterogeneous* players means that the latency functions are player-specific: the same link under the same congestion might give different latencies to different players using it. For the nonatomic routing with heterogeneous players, Milchtaich [8] characterized undirected networks such that, given any strictly increasing latency functions, each player's latency is the same at all NE. These networks are either nearly parallel or consist of two or more nearly parallel networks connected in series. In contrast to nonatomic routing, an *atomic routing* game has only a finite number of players, each controlling a noneligible part of traffic. The most studied scenarios are *unsplittalbe routing* where each player routes his flow through a single path from his origin to his destination, and *routing with unit demands* where each player controls a unit of flow. Milchtaich [9] identified some sufficient conditions for directed networks that guarantee the existence of at least one pure NE in unsplittable atomic routing with unit demands for either player-specific latency functions or weighted players.

As far as multicommodity networks are concerned, Holzman and Monderer [6] studied the unsplittable atomic routing game with unit demands that is played on a directed network with two distinguished vertices u and v, where every arc belongs to at least one u-v path. Under this topological constraint, the authors [6] proved that the class of extension-parallel networks is exact the one guaranteeing the existence of strong equilibrium. This is a generalization of the previous result on the single-commodity case [5]. Epstein et al. [3] characterized the so-called *efficient* undirected networks in which all NE of each unsplittable routing with unit demands are socially optimal, i.e., they are among routings that minimize the maximum path latency between any origin-destination pair. It was shown that the efficient multicommodity networks are either trees or two vertices joined by parallel edges, while the efficient single-commodity networks are exactly those with linearly independent routes. The authors [3] also obtained characterizations of efficient undirected networks for the routing game where both individual players and the network (society) wish to minimize their own maximum edge latencies.

Organization. In Sect. 2, we briefly discuss our proofs for single-commodity networks and multicommodity networks. In Sect. 3, we give concluding remarks. Due to the space limit, the technical details, as well as discussions on corollaries and future research, are deferred to the full version of the paper.

2 Proofs

We first study single-commodity network $G = (V, E)$ with origin-destination pair (s, t). It can be assumed w.l.o.g. that G is connected.

For the undirected case, the characterization for irredundant networks can be easily extended to all single-commodity networks. This particularly provides a polynomial time algorithm for determining whether a given undirected single-commodity network is paradox-free or not.

Proof of Theorem 1 (Directed Case). We further assume that G is an irredundant directed network, meaning that each arc of G is contained in some s-t path. Let $\mathbb{u}(G)$ denote the underlying graph of G. Let $P = v_1 v_2 \dots v_h$ be a v_1-v_h path in G. For any $1 \leq i < j \leq h$, we say that v_i *precedes* v_j in P, or equivalently v_j *follows* v_i in P; such a relation is written as $v_i \prec_P v_j$. If $1 \leq i \leq j \leq h$, we write $v_i \preceq_P v_j$, and use $P[v_i, v_j]$ to denote the subpath of P from v_i to v_j. For convenience, we set $P(v_i, v_j] = P[v_i, v_j] \setminus v_i$, $P[v_i, v_j) = P[v_i, v_j] \setminus v_j$ and $P(v_i, v_j) = P[v_i, v_j] \setminus \{v_i, v_j\}$.

Sufficiency Proof. Suppose that G is s-t series-parallel. Then so is $\mathbb{u}(G)$ by Definition 3. It can be seen that $\{\mathbb{u}(P) : P \in \mathcal{P}\}$ is exactly the set of s-t paths in $\mathbb{u}(G)$. Suppose for a contradiction that there exist $H \subseteq G$, traffic demand r and nonnegative, continuous, nondecreasing latency functions ℓ such that $\ell(H, r) < \ell(G, r)$. For each edge e of $\mathbb{u}(G)$, set $\underline{\ell}_e(\cdot) = \ell_{e'}(\cdot)$ where e is the underlying edge of the arc $e' \in E$. Then $\underline{\ell}(\mathbb{u}(H), r) = \ell(H, r) < \ell(G, r) = \underline{\ell}(\mathbb{u}(G), r)$, exhibiting a Braess's paradox in $\mathbb{u}(G)$. We deduce from Theorem 1 (Undirected Case) that $\mathbb{u}(G)$ is not s-t series-parallel. The contradiction proves the sufficiency.

Necessity Proof. Our approach is to derive Braess's paradox from directed networks that are not two-terminal series-parallel. The contradictory method was adopted by Milchtaich [10], who used the property that the graph has two s-t paths which goes through an edge in opposite directions. Such a property is lost when considering digraphs. This is main hurdle to extending Milchtaich's proof to digraphs.

Our first step is to use the existence of a cycle to derive a special pair of vertices u and v as in the following lemma.

Lemma 1. *If there is a cycle in G, then there exist s-t paths $P, Q \in \mathcal{P}$ and distinct vertices $u, v \in V(P) \cap V(Q)$ such that u precedes v in P and u follows v in Q, i.e., $u \prec_P v$ and $v \prec_Q u$.* □

Consequently, such a pair of special vertices u and v helps us to construct (in Lemma 2) an s-t paradox defined as follows.

Definition 5. *We call G an s-t* paradox *if $G = P_1 \cup P_2 \cup P_3$ is the union of three paths P_1, P_2 and P_3 with the following properties:*

(i) P_1 is an s-t path going through distinct vertices a, u, v, b such that $s \preceq a \prec_{P_1} u \prec_{P_1} v \prec_{P_1} b \preceq t$;
(ii) P_2 is an a-v path with $V(P_2) \cap V(P_1) = \{a, v\}$;
(iii) P_3 is a u-b path with $V(P_3) \cap V(P_1) = \{u, b\}$ and $V(P_3) \cap V(P_2) = \emptyset$.

The s-t paradox defined above is often denoted as (P_1, P_2, P_3). See Fig. 2 for an illustration.

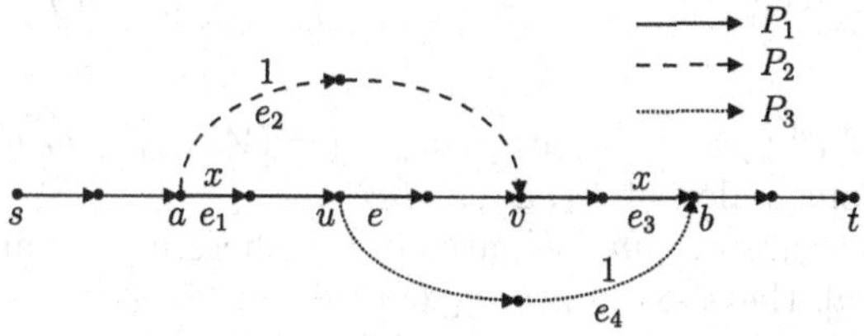

Fig. 2. An s-t paradox (P_1, P_2, P_3).

Lemma 2. *If there is a cycle in G, then G contains an s-t paradox.* □

Observe that the four-node Braess's paradox G_a in Fig. 1(a) is an s-t paradox $(suvt, sv, ut)$ with $s = a$ and $t = b$. More generally, we have

Lemma 3. *If G contains an s-t paradox, then G is paradox-ridden.*

Proof. Let $G' = (V', E') = P_1 \cup P_2 \cup P_3$ be an s-t paradox (P_1, P_2, P_3) contained in G. In the subdigraph G', let e_1, e_2 be the two outgoing arc from a with $e_1 \in P_1$ and $e_2 \in P_2$, and let e_3, e_4 the two incoming arcs to b with $e_3 \in P_1$ and $e_4 \in P_3$. We define routing instance $(G, 1, \ell)$ by $\ell_e(x) = x$ if $e \in \{e_1, e_3\}$, $\ell_e(x) = 1$ if $e \in \{e_2, e_4\}$, $\ell_e(x) = 0$ if $e \in E' - \{e_1, e_2, e_3, e_4\}$, and $\ell_e(x) = \infty$ if $e \in E - E'$. See Fig. 2 for an illustration. The unique equilibrium in $(G, 1, \ell)$ sends the one-unit flow all through path P_1 and suffers from a latency $\ell(G, 1) = \ell_{e_1}(1) + \ell_{e_3}(1) = 2$. Let subnetwork H of $(G, 1)$ be obtained from G by removing an arc e on $P_1[u, v]$. Then the unique equilibrium in (H, r, ℓ) splits the one-unit flow equally at a, sending half via path $P_2[a, v] \cup P_1[v, t]$ and the other half via path $P_1[a, u] \cup P_3[u, b] \cup P_1[b, t]$. This incurs a latency $\ell(H, 1) = 1 + 0.5 = 1.5$, showing a Braess's paradox. □

Finally, we restrict our attention to acyclic digraphs. Using the existences of s-t path through reversed arcs in digraphs, we derive s-t paradoxes directly via an inductive argument, instead of relying on those special pair of vertices u, v provided by Lemma 1.

For any arc subset $K \subseteq E$, let $G\langle K\rangle$ be the digraph obtained from G by reversing all arcs in K. The set of reversed arcs is written as $\bar{K}$.

Theorem 3. *If G is not s-t series-parallel, then G is paradox-ridden.*

Proof. By Lemma 3, suppose on the contrary that G contains no s-t paradox. It follows from Lemma 2 that G is acyclic. Therefore we have

(i) for any $u, v, w \in V$ and any u-v path P_{uv} and v-w path P_{vw} in G, $P_{uv} \cup P_{vw}$ is a u-w path in G.

By Definitions 3, since digraph G is not s-t series-parallel, its underlying graph has a pair of s-t paths which go through some edge in opposite directions. Recall that each arc of G is contained in some s-t path in G. We deduce that there is a nonempty subset K of E such that $G\langle K\rangle$ contains an s-t path P going through some arc(s) in $\bar{K}$. Take such a K with minimum $|K| > 0$. The minimality of K implies that $\bar{K} \subseteq P$, and $\bar{K}$ induces a number of, say m, subpaths of P from v_i to w_i, $i = 1, 2, \ldots, m$, where $v_i \prec_P w_i \prec_P v_{i+1} \prec_P w_{i+1}$ for all $i = 1, 2, \ldots, m-1$. Notice that

(ii) the reverse of $P[v_i, w_i]$ is a path $\bar{P}[w_i, v_i] = (V_i, E_i)$ in G for each $i \in [m]$;
(iii) K is the disjoint union of $E_1, E_2, \ldots, E_m$;
(iv) $P[w_i, v_{i+1}]$, $i = 0, 1, \ldots, m$ are paths in G, where $w_0 = s$ and $v_{m+1} = t$;
(v) for any $i \in [m]$, there exists a v_i-v_1 path Q_i in G.

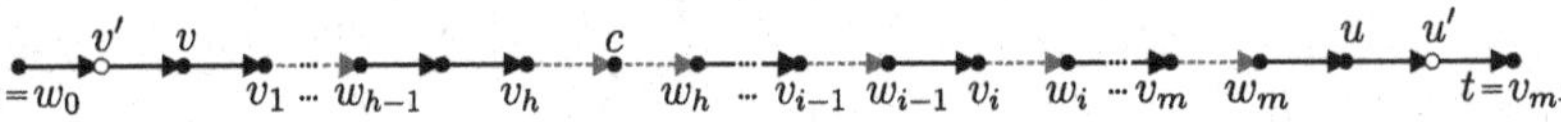

Fig. 3. The path P in $G\langle K\rangle$, where arcs in G and $\bar{K}$ are drawn as solid and dotted ones, respectively.

Statements (ii)–(iv) are straightforward observations. See Fig. 3 for an illustration. We prove (v) by induction on i. The base case $i = 1$ is trivial. Suppose that $2 \le i \le m$ and there is a v_h-v_1 path Q_h in G for each $h \in [i-1]$.

Since G is an irredundant network, it contains a v_i-t path Q. If Q and $P[s, v_i)$ are vertex-disjoint, then $P[s, v_i] \cup Q$ is an s-t path in $G\langle \cup_{h=1}^{i-1} E_h\rangle$ going through $\cup_{h=1}^{i-1} \bar{E}_h$. However $\emptyset \neq \cup_{h=1}^{i-1} E_h \subsetneq \cup_{h=1}^{m} E_h$ and (iii) imply a contradiction to the minimality of K. Hence

- $P(s, v_i)$ and $Q(v_i, t)$ have some vertices in common, say c.

Observe from (iv) and (i) that $P[w_{i-1}, v_i] \cup Q$ is a path in G, implying that $P[w_{i-1}, v_i)$ and Q are vertex-disjoint, and

- $c \in P(s, w_{i-1}) \cap Q(v_i, t)$.

Note that $c \in P[w_{h-1}, w_h)$ for some $h \leq i-1$. By (iv) and (ii), the subdigraph $P[w_{h-1}, v_h] \cup \bar{P}[w_h, v_h]$ of G contains a c-v_h path, written as R. Consider the concatenation of R and the v_h-v_1 path Q_h (whose existence is guaranteed by the inductive hypothesis). We see from (i) that $R \cup Q_h$ is a c-v_1 path in G. In turn we concatenate the v_i-c path $Q[v_i, c]$ and the c-v_1 path $R \cup Q_h$ into a v_i-v_1 path in G, which establishes (v).

Applying (ii) and (v) with $i = m$, we deduce from (i) that $S_{w_m v_1} = \bar{P}[w_m, v_m] \cup Q_m$ is a w_m-v_1 path in G. Since G is an irredundant network, it contains an s-w_m path S_{sw_m} and a v_1-t path $S_{v_1 t}$. By (i), $S = S_{sw_m} \cup S_{w_m v_1} \cup S_{v_1 t}$ is an s-t path in G satisfying $s \prec_S w_m \prec_S v_1 \prec_S t$. Observe that $v_1, w_m \in P \cap S$. This enables us to take vertices $u, v \in P \cap S$ such that

- $u \prec_S v$, $v \prec_P u$,
- $P[v_1, w_m] \subseteq P[v, u]$, and
- $P[v, u]$ is as long as possible.

Notice from $s \preceq_S u \prec_S v \preceq_S t$ that $v \neq s$ and $u \neq t$, giving $s \in P[s, v)$ and $t \in P(u, t]$. Recall from (iv) that $P[s, v_1] \cup P[w_m, t] = P[w_0, v_1] \cup P[w_m, v_{m+1}] \subseteq G$. From $P[v_1, w_m] \subseteq P[v, u]$, we derive

(vi) $P[s, v] \cup P[u, t] \subseteq G$.

The maximality of $P[v, u]$ enforces that

(vii) $P[s, v) \cap S[u, t] = \emptyset$ and $P(u, t] \cap S[s, v] = \emptyset$.

Otherwise we should have taken v to be some vertex $v' \in P[s, v) \cap S(u, t)$ with $u \prec_S v'$, or u to be some vertex $u' \in P(u, t] \cap S(s, v)$ with $u' \prec_S v$. See Fig. 3 for the positions of the contradictory v' and u' on P.

Consider a traverse of P from s to t. Let a be the last vertex of $P[s, v)$ with $a \in S$, meaning that $P(a, v) \cap S = \emptyset$ (such a vertex a exists because $s \in P[s, v) \cap S$). Let b be the first vertex of $P(u, t]$ with $b \in S$, meaning $P(u, b) \cap S = \emptyset$ (such a vertex b exists because $t \in P(u, t] \cap S$). It follows from (vii) that $a \in S[s, u)$ and $b \in S(v, t]$, giving $a \prec_S u \prec_S v \prec_S b$. (Notice from $s \prec_P v \prec_P u \prec_P t$ that $s \neq u$ and $v \neq t$.) Thus, by (vi), we see that $(S, P[a, v], P[u, b])$ is an s-t paradox in G. The contradiction completes the proof of the theorem. □

Proof of Theorem 2. Our goal is to prove the series-parallel and coincident conditions in the theorem are necessary and sufficient for a multicommodity network to be paradox-free.

By virtue of the block chain structure of series-parallel networks, the sufficiency proof, which builds on the results for single-commodity (Theorem 1), turns out to be easier. For the harder necessity proof, the main focus is on establishing the coincident condition for 2-commodity networks.

3 Conclusion

The main result of this paper is a graphical characterization for all irredundant networks to be paradox-free; the series-parallel and coincident conditions are shown to be sufficient and necessary for the paradox-freeness. In our proofs, the constructions of Braess's paradoxes only use linear latency functions of form $\ell_e(x) = a_e x + b_e$ for constants $a_e, b_e \geq 0$. An immediate corollary is that our results all hold even if the paradox-freeness is defined with respect to this kind of linear latency functions.

References

1. Beckmann, M.J., McGuire, C.B., Winsten, C.B.: Studies in the Economics of Transportation. Yale University Press, New Haven (1956)
2. Braess, D.: Über ein paradoxon aus der verkehrsplanung. Unternehmensforschung **12**(1), 258–268 (1968)
3. Epstein, A., Feldman, M., Mansour, Y.: Efficient graph topologies in network routing games. Game. Econ. Behav. **66**(1), 115–125 (2009)
4. Fotakis, D., Kaporis, A.C., Lianeas, T., Spirakis, P.G.: On the hardness of network design for bottleneck routing games. Theor. Comput. Sci. **521**, 107–122 (2014)
5. Holzman, R., Yone (Lev-tov), N.L.: Network structure and strong equilibrium in route selection games. Math. Soc. Sci. **46**(2), 193–205 (2003)
6. Holzman, R., Monderer, D.: Strong equilibrium in network congestion games: increasing versus decreasing costs. Int. J. Game Theory **44**, 1–20 (2014)
7. Lin, H., Roughgarden, T., Tardos, É., Walkover, A.: Stronger bounds on braess's paradox and the maximum latency of selfish routing. SIAM J. Discrete Math. **25**(4), 1667–1686 (2011)
8. Milchtaich, I.: Topological conditions for uniqueness of equilibrium in networks. Math. Oper. Res **30**(1), 225–244 (2005)
9. Milchtaich, I.: The equilibrium existence problem in finite network congestion games. In: Spirakis, P.G., Mavronicolas, M., Kontogiannis, S.C. (eds.) WINE 2006. LNCS, vol. 4286, pp. 87–98. Springer, Heidelberg (2006)
10. Milchtaich, I.: Network topology and the efficiency of equilibrium. Game. Econ. Behav. **57**(2), 321–346 (2006)
11. Roughgarden, T.: On the severity of braess's paradox: designing networks for selfish users is hard. J. Compu. Syst. Sci. **72**(5), 922–953 (2006)
12. Roughgarden, T., Tardos, É.: How bad is selfish routing? J. ACM **49**(2), 236–259 (2002)
13. Tutte, W.T.: Graph Theory. Electronic Library of Mathematics. China Machine Press, Beijing (2004)

"Beat-Your-Rival" Routing Games

Gideon Blocq(✉) and Ariel Orda

Technion, Israel Institute of Technology, Haifa, Israel
gideon@tx.technion.ac.il, ariel@ee.technion.ac.il

Abstract. In the traditional setting of routing games, the standard assumption is that selfish agents are unconcerned with the performance of their competitors in the network. We propose an extension to this setting by modeling agents to consider a combination of their own performance as well as that of their *rivals*. Per agent, we parameterize this trade-off, thereby allowing agents to be partially selfish and partially malicious.

We consider two types of routing games based on the structure of the agents' performance objectives, namely *bottleneck routing games* and *additive routing games*. For *bottleneck routing games*, the performance of an agent is determined by its worst-case link performance, and for *additive routing games*, performance is determined by the sum of its link performances. For the bottleneck routing scenario we establish the existence of a Nash equilibrium and show that the *Price of Stability* is equal to 1. We also prove that the *Price of Anarchy* is unbounded. For *additive routing games*, we focus on the fundamental load balancing game of routing over parallel links. For an interesting class of agents, we prove the existence of a Nash equilibrium. Specifically, we establish that a special case of the *Wardrop equilibrium* is likewise a *Nash equilibrium*. Moreover, when the system consists of two agents, this Nash equilibrium is unique, and for the general case of N agents, we present an example of its non-uniqueness.

1 Introduction

To date, game theoretic models have been employed in virtually all networking contexts. These include control tasks at the network layer, such as flow control (e.g., [16]), and routing (e.g., [1,5,18,20,24,25] and references therein), as well as numerous studies on control tasks at the link and MAC layers. A fundamental assumption in all of these referenced studies is that the selfish agents compete over resources in the network and aim to optimize their own performance; agents do not care (either way) about the performance of their competitors. However, and typically in the context of routing, scenarios exist in which this assumption is not warranted.

For example, consider the scenario where two Content Providers, A and B, offer video-on-demand services in a network. Both A and B compete over the network resources, however only Content Provider A aspires to minimize its own latency. Due to business considerations, Content Provider B aims at offering its

M. Hoefer (Ed.): SAGT 2015, LNCS 9347, pp. 231–243, 2015.
DOI: 10.1007/978-3-662-48433-3_18

clients a performance that is equal or better than A's performance. Thus, the objective of B is not solely to maximize its performance.

In light of examples like the one above, previous research in routing games has extended the classical model of "performance-maximizing" or "selfish" agents, and focused on different scenarios, e.g., settings where certain agents may act *maliciously* towards other agents [4,7,22]. Such malicious behavior could be due to a range of reasons, e.g., hackers or rivaling companies that aim to degrade network quality. In contrast, other studies in routing games consider agents to have an altruistic component to their objective [3,9,15].

In order to best model real-life scenarios, each agent's objective should lie somewhere in the range between *malicious*, *selfish* and *altruistic*, as depicted in Fig. 1. This direction has been proposed in [11], where each agent i has a parameter that captures how important the social performance is to i. In this setting, a malicious agent aims to minimize the social performance, an altruistic agent aims to maximize it and a selfish agent does not take the social performance into account at all. However, [11] focuses on a non-atomic game, i.e., a game with an infinite amount of agents, where each agent controls a negligible amount of flow. Following a similar course, in [3,9,10,15], agents are of finite size, and their objectives are parameterized to lie somewhere between *selfish* and *altruistic*, yet *malicious* objectives are not taken into account.

In this study, we intend to investigate agents of finite size whose objectives lie in the range between *malicious* and *selfish*. Per agent i, we parameterize this trade-off through a coefficient $\alpha^i \in [0,1]$, where $\alpha^i = 1$ corresponds to a selfish agent and $\alpha^i = 0$ to a malicious agent. However, unlike [11], we represent agent i's cost as a combination of its own performance and that of its *rival*. We define the rival of an agent i as the agent $j \neq i$ with the current best performance in the system. Note that an agent's rival is not fixed, but is dependent on the current performance of all the agents in the system. In our setting, a totally malicious agent aims to minimize the performance of its rival, while a totally selfish agent does not take its rival's performance into account.

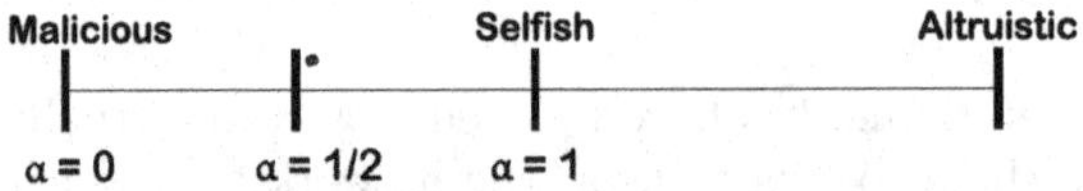

Fig. 1. The range of agents' objectives.

We consider two types of routing games based on the structure of the agents' performance objectives. The first game considers agents with *bottleneck objectives* (also known as Max-Min or Min-Max objectives), i.e., their performance is determined by the worst component (link) in the network [5,8,12]. *Bottleneck routing games* have been shown to emerge in many practical scenarios. For example, in wireless networks, the weakest link in a transmission is determined by the node with the least remaining battery power. Hence, each agent would route traffic so as to maximize the smallest battery lifetime along its routing topology.

Additionally, bottleneck routing games arise in congested networks where it is desirable to move traffic away from congested hot spots. For further discussion and additional examples see [5]. The second type of game considers agents with *additive* performance measures, e.g., delay or packet loss. Much of the current literature on networking games has focused on such games, e.g., [1,14,17–20,25], albeit in the traditional setting of selfish agents.

In [5] and [20], the existence of a Nash equilibrium has been established respectively, for bottleneck and additive routing games with selfish agents. We note that a major complication in proving the existence of a Nash equilibrium for agents with a malicious component, i.e., $\alpha^i < 1$, is the inherent lack of convexity of the objective functions. Thus, we cannot rely on the proofs of existence from the referenced works, and need to establish proofs of our own that do not require (quasi-)convexity of the performance functions.

For both types of games, many studies have attempted to bound the Price of Anarchy (PoA) [17] and the Price of Stability (PoS) [2]. The PoA and PoS quantify the deficiency of the network from a social perspective, at respectively, the worst and best Nash equilibrium. Due to the ever-growing work in this context, it is beyond the scope of this writing to do justice and present an exhaustive survey of previous work on routing games with selfish agents. We refer the reader to the above cited papers and to the references therein for a broader review of the literature.

1.1 Our Contribution

We focus our study on the atomic splittable routing model [5,20], in which each agent sends its non-negligible demand to its destination by splitting it over a set of paths in the network. All agents share the same source and destination, and each agent i has a coefficient α^i, which captures the importance of its rival's performance. We first consider agents with *bottleneck performance measures*, and for which $\alpha^i \in [1/2, 1]$. Intuitively, this range of α^i implies that they care more about their own performance than that of their rivals'. We prove that the Price of Stability is equal to 1, i.e., there always exists a system optimal Nash equilibrium. Moreover, we establish that the Price of Anarchy is unbounded.

We then consider agents with *additive* performance objectives and focus on the fundamental load balancing game of routing over parallel links. Beyond being a basic framework of routing, this is the *generic framework of load balancing* among servers in a network. It has been the subject of numerous studies in the context of non-cooperative networking games, e.g., [14,17,18,20,23,26], to name a few. We consider agents that view their own performance and that of their rivals with equal importance, i.e., for all i, $\alpha^i = 1/2$. We establish the existence of a Nash equilibrium and show that the Wardrop equilibrium (which necessarily exists and is unique [13]) is also a Nash equilibrium. Moreover, for a system with two agents, we prove the Nash equilibrium's uniqueness and for the general case of N agents, we provide an example of its non-uniqueness. Finally, we present an example of a system with agents for which $\alpha^i \in [0, 1]$ and show that for both *bottleneck* and *additive* routing games, no Nash equilibrium necessarily exists. Due to space limitations, some proofs are omitted and can be found (online) in [6].

2 Model and Game Theoretic Formulations

2.1 Model

We consider a set $\mathcal{N} = \{1, 2, \ldots, N\}$ of selfish "users" (or, "players", "agents"), which share a communication network modeled by a directed graph $G(V, E)$. We denote by $\mathcal{P}$ the set of all paths in the network. Each user $i \in \mathcal{N}$ has a traffic demand r^i and all users share a common source S and common destination T. Denote the total demand of all the users by R, i.e., $R = \sum_{i \in \mathcal{N}} r^i$. For every i, we denote by $-i$ the set of all users in the system, excluding i. A user ships its demand from S to T by splitting it along the paths in $\mathcal{P}$, i.e., user i decides what fraction of r^i should be sent on through each path. We denote by f_p^i, the flow that user $i \in \mathcal{N}$ sends on path $p \in \mathcal{P}$. User i can fix any value for f_p^i, as long as $f_p^i \geq 0$ (non-negativity constraint) and $\sum_{p \in \mathcal{P}} f_p^i = r^i$ (demand constraint); this assignment of traffic to paths, $\mathbf{f^i} = \{f_p^i\}_{p \in \mathcal{P}}$ shall also be referred to as the *routing strategy* of user i. The *(routing strategy) profile* $\mathbf{f}$ is the vector of all user routing strategies, $\mathbf{f} = (\mathbf{f^1}, \mathbf{f^2}, \ldots, \mathbf{f^N})$. We say that a profile $\mathbf{f}$ is feasible if it is composed of feasible routing strategies and we denote by $\mathbf{F}$ the set of all feasible profiles. Turning our attention to a path $p \in \mathcal{P}$, let f_p be the total flow on that path i.e., $f_p = \sum_{i \in \mathcal{N}} f_p^i$; also denote by f_e^i the flow that i sends on link $e \in E$, i.e., $f_e^i = \sum_{p | e \in p} f_p^i$. Similarly, the total flow on link $e \in E$ is denoted by $f_e = \sum_{i \in \mathcal{N}} f_e^i$. We associate with each link a performance function $T_e(\cdot)$, which corresponds to the *cost per unit of flow* through link e and only depends on the total flow f_e. Furthermore, we impose the following assumptions on $T_e(f_e)$:

A1 $T_e : [0, \infty) \rightarrow [0, \infty]$.
A2 $T_e(f_e)$ is continuous and strictly increasing in f_e.

The performance measure of a user $i \in \mathcal{N}$ is given by a cost function $H^i(\mathbf{f})$, which we shall refer to as the *selfish* cost of i. In bottleneck routing games, $H^i(\mathbf{f})$ corresponds to the performance of the worst-case link, and in additive routing games it corresponds to the sum of all link performances in the system. We define the *rival* of i at $\mathbf{f}$, as the user with the lowest selfish cost at $\mathbf{f}$, i.e., $\min_{j \neq i} H^j(\mathbf{f})$. The aim of each user is to minimize the weighted difference between its own cost and the cost of its *rival* in the network. Thus, the aim of i is to minimize

$$J^i(\mathbf{f}) \equiv \alpha^i H^i(\mathbf{f}) - (1 - \alpha^i) \min_{j \neq i} \{H^j(\mathbf{f})\}. \tag{1}$$

Note that $J^i(\mathbf{f})$ is not necessarily convex in its user flows.

2.2 Bottleneck Routing Cost Function

Following [5], we define the bottleneck of a user $i \in \mathcal{N}$, $b^i(\mathbf{f})$, as the worst performance of any link in the network that i sends a positive amount of flow on,

$$b^i(\mathbf{f}) = \max_{e \in E | f_e^i > 0} T_e(f_e).$$

The *selfish cost* of user i is equal to its bottleneck, $H^i(\mathbf{f}) = b^i(\mathbf{f}) = \max_{e\in E|f_e^i>0} T_e(f_e)$. Thus, we consider users whose cost functions contain the following form,

$$J^i(\mathbf{f}) = \alpha^i \max_{e\in E|f_e^i>0} \{T_e(f_e)\} - (1-\alpha^i) \min_{j\neq i} \max_{l\in E|f_l^j>0} \{T_l(f_l)\}. \tag{2}$$

In other words, user i aims to minimize the weighted difference between its bottleneck and that of its best-off competitor. We define the bottleneck of a path $p \in \mathcal{P}$ with $f_p > 0$ as $b_p(\mathbf{f}) = \max_{e\in p} T_e(f_e)$ and we define the bottleneck of the system as

$$b(\mathbf{f}) = \max_{e\in E|f_e>0} T_e(f_e).$$

We equate the "welfare" of the system to its bottleneck and denote by $\mathbf{f}^* = (\mathbf{f}^*)_{e\in E}$, the optimal vector of link flows. Thus, the social optimum equals $b(\mathbf{f}^*) = \min_{\mathbf{f}\in\mathbf{F}} b(\mathbf{f})$.

2.3 Additive Routing Cost Functions

An important class of problems is when users are interested additive performance measures, e.g., delay or packet loss. In this case, T_e may correspond to the total delay of link e. For *additive routing games*, we consider the framework of routing in a "parallel links" network. Thus, $G(V, E)$ corresponds to a graph with parallel "links" (e.g., communication links, servers, etc.) $\mathcal{L} = \{1, 2, \ldots, L\}$, $L > 1$, and a users ships its demand by splitting it over the links $\mathcal{L}$. In particular, we consider users whose selfish cost functions are of the following form:

$$H^i(\mathbf{f}) = \frac{1}{r^i} \sum_{l\in\mathcal{L}} f_l^i T_l(f_l). \tag{3}$$

Thus, $H^i(\mathbf{f})$ corresponds to the average sum of the link costs. From (1) we get that

$$J^i(\mathbf{f}) = \alpha^i \sum_{l\in\mathcal{L}} \frac{f_l^i}{r^i} T_l(f_l) - (1-\alpha^i) \min_{j\neq i} \left\{ \sum_{l\in\mathcal{L}} \frac{f_l^j}{r^j} T_l(f_l) \right\}. \tag{4}$$

2.4 Nash Equilibrium

A profile $\mathbf{f}$ is said to be a Nash equilibrium if, given $\mathbf{f^{-i}}$, no user finds it beneficial to deviate from its routing strategy $\mathbf{f^i}$. More formally, $\mathbf{f}$ is a Nash equilibrium if, for all $i \in \mathcal{N}$ and any feasible routing strategy $\mathbf{\bar{f}^i} \neq \mathbf{f^i}$, the following condition holds

$$J^i(\mathbf{f^i}, \mathbf{f^{-i}}) \leq J^i(\mathbf{\bar{f}^i}, \mathbf{f^{-i}}). \tag{5}$$

In order to quantify the degradation of a Nash equilibrium, we turn towards the *Price of Anarchy* [17] (the *Price of Stability* [2]), which is defined as the ratio between the *worst* (*best*) Nash equilibrium, and the social optimum.

3 Bottleneck Routing Games

We start by establishing the existence of a Nash equilibrium in our bottleneck routing game. Note that the user cost function in (2) is not continuous, as pointed out in [5]. Moreover, $J^i(\mathbf{f})$ is not necessarily quasi-convex in f_l^i. Consequently, we need to construct an existence proof that does not rely on the continuity or the quasi-convexity of the cost functions. We establish the existence of a Nash equilibrium by constructing a feasible strategy profile for all users, such that no user wishes to unilaterally deviate from its routing strategy. We first provide the following definition.

Definition 1. *A profile, $\mathbf{f}$, is referred to as balanced, if for any two paths $p_1, p_2 \in \mathcal{P}$ with $f_{p_1} > 0$, it holds that, $b_{p_1}(\mathbf{f}) \le \max_{e \in p_2}\{T_e(f_e)\}$.*

Thus, at a *balanced* flow profile, for any two paths $p_1, p_2 \in \mathcal{P}$ with positive flow, their bottlenecks are equal, $b_{p_1}(\mathbf{f}) = b_{p_2}(\mathbf{f})$.

Definition 2. *A profile, $\mathbf{f}$, is referred to as proportional, if for any path $p \in \mathcal{P}$, and for any user $i \in \mathcal{N}$, $f_p^i = \frac{r^i}{R} f_p$.*

To demonstrate that a proportional profile is feasible, it needs to satisfy **(i)** the non-negativity constraint and **(ii)** the demand constraint of all users. Consider a user $i \in \mathcal{N}$. It follows that $f_p^i = \frac{r^i}{R} f_p \ge 0$, thus the non-negativity constraint is satisfied. Furthermore, $\sum_{p \in \mathcal{P}} f_p^i = \frac{r^i}{R} \sum_{p \in \mathcal{P}} f_p = r^i$, thus the demand constraint is also satisfied. In order to construct a feasible Nash equilibrium, we first establish following lemma.

Lemma 1. *Consider a bottleneck routing game. Any system optimal strategy profile is balanced.*

Proof. See [6]. □

We continue to construct a feasible profile, which is also a Nash equilibrium. Specifically, we focus on a profile that is proportional and system optimal.

Theorem 1. *Consider a bottleneck routing game, where for any user i, $\alpha^i \in [1/2, 1]$. Each system optimal proportional profile is a Nash equilibrium.*

Proof. Consider a system optimal, proportional profile, $\mathbf{f}$. As a result of Lemma 1, $\mathbf{f}$ is balanced, thus for all $i \in \mathcal{N}$, $b^i(\mathbf{f}) = b(\mathbf{f})$. Therefore, for any user i,

$$J^i(\mathbf{f}) = \alpha^i b^i(\mathbf{f}) - (1 - \alpha^i) \min_{j \ne i}\{b^j(\mathbf{f})\} = (2\alpha^i - 1) \cdot b(\mathbf{f}). \tag{6}$$

Assume by contradiction that $\mathbf{f}$ is not a Nash equilibrium. In other words, there exists a user i, which can send its flow according to $\overline{\mathbf{f}}^{\mathbf{i}} \ne \mathbf{f}^{\mathbf{i}}$ and by doing so, decreases its cost. Moreover, consider the case that the cost of the bottleneck link of i's rival, has increased due to i's deviation, i.e.,

$$\min_{j \ne i}\{b^j(\overline{\mathbf{f}}^{\mathbf{i}}, \mathbf{f}^{-\mathbf{i}})\} > \min_{j \ne i}\{b^j(\mathbf{f})\}. \tag{7}$$

Denote the bottleneck link of i's rival at $(\bar{\mathbf{f}}^{\mathbf{i}}, \mathbf{f}^{-\mathbf{i}})$ as n, thus

$$\min_{j \neq i}\{b^j(\bar{\mathbf{f}}^{\mathbf{i}}, \mathbf{f}^{-\mathbf{i}})\} \equiv T_n(\bar{f}_n^i + f_n^{-i}). \tag{8}$$

Since $\mathbf{f}$ is balanced, from (8) it follows that,

$$T_n(\bar{f}_n^i + f_n^{-i}) > \min_{j \neq i}\{b^j(\mathbf{f})\} = b(\mathbf{f}) \geq T_n(f_n). \tag{9}$$

From (9) and Assumption A2 it follows that $\bar{f}_n^i > 0$. Therefore from (6), (7) and (9),

$$\begin{aligned} J^i(\bar{\mathbf{f}}^{\mathbf{i}}, \mathbf{f}^{-\mathbf{i}}) &= \alpha^i b^i(\bar{\mathbf{f}}^{\mathbf{i}}, \mathbf{f}^{-\mathbf{i}}) - (1 - \alpha^i) \min_{j \neq i}\{b^j(\bar{\mathbf{f}}^{\mathbf{i}}, \mathbf{f}^{-\mathbf{i}})\} \\ &= \alpha^i b^i(\bar{\mathbf{f}}^{\mathbf{i}}, \mathbf{f}^{-\mathbf{i}}) - (1 - \alpha^i) T_n(\bar{f}_n^i + f_n^{-i}) \\ &\geq (2\alpha^i - 1) T_n(\bar{f}_n^i + f_n^{-i}) \geq (2\alpha^i - 1) b(\mathbf{f}) = J^i(\mathbf{f}). \end{aligned}$$

The last inequality follows from (9) and from $\alpha^i \in [1/2, 1]$. Therefore, $J^i(\bar{\mathbf{f}}^{\mathbf{i}}, \mathbf{f}^{-\mathbf{i}}) \geq J^i(\mathbf{f})$, which is a contradiction. We now consider the case where $\mathbf{f}$ is not a Nash equilibrium and

$$\min_{j \neq i}\{b^j(\bar{\mathbf{f}}^{\mathbf{i}}, \mathbf{f}^{-\mathbf{i}})\} \leq \min_{j \neq i}\{b^j(\mathbf{f})\}. \tag{10}$$

Since $\mathbf{f}$ is system optimal, it holds that

$$b(\bar{\mathbf{f}}^{\mathbf{i}}, \mathbf{f}^{-\mathbf{i}}) \geq b(\mathbf{f}). \tag{11}$$

Denote the bottleneck link of the system, at $(\bar{\mathbf{f}}^{\mathbf{i}}, \mathbf{f}^{-\mathbf{i}})$ as s and consider the case where $\bar{f}_s^i = 0$. By definition $f_s^{-i} > 0$, otherwise s cannot be the system's bottleneck. Since, $f_s^{-i} > 0$ it follows that $f_s > 0$ and $f_s^i = \frac{r^i}{R} f_s > 0$. Consequently, from (11) and Assumption A2, $T_s(f_s^i + f_s^{-i}) > T_s(\bar{f}_s^i + f_s^{-i}) \geq b(\mathbf{f})$, which is a contradiction to s being the system's bottleneck. Therefore, $\bar{f}_s^i > 0$ and

$$b(\bar{\mathbf{f}}^{\mathbf{i}}, \mathbf{f}^{-\mathbf{i}}) \equiv T_s(\bar{f}_s^i + f_s^{-i}) = b^i(\bar{\mathbf{f}}^{\mathbf{i}}, \mathbf{f}^{-\mathbf{i}}). \tag{12}$$

Finally, from (6), (10), (11) and (12)

$$\begin{aligned} J^i(\bar{\mathbf{f}}^{\mathbf{i}}, \mathbf{f}^{-\mathbf{i}}) &= \alpha^i b^i(\bar{\mathbf{f}}^{\mathbf{i}}, \mathbf{f}^{-\mathbf{i}}) - (1 - \alpha^i) \min_{j \neq i}\{b^j(\bar{\mathbf{f}}^{\mathbf{i}}, \mathbf{f}^{-\mathbf{i}})\} \\ &\geq \alpha^i b^i(\bar{\mathbf{f}}^{\mathbf{i}}, \mathbf{f}^{-\mathbf{i}}) - (1 - \alpha^i) \min_{j \neq i}\{b^j(\mathbf{f})\} \\ &= \alpha^i b(\bar{\mathbf{f}}^{\mathbf{i}}, \mathbf{f}^{-\mathbf{i}}) - (1 - \alpha^i) \min_{j \neq i}\{b^j(\mathbf{f})\} \\ &\geq \alpha^i b(\mathbf{f}) - (1 - \alpha^i) \min_{j \neq i}\{b^j(\mathbf{f})\} = (2\alpha^i - 1) \cdot b(\mathbf{f}) = J^i(\mathbf{f}), \end{aligned}$$

which is a contradiction. Thus, any system optimal proportional balanced flow is a Nash equilibrium. □

Theorem 1 illustrates that in any bottleneck routing game where for each user i, $\alpha^i \in [1/2, 1]$, there exists a Nash equilibrium. Moreover, there always exists a Nash equilibrium, which is system optimal[1]. This brings us to the following conclusion.

Corollary 1. *Consider a bottleneck routing game, where for any user i, $\alpha^i \in [1/2, 1]$. The Price of Stability is equal to 1.*

Even though Theorem 1 establishes the existence of desirable equilibria from a system's perspective, it might also happen that the selfishness of the users degrades the system substantially. This deficiency is captured by the Price of Anarchy.

Theorem 2. *Consider a bottleneck routing game, where for any user i, $\alpha^i \in [1/2, 1]$. The Price of Anarchy is unbounded.*

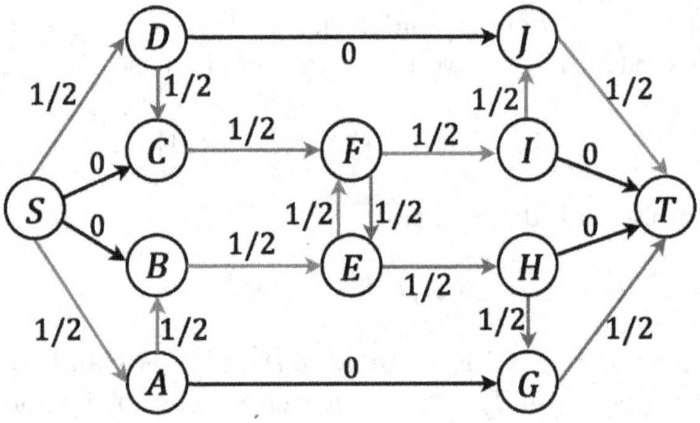

Fig. 2. Example of a network with an unbounded PoA.

Proof. We establish the theorem through the following example.

Example 1. Consider the network $G = (V, E)$ as depicted in Fig. 2. Further, consider two users i and j, each with a flow demand of $r^i = r^j = \frac{R}{2}$ and $\alpha^i = \alpha^j \equiv \alpha \in [1/2, 1]$. For any edge $e \in E$, the cost per unit of flow is equal to $T_e(f_e) = e^{f_e} - 1$. We focus on a specific profile $\mathbf{f}$, in which user i sends its total demand on a single path, namely $\{S, A, B, E, F, I, J, T\}$, and, user j sends its demand on the path $\{S, D, C, F, E, H, G, T\}$. The labels on the edges in Fig. 2 correspond to the portion of the total flow that transverses on that edge at $\mathbf{f}$, i.e., f_e/R. Thus,

$$J^i(\mathbf{f}) = \alpha b^i(\mathbf{f}) - (1 - \alpha) b^j(\mathbf{f}) = (2\alpha - 1) \cdot (e^{R/2} - 1).$$

[1] In [5] a similar theorem was proven for a more general topology. However, they only considered selfish users (i.e., $\forall i,\ \alpha^i = 1$).

It is straightforward that $J^i(\mathbf{f}) = J^j(\mathbf{f})$. Now assume by contradiction that $\mathbf{f}$ is not a Nash equilibrium. Hence, there exists a different routing strategy for user i, $\overline{\mathbf{f}}^{\mathbf{i}} \neq \mathbf{f}^{\mathbf{i}}$ at which user i can decrease its cost. If i places a positive flow on either (S,D), (S,C) or (A,G), it is immediate that $b^i(\overline{\mathbf{f}}^{\mathbf{i}}, \mathbf{f}^{\mathbf{j}}) = b^j(\overline{\mathbf{f}}^{\mathbf{i}}, \mathbf{f}^{\mathbf{j}}) > b^i(\mathbf{f})$ and $J^i(\overline{\mathbf{f}}^{\mathbf{i}}, \mathbf{f}^{\mathbf{j}}) > J^i(\mathbf{f})$.

Thus, if i wishes to refrain from increasing its cost, it will send all its flow on (B,E) and its bottleneck will be at least $T_{(B,E)}(\frac{R}{2}) = e^{R/2} - 1$. It follows that at $(\overline{\mathbf{f}}^{\mathbf{i}}, \mathbf{f}^{\mathbf{j}})$, there cannot exist an edge on which both i and j send a positive amount of flow, otherwise i increases its cost. Thus, the bottleneck of j stays the same. Hence,

$$\begin{aligned} J^i(\overline{\mathbf{f}}^{\mathbf{i}}, \mathbf{f}^{\mathbf{j}}) = \alpha b^i(\overline{\mathbf{f}}^{\mathbf{i}}, \mathbf{f}^{\mathbf{j}}) - (1-\alpha) b^j(\overline{\mathbf{f}}^{\mathbf{i}}, \mathbf{f}^{\mathbf{j}}) &= \alpha b^i(\overline{\mathbf{f}}^{\mathbf{i}}, \mathbf{f}^{\mathbf{j}}) - (1-\alpha) b^j(\mathbf{f}) \\ &\geq (2\alpha - 1)\cdot(e^{R/2} - 1) = J^i(\mathbf{f}), \end{aligned}$$

which is a contradiction. Because the users i and j are symmetric, the above analysis also holds for j. Therefore $\mathbf{f}$ is a Nash equilibrium. The bottleneck of the system at $\mathbf{f}$ is equal to $b(\mathbf{f}) = e^{R/2} - 1$.

On the other hand, at the system optimum, $\mathbf{f}^*$, an amount of flow, $R/4$, is sent through the following four paths: **1:** $\{S, A, G, T\}$, **2:** $\{S, B, E, H, T\}$, **3:** $\{S, C, F, I, T\}$, **4:** $\{S, D, J, T\}$. The system bottleneck at $\mathbf{f}^*$ is equal to $b(\mathbf{f}^*) = e^{R/4} - 1$. As a result, the Price of Anarchy in our example is lower bounded by

$$PoA = \frac{e^{R/2} - 1}{e^{R/4} - 1} \geq \frac{e^{R/2} - 1}{e^{R/4}} = e^{R/4} - \frac{1}{e^{R/4}} \geq e^{R/4} - 1.$$

Since R can be any positive number, the PoA is unbounded. □

4 Additive Routing Games

In this section we consider additive performance measures, such as delay, jitter and packet loss. Similar to bottleneck routing games, we first need to prove the existence of a Nash equilibrium. As mentioned in Sect. 3, due to the lack of quasi-convexity we cannot rely on well-known existence proofs for convex-games, such as the one given in [21]. We establish the existence of a Nash equilibrium by constructing a feasible strategy profile for all users, such that no user wishes to unilaterally deviate from its routing strategy. Moreover, we consider the specific case where for all $i \in \mathcal{N}$, $\alpha^i = 1/2$.[2] In other words, each user views its own performance and that of its rival, with equal importance. From (3), the cost of user i turns into

$$J^i(\mathbf{f}) = \sum_{l \in \mathcal{L}} \frac{f_l^i}{r^i} T_l(f_l) - \min_{j \neq i} \left\{ \sum_{l \in \mathcal{L}} \frac{f_l^j}{r^j} T_l(f_l) \right\}. \tag{13}$$

Note that we disregard $\alpha_i \equiv \alpha = 1/2$ from our equilibrium analysis, since it multiplies all users' costs by the same constant. We now bring the following definition from [27].

[2] An existence and uniqueness proof for selfish users is given in [20].

Definition 3. *A profile,* $\mathbf{f}$*, is a Wardrop equilibrium if for any two links* $l, n \in \mathcal{L}$ *with* $f_l > 0$*,* $T_l(f_l) \leq T_n(f_n)$*.*

In any additive routing game, there exists a Wardrop equilibrium. Moreover, it is unique with respect to the aggregated link flows f_l, [13,27]. We focus on a specific Wardrop equilibrium, which is also proportional in the sense of Definition 2. Hence, it is also unique with respect to the individual user flows.

Theorem 3. *Consider an additive routing game as described in Sect. 2, where for all users* i*,* $\alpha^i = 1/2$*. There exists a Nash equilibrium. In particular, it is equal to the proportional Wardrop equilibrium.*

Proof. We consider the unique proportional Wardrop equilibrium, $\mathbf{f}$, and prove that no user wishes to unilaterally deviate from $\mathbf{f}$. Assume by contradiction that $\mathbf{f}$ is not a Nash equilibrium. Hence, there exists a user i and a routing strategy, $\bar{\mathbf{f}}^{\mathbf{i}} \neq \mathbf{f}^{\mathbf{i}}$ such that $J^i(\bar{\mathbf{f}}^{\mathbf{i}}, \mathbf{f}^{-\mathbf{i}}) < J^i(\mathbf{f})$. We split the set of links $\mathcal{L}$, into three subsets: $\mathcal{L}^+ = \{l \in \mathcal{L} | \bar{f}_l^i > f_l^i\}$, $\mathcal{L}^- = \{l \in \mathcal{L} | \bar{f}_l^i < f_l^i\}$ and $\mathcal{L}^0 = \{l \in \mathcal{L} | \bar{f}_l^i = f_l^i\}$. Since $\bar{\mathbf{f}}^{\mathbf{i}} \neq \mathbf{f}^{\mathbf{i}}$, it follows that $\mathcal{L}^+$ and $\mathcal{L}^-$ are not empty. For any link $l \in \mathcal{L}^+$, denote $\epsilon_l \equiv \bar{f}_l^i - f_l^i$ and for any link $l \in \mathcal{L}^-$, denote $\delta_l \equiv f_l^i - \bar{f}_l^i$. Since r^i is constant, the differences in $\mathcal{L}^+$ and $\mathcal{L}^-$ are equal and $\sum_{l \in \mathcal{L}^+} \epsilon_l = \sum_{l \in \mathcal{L}^-} \delta_l$.

Because $\mathbf{f}$ is a proportional profile, it holds that for any two users $i, k \in \mathcal{N}$ and for any link $l \in \mathcal{L}$, $f_l^i / r^i = f_l^k / r^k$. Thus, for any link $l \in \mathcal{L}^0$ and any user $k \in \mathcal{N}$,

$$\left[\frac{\bar{f}_l^i}{r^i} - \frac{f_l^k}{r^k}\right] = \left[\frac{f_l^i}{r^i} - \frac{f_l^k}{r^k}\right] = 0. \tag{14}$$

Equation (14) holds for any $k \in \mathcal{N}$, hence also for i's rival at $(\bar{\mathbf{f}}^{\mathbf{i}}, \mathbf{f}^{-\mathbf{i}})$. Denote i's rival at $(\bar{\mathbf{f}}^{\mathbf{i}}, \mathbf{f}^{-\mathbf{i}})$ as j. Combining (14) with (13), we get

$$\begin{aligned}
J^i(\bar{\mathbf{f}}^{\mathbf{i}}, \mathbf{f}^{-\mathbf{i}}) &= \sum_{l \in \mathcal{L}^+} \left[\frac{\bar{f}_l^i}{r^i} - \frac{f_l^j}{r^j}\right] T_l(\bar{f}_l^i + f_l^{-i}) + \sum_{l \in \mathcal{L}^-} \left[\frac{\bar{f}_l^i}{r^i} - \frac{f_l^j}{r^j}\right] T_l(\bar{f}_l^i + f_l^{-i}) \\
&+ \sum_{l \in \mathcal{L}^0} \left[\frac{\bar{f}_l^i}{r^i} - \frac{f_l^j}{r^j}\right] T_l(\bar{f}_l^i + f_l^{-i}) \qquad (15) \\
&= \sum_{l \in \mathcal{L}^+} \left[\frac{f_l^i + \epsilon_l}{r^i} - \frac{f_l^j}{r^j}\right] T_l(f_l + \epsilon_l) + \sum_{l \in \mathcal{L}^-} \left[\frac{f_l^i - \delta_l}{r^i} - \frac{f_l^j}{r^j}\right] T_l(f_l - \delta_l) \\
&= \sum_{l \in \mathcal{L}^+} \frac{\epsilon_l}{r^i} T_l(f_l + \epsilon_l) - \sum_{l \in \mathcal{L}^-} \frac{\delta_l}{r^i} T_l(f_l - \delta_l) > \sum_{l \in \mathcal{L}^+} \frac{\epsilon_l}{r^i} T_l(f_l) - \sum_{l \in \mathcal{L}^-} \frac{\delta_l}{r^i} T_l(f_l).
\end{aligned}$$

The last inequality follows from Assumption A2. Since $\mathbf{f}$ is a Wardrop equilibrium, we make two observations, namely
(1): $\forall l \in \mathcal{L}^-$, $\delta_l > 0$, thus $f_l > 0$. Therefore, from Definition 3, for any two links $l, n \in \mathcal{L}^-$, $T_l(f_l) = T_n(f_n)$.
(2): From Definition 3 it follows that for any link $l \in \mathcal{L}^+$ and any link $n \in \mathcal{L}^-$, $T_l(f_l) \geq T_n(f_n)$.

Consider a link $e \in \mathcal{L}^-$. Consequently, Eq. (15) turns into

$$J^i(\bar{\mathbf{f}}^{\mathbf{i}}, \mathbf{f}^{-\mathbf{i}}) > \sum_{l \in \mathcal{L}^+} \frac{\epsilon_l}{r^i} T_e(f_e) - \sum_{l \in \mathcal{L}^-} \frac{\delta_l}{r^i} T_e(f_e) = \frac{1}{r^i} T_e(f_e) \cdot \left[\sum_{l \in \mathcal{L}^+} \epsilon_l - \sum_{l \in \mathcal{L}^-} \delta_l \right] = 0.$$

On the other hand, because $\mathbf{f}$ is proportional, it follows from (13) that

$$J^i(\mathbf{f}) = \sum_{l \in \mathcal{L}} \frac{f_l^i}{r^i} T_l(f_l) - \min_{k \neq i} \{ \sum_{l \in \mathcal{L}} \frac{f_l^k}{r^k} T_l(f_l) \} = \sum_{l \in \mathcal{L}} \frac{f_l^i}{r^i} T_l(f_l) - \sum_{l \in \mathcal{L}} \frac{f_l^i}{r^i} T_l(f_l) = 0.$$

Thus, $J^i(\bar{\mathbf{f}}^{\mathbf{i}}, \mathbf{f}^{-\mathbf{i}}) > J^i(\mathbf{f})$, which is a contradiction. Hence, $\mathbf{f}$ is a Nash equilibrium.

Now that we have proven the existence of a Nash equilibrium, we continue to investigate its uniqueness. We focus on a special case in which the network has two users, i.e., $N = 2$, and we denote these two users as i and j. It follows from (13) that $J^j(\mathbf{f}) = -J^i(\mathbf{f})$. In order to prove the Nash equilibrium's uniqueness, we use of the following lemma.

Lemma 2. *Consider an additive routing game as described in Sect. 2, where $N = 2$ and $\alpha^i = \alpha^j = 1/2$. At any Nash equilibrium $\mathbf{f}$, $J^i(\mathbf{f}) = J^j(\mathbf{f}) = 0$.*

Proof. See [6]. □

We are now ready to prove our theorem.

Theorem 4. *For $N = 2$, the proportional Wardrop equilibrium is the unique Nash equilibrium, i.e., the Nash equilibrium is unique in the users' individual flows.*

Proof. See [6]. □

An immediate consequence of Theorem 4 is that, the PoA of two-user systems is bounded by well-known bounds on the Wardrop equilibrium, e.g., see [19,25].

Although Theorem 4 holds for a network with two users, in the general case of N-players, it does not hold. Indeed, in [6] we provide an example of a network with N users and multiple Nash equilibria. Finally, in [6], we provide an example of a network with users for which $\alpha^i \in [0, 1]$, in which no Nash equilibrium exists for either bottleneck routing games or additive routing games.

5 Conclusions

In this study we investigated routing games where the cost of each agent is represented as a combination of its own performance and that of its *rival*. We established the existence of Nash equilibria in games with bottleneck performance measures and games with additive performance measures. For bottleneck routing games and agents with $\alpha^i \in [1/2, 1]$, namely, games where agents care more about their own performance than that of their rivals', we established that the Price of Stability is equal to 1, i.e., a system optimal Nash equilibrium always exists. Moreover, we provide an example in which the Price of Anarchy is unbounded. For additive routing games, we focused on the fundamental load balancing game of routing over parallel links and on agents with $\alpha^i = 1/2$, namely, games where agents view their own performance and that of their rivals with equal importance. We proved that the proportional Wardrop equilibrium (which exists and is unique) is also a Nash equilibrium. Moreover, for a two-player system, we established the uniqueness of the Nash equilibrium. In this case, the PoA can be bounded by well-known bounds on the Wardrop equilibrium. We also provided an example of the non-uniqueness of the Nash equilibrium for a system with N-players, and an example of its non-existence for agents with $\alpha^i \in [0,1]$. In future research, it would be interesting to consider networks with multiple sources and destination pairs. Lastly, establishing the existence of a Nash equilibrium for additive games and agents with $\alpha^i \in [1/2,1]$, remains an open problem.

Acknowledgments. This research was supported by the European Union through the CONGAS project (http://www.congasproject.eu/) in the 7th Framework Programme. Gideon Blocq is supported by the Google Europe Fellowship in Computer Networking.

References

1. Altman, E., Basar, T., Jiménez, T., Shimkin, N.: Competitive routing in networks with polynomial cost. In: Proceedings of INFOCOM 2000, pp. 1586–1593 (2000)
2. Anshelevich, E., Dasgupta, A., Kleinberg, J.M., Tardos, É., Wexler, T., Roughgarden, T.: The price of stability for network design with fair cost allocation. SIAM J. Comput. **38**(4), 1602–1623 (2008)
3. Azad, A.P., Altman, E., Azouzi, R.E.: Routing games : from egoism to altruism. In: Proceedings of WiOpt 2010, pp. 528–537 (2010)
4. Babaioff, M., Kleinberg, R., Papadimitriou, C.H.: Congestion games with malicious players. Games Econ. Behav. **67**(1), 22–35 (2009)
5. Banner, R., Orda, A.: Bottleneck routing games in communication networks. IEEE J. Sel. Areas Commun. **25**(6), 1173–1179 (2007)
6. Blocq, G., Orda, A.: "Beat-Your-Rival" routing games. Technical report, Department of Electrical Engineering, Technion, Haifa, Israel (2015). http://webee.technion.ac.il/Sites/People/ArielOrda/Info/Other/BO15.pdf
7. Blocq, G., Orda, A.: Worst-case coalitions in routing games. CoRR abs/1310.3487 (2013)
8. Busch, C., Magdon-Ismail, M.: Atomic routing games on maximum congestion. Theoret. Comput. Sci. **410**(36), 3337–3347 (2009)
9. Caragiannis, I., Kaklamanis, C., Kanellopoulos, P., Kyropoulou, M., Papaioannou, E.: The impact of altruism on the efficiency of atomic congestion games. In: Wirsing, M., Hofmann, M., Rauschmayer, A. (eds.) TGC 2010, LNCS, vol. 6084, pp. 172–188. Springer, Heidelberg (2010)
10. Chen, P., de Keijzer, B., Kempe, D., Schäfer, G.: Altruism and its impact on the price of anarchy. ACM Trans. Econ. Comput. **2**(4), 17:1–17:45 (2014)

11. Chen, P., Kempe, D.: Altruism, selfishness, and spite in traffic routing. In: Proceedings EC 2008, pp. 140–149 (2008)
12. Cole, R., Dodis, Y., Roughgarden, T.: Bottleneck links, variable demand, and the tragedy of the commons. Networks **60**(3), 194–203 (2012)
13. Correa, J.R., Moses, N.E.S.: Wardrop equilibria. Wiley Encyclopedia of Operations Research and Management Science (2010)
14. Harks, T.: Stackelberg strategies and collusion in network games with splittable flow. In: Bampis, E., Skutella, M. (eds.) WAOA 2008. LNCS, vol. 5426, pp. 133–146. Springer, Heidelberg (2009)
15. Hoefer, M., Skopalik, A.: Altruism in atomic congestion games. ACM Trans. Econ. Comput. **1**(4), 21 (2013)
16. Korilis, Y.A., Lazar, A.A.: On the existence of equilibria in noncooperative optimal flow control. J. ACM **42**(3), 584–613 (1995)
17. Koutsoupias, E., Papadimitriou, C.: Worst-case equilibria. In: Meinel, C., Tison, S. (eds.) STACS 1999. LNCS, vol. 1563, p. 404. Springer, Heidelberg (1999)
18. La, R.J., Anantharam, V.: Optimal routing control: repeated game approach. IEEE Trans. Autom. Control **47**, 437–450 (2002)
19. Nisan, N., Roughgarden, T., Tardos, E., Vazirani, V.V.: Algorithmic Game Theory. Cambridge University Press, New York (2007)
20. Orda, A., Rom, R., Shimkin, N.: Competitive routing in multiuser communication networks. IEEE/ACM Trans. Networking **1**, 510–521 (1993)
21. Rosen, J.B.: Existence and uniqueness of equilibrium points for concave n-person games. Econometrica **33**(3), 520–534 (1965)
22. Roth, A.: The price of malice in linear congestion games. In: Papadimitriou, C., Zhang, S. (eds.) WINE 2008. LNCS, vol. 5385, pp. 118–125. Springer, Heidelberg (2008)
23. Roughgarden, T.: Stackelberg scheduling strategies. In: Proceedings of STOC 2001, pp. 104–113 (2001)
24. Roughgarden, T.: Algorithmic game theory. Commun. ACM **53**(7), 78–86 (2010)
25. Roughgarden, T., Tardos, E.: How bad is selfish routing? J. ACM **49**, 236–259 (2002)
26. Wan, C.: Coalitions in nonatomic network congestion games. Math. Oper. Res. **37**(4), 654–669 (2012)
27. Wardrop, J.G.: Some theoretical aspects of road traffic research. In: Proceedings of the Institute of Civil Engineers, Part II, vol. 1, pp. 325–378 (1952)

Characterization and Computation of Equilibria for Indivisible Goods

Simina Brânzei[1], Hadi Hosseini[2](✉), and Peter Bro Miltersen[1]

[1] Aarhus University, Aarhus, Denmark
[2] University of Waterloo, Waterloo, Canada
h5hossei@uwaterloo.ca

Abstract. We consider the problem of allocating indivisible goods using the leading notion of fairness in economics: *the competitive equilibrium from equal incomes*. Focusing on two major classes of valuations, namely *perfect substitutes* and *perfect complements*, we establish the computational properties of algorithms operating in this framework. For the class of valuations with perfect complements, our algorithm yields a surprisingly succinct characterization of instances that admit a competitive equilibrium from equal incomes.

1 Introduction

The systematic study of economic mechanisms began in the 19th century with the pioneering work of Irving Fisher [4] and Léon Walras [15], who proposed the Fisher market and the exchange economy as answers to the question: *"How does one allocate scarce resources among the participants of an economic system?"*. These models of a competitive economy are central in mathematical economics and have been studied ever since in an extensive body of literature [11].

The high level scenario is that of several economic players arriving at the market with an initial endowment of resources and a utility function for consuming goods. The problem is to compute prices and an allocation for which an optimal exchange takes place: each player is maximally satisfied with the bundle acquired, given the prices and his initial endowment. Such allocation and prices form a *market equilibrium* and, remarkably, are guaranteed to exist under mild assumptions when goods are divisible [1].

In real scenarios, however, goods often come in discrete quantities; for example, clothes, furniture, houses, or cars may exist in multiple copies, but cannot be infinitely divided. Scarce resources, such as antique items or art collection pieces are even rarer – often unique (and thus *indivisible*). The problem of allocating discrete or indivisible resources is much more challenging because the theoretical guarantees from the divisible case do not always carry over; however, it can be tackled as well using market mechanisms [3,5,6,12]. In this paper, we are concerned with the question of allocating indivisible resources using the leading fairness concept from economics: the *competitive equilibrium from equal incomes* (CEEI).

M. Hoefer (Ed.): SAGT 2015, LNCS 9347, pp. 244–255, 2015.
DOI: 10.1007/978-3-662-48433-3_19

The competitive equilibrium from equal incomes solution embodies the ideal notion of fairness [8,9,12,14] and is a special case of the Fisher market model [13]. Informally, there are m goods to be allocated among n buyers, each of which is endowed with one unit of an artificial currency that they can use to acquire goods. The buyers declare their preferences over the goods, after which the equilibrium prices and allocation are computed. When the goods are divisible, a competitive equilibrium from equal incomes is guaranteed to exist for very general conditions and each equilibrium allocation satisfies the desirable properties of envy-freeness and efficiency.

In recent years, the competitive equilibrium from equal incomes has been studied for the allocation of discrete and indivisible resources in a series of papers. Bouveret and Lemaître [3] considered it for allocating indivisible goods, together with notions of fairness such as proportionality, envy-freeness, and maximin fairness. Budish [5] analyzed the allocation of multiple discrete goods for the course assignment problem[1] and designed an approximate variant of CEEI that is guaranteed to exist for any instance. In this variant, buyers have permissible bundles of goods and the approximation notion requires randomization to perturb the budgets of the buyers while relaxing the market clearing condition. In follow-up work, Othman, Papadimitriou, and Rubinstein [12] analyzed the computational complexity of this variant, showing that computing the approximate solution proposed by Budish is PPAD-complete, and that it is NP-hard to distinguish between an instance where an exact CEEI exists and the one in which there is no approximate-CEEI tighter than guaranteed in Budish [5].

A key requirement in the variants of the CEEI solution concept is the envy-freeness condition among all participants of the market. In this vein, Moulin [10] analyzes a variant of CEEI where the buyers have intrinsic value for money and goods are allocated to those with the highest valuations (i.e. higher than the competitive price), while the unallocated buyers receive a fair compensation instead. Moulin's definition can be viewed as a special case of the Arrow-Debreu market model and is suitable for markets with extremely uneven supply and demand, such as assigning seats in overbooked planes.

In this paper, we study the competitive equilibrium from equal incomes for two major classes of valuations, namely *perfect substitutes* and *perfect complements*, following the definition of Bouveret and Lemaître [3] and the exact notion in Budish [5]. Perfect substitutes represent goods that can replace each other in consumption, such as Pepsi and Coca-Cola, and are modeled mathematically through *additive utilities*. This is the setting examined by Bouveret and Lemaître [3] as well. Perfect complements represent goods that have to be consumed together, such as a left shoe and a right shoe, and are modeled mathematically through *Leontief utilities*. For indivisible goods, Leontief utilities are in fact equivalent to the class of single-minded buyers, which have been studied extensively in the context of auctions [11].

[1] Given a set of students and courses to be offered at a university, how should the courses be scheduled given that the students have preferences over their schedules and the courses have capacity constraints on enrollment?

We study the computation of competitive equilibria for indivisible goods and establish polynomial time algorithms and hardness results (where applicable). Our algorithm for Leontief utilities gives a very succinct characterization of markets that admit a competitive equilibrium from equal incomes for indivisible resources. The computational results of Othman, Papadimitriou, and Rubinstein [12] are orthogonal to our setting since they refer to combinatorial valuations. Also, since in our model, buyers have no value for money, the literature on markets with quasi-linear utilities is not directly relevant.

2 Competitive Equilibrium from Equal Incomes

We begin by formally introducing the competitive equilibrium from equal incomes; the model is equivalent to a Fisher market with identical budgets [11]. Formally, there is a set $N = \{1, \ldots, n\}$ of buyers and a set $M = \{1, \ldots, m\}$ of goods which are brought by a seller. In general, the goods can be either infinitely divisible or discrete and, without loss of generality, there is exactly one unit from every good $j \in M$. Each buyer i is endowed with:

- A utility function $u_i : [0,1]^m \to \mathbb{R}_{\geq 0}$ for consuming the goods, which maps each vector $\mathbf{x} = \langle x_1, \ldots, x_m \rangle$ of resources to a real value, where x_j denotes the amount received by buyer i from good j and $u_i(\mathbf{x})$ represents i's utility for bundle $\mathbf{x}$.
- An initial budget $B_i = 1$, which can be viewed as (artificial) currency to acquire goods, but has no intrinsic value to the buyer. However, the currency does have intrinsic value to the seller.

Each buyer in the market wants to spend its entire budget to acquire a bundle of items that maximizes its utility, while the seller aims to sell all the goods (which it has no intrinsic value for) and extract the money from the buyers.

A market outcome is defined as a tuple $(\mathbf{x}, \mathbf{p})$, where $\mathbf{p}$ is a vector of prices for the m items and $\mathbf{x} = \langle \mathbf{x}_1, \ldots, \mathbf{x}_n \rangle$ is an allocation of the m items, where p_j denotes the price of item j, $\mathbf{x}_i$ is the bundle received by buyer i, and x_{ij} is the amount of item j received by i. A market outcome that maximizes the utility of each buyer subject to its budget constraint and clears the market is called a *market equilibrium* [11]. Formally, $(\mathbf{x}, \mathbf{p})$ is a market equilibrium if and only if:

- For each buyer $i \in N$, the bundle $\mathbf{x}_i$ maximizes buyer i's utility given the prices $\mathbf{p}$ and budget $B_i = 1$.
- Each item $j \in M$ is completely sold or has price zero: $(\sum_{i=1}^{n} x_{ij} - 1)\, p_j = 0$.
- All the buyers exhaust their budgets; that is, $\sum_{j=1}^{m} p_j \cdot x_{ij} = 1$, for all $i \in N$.

Every competitive equilibrium from equal incomes $(\mathbf{x}, \mathbf{p})$ is envy-free; if buyer i would strictly prefer another buyer j's bundle $\mathbf{x}_j$, then i could simply purchase $\mathbf{x}_j$ instead of $\mathbf{x}_i$ since they have the same buying power, which is in contradiction with the equilibrium property.

A market with divisible goods is guaranteed to have a competitive equilibrium under mild conditions [1]. Moreover, for the general class of *Constant Elasticity of Substitution* valuations, the equilibrium can be computed using a remarkable convex program due to Eisenberg and Gale [7], as one of the few algorithmic results in general equilibrium theory. The classes of valuations studied in this paper – perfect complements and perfect substitutes – belong to the constant elasticity of substitution family.

In the following sections we study these classes in detail in the context of allocating indivisible resources.

3 Perfect Complements

Let $\mathcal{M} = (N, M, \mathbf{v})$ denote a market with perfect complements, represented through Leontief utilities; recall N is the set of buyers, M the set of items, and $\mathbf{v}$ a matrix of constants, such that $v_{i,j}$ is the value of buyer i for consuming one unit of good j. The utility of buyer i for a bundle $\mathbf{x} = \langle x_1, \ldots, x_m \rangle \in [0,1]^m$ is:

$$u_i(\mathbf{x}) = \min_{j=1}^{m} \left(\frac{x_j}{v_{i,j}} \right) \tag{1}$$

In our model the goods are indivisible, and so $x_{i,j} \in \{0,1\}$, for all i, j. By examining Eq. 1, it can be observed that buyer i's utility for a bundle depends solely on whether the buyer gets all the items that it values positively (or not). To capture this we define the notion of *demand set*.

Definition 1 (Demand Set). *Given a CEEI market with indivisible goods and Leontief utilities, let the* demand set *of buyer i be the set of items that i has a strictly positive value for; that is, $D_i = \{j \in M \mid v_{i,j} > 0\}$.*

Now we can introduce the precise utility equation for indivisible goods with Leontief valuations.

Definition 2 (Leontief Utility for Indivisible Goods). *Given a market with Leontief utilities and indivisible goods, the utility of a buyer i for a bundle $\mathbf{x} = \langle x_1, \ldots, x_m \rangle \in \{0,1\}^m$ is:*

$$u_i(\mathbf{x}) = \begin{cases} \min_{j \in D_i} \left(\frac{1}{v_{i,j}} \right), & \text{if } D_i \subseteq \mathbf{x}, \\ 0, & \text{otherwise} \end{cases}$$

where D_i represents buyer i's demand set.

We illustrate this utility class with an example. Note that valuations can be arbitrary (i.e. not necessarily normalized).

Example 1. Let $\mathcal{M}$ be a market with buyers $N = \{1, 2, 3\}$, items $M = \{1, 2, 3, 4\}$, and values: $v_{1,1} = 1$, $v_{2,2} = 2$, $v_{2,4} = 3$, $v_{3,1} = 0.5$, $v_{3,2} = 2.5$, $v_{3,3} = 5$, and $v_{i,j} = 0$, for all other i, j. Recall the demand set of each buyer consists of the items it values strictly positively, and so: $D_1 = \{1\}$, $D_2 = \{2, 4\}$, $D_3 = \{1, 2, 3\}$. Then the utility of buyer 1 for a bundle $S \subseteq M$ is: $u_1(S) = 0$ if $D_1 \not\subseteq S$, and $u_1(S) = \min_{j \in D_1} \{1/v_{1,j}\} = 1/v_{1,1} = 1$ otherwise. Similarly, $u_2(S) = 0$ if $D_2 \not\subseteq M$ and $u_2(S) = \min \left\{ \frac{1}{v_{2,2}}, \frac{1}{v_{2,4}} \right\} = 1/3$ otherwise.

Next, we examine the computation of allocations that are fair according to the CEEI solution concept. The main computational problems that we consider are: *Given a market, determine whether a competitive equilibrium exists and compute it when possible.* Depending on the scenario at hand, an allocation of the resources to the buyers may have already been made (or the seller may have already set prices for the items). The questions then are to determine whether an equilibrium exists at those prices or allocations. Our algorithm for computing a competitive equilibrium for Leontief utilities with indivisible goods yields a characterization of when a market equilibrium is guaranteed to exist.

Theorem 1. *Given a market $\mathcal{M} = (N, M, \mathbf{v})$ with Leontief utilities, indivisible goods, and a tuple $(\mathbf{x}, \mathbf{p})$, where $\mathbf{x}$ is an allocation and $\mathbf{p}$ a price vector, it can be decided in polynomial time if $(\mathbf{x}, \mathbf{p})$ is a market equilibrium for $\mathcal{M}$.*

Proof. It is sufficient to verify that these conditions hold: (*i*) Each buyer i exhausts their budget: $\sum_{j \in \mathbf{x}_i} p_j = 1$; (*ii*) Each item is either allocated or has a price of zero; (*iii*) No buyer i can afford a better bundle: if $u_i(\mathbf{x}) = 0$, then $\sum_{j \in D_i} p_j > 1$. All the conditions can be verified in $O(mn)$.

Theorem 2. *Given a market $\mathcal{M} = (N, M, \mathbf{v})$ with Leontief utilities, indivisible goods, and a price vector $\mathbf{p}$, it is NP-complete to decide if there exists an allocation $\mathbf{x}$ such that $(\mathbf{x}, \mathbf{p})$ is a market equilibrium for $\mathcal{M}$.*

The proof of Theorem 2 uses the PARTITION problem and is included in the full version of the paper, together with the other omitted proofs.

Theorem 3. *Given a market $\mathcal{M} = (N, M, \mathbf{v})$ with Leontief utilities and an allocation $\mathbf{x}$, it can be decided in polynomial time if there exists price vector $\mathbf{p}$ such that $(\mathbf{x}, \mathbf{p})$ is a market equilibrium for $\mathcal{M}$.*

Proof. This problem can be solved using linear programming (see Algorithm 1). At a high level, one needs to check that the allocation $\mathbf{x}$ is feasible, that each item is either sold or has a price of zero, and that (i) each buyer spends all their money and (ii) whenever a buyer does not get their demand set, the bundle is too expensive. Since the number of constraints is polynomial in the number of buyers and items, the algorithm runs in polynomial time.

Finally, we investigate the problem of computing both market equilibrium allocation and prices, which yields the following characterization.

Theorem 4. *Given a market $\mathcal{M} = (N, M, \mathbf{v})$ with Leontief utilities and indivisible items, a competitive equilibrium from equal incomes exists if and only if the following hold:*

- *There are at least as many items as buyers ($m \geq n$)*
- *No two buyers have identical demand sets of size one.*

Moreover, an equilibrium can be computed in polynomial time if it exists.

```
input: Market M with Leontief valuations; allocation x
output: price vector p such that (x, p) is a market equilibrium for M, or NULL if
none exists
A ← ∅ // Set of items allocated by x
for i = 1 to n do
    A ← A ∪ x_i
    for j = i + 1 to n do
        if (x_i ∩ x_j ≠ ∅) then
            return NULL // The allocation x is unfeasible
        end if
    end for
end for
C ← ∅ // Initialize the set of constraints
for j ∈ M \ A do
    C ← C ∪ {p_j ≤ 0} // Price the unsold items at zero
end for
for i ∈ {1, ..., n} do
    C ← C ∪ {Σ_{j∈x_i} p_j ≤ 1, −Σ_{j∈x_i} p_j ≤ −1}
    if (D_i ⊈ x_i) then
        C ← C ∪ {−Σ_{j∈x_i} p_j ≤ −1 − ε} // If i doesn't get D_i, the bundle is expensive
    end if
end for
return SOLVE(max ε, C, p ≥ 0) //Linear program solver: maximize ε subject to non-
negative prices and the constraints C
```

Algorithm 1. COMPUTE-EQUILIBRIUM-PRICES($\mathcal{M}, \mathbf{x}$)

Proof. Clearly the two conditions are necessary; if there are fewer items than buyers, then the budgets can never be exhausted, while if there exist two buyers whose demand sets are identical and consist of exactly the same item, at least one of them will be envious under any pair of feasible allocation and prices.

To see that the conditions are also sufficient, consider the allocation produced by Algorithm 2. At a high level, the algorithm first sorts the buyers in increasing order by the sizes of their demand sets, breaking ties lexicographically. Then each buyer i in this order is given one item, j, selected from the unallocated items in the buyer's demand set (if possible), and an arbitrary un-allocated item otherwise. Finally, the last buyer (i.e. with the largest demand set) additionally gets all the items that remained unallocated at the end of this iteration (if any).

The prices are set as follows. For each buyer i among the first $n-1$ allocated, the items in its bundle, $\mathbf{x}_i$, are priced equally, at $1/|\mathbf{x}_i|$. For the last allocated buyer, L, the items in $\mathbf{x}_L \cap D_L$ are priced high (at $(1-\epsilon)/|\mathbf{x}_L \cap D_L|$), while the unwanted items, in $\mathbf{x}_L \setminus D_L$, are priced low (at $\epsilon/|\mathbf{x}_L \setminus D_L|$).

Now we verify that the allocation and prices $(\mathbf{x}, \mathbf{p})$ computed by Algorithm 2 represent indeed a market equilibrium for $\mathcal{M}$ (if one exists):

- BUDGETS EXHAUSTED: Each buyer gets a non-empty bundle priced at 1.
- ITEMS SOLD: Each item is allocated by the algorithm.
- OPTIMALITY FOR EACH BUYER: We show that each buyer i either gets its demand set or cannot afford it using a few cases:

- *Case 1*: ($|D_i| = 1$). Since there are no two identical demand sets with the size of one, buyer i gets the unique item in its demand set, and this allocation maximizes i's utility.
- *Case 2*: ($|D_i| \geq 2$) and i is not the last buyer. Then if i gets an item from its demand set, since $|D_i| \geq 2$ and all items are positively priced, the bundle D_i is too expensive: $\mathbf{p}(D_i) > 1$. Otherwise, i gets an item outside of its demand set. Then all the items in D_i were allocated to the previous buyers. Since $|D_i| \geq 2$ and each previously allocated item has price 1, D_i is too expensive: $\mathbf{p}(D_i) > 1$.
- *Case 3*: ($|D_i| \geq 2$) and i is the last buyer. If i does not get all its demand, then some item in D_i was given to an earlier buyer at price 1. From $|D_i| \geq 2$, there is at least one other desired item in D_i positively priced, thus $\mathbf{p}(D_i) > 1$. Thus, Algorithm 2 computes an equilibrium, which completes the proof.

To gain more intuition, we illustrate Algorithm 2 on an example.

input: Market $\mathcal{M}$ with Leontief valuations; (optional) set $\mathcal{A}$ of pre-allocated items
// if $\mathcal{A}$ is not given, default to $\mathcal{A} \leftarrow \emptyset$
output: Equilibrium allocation and prices $(\mathbf{x}, \mathbf{p})$, or NULL if none exist
if $((m < n)$ or $(\exists\, i, j \in N$ with $D_i = D_j$ and $|D_i| = 1)$ **then**
  return NULL *// No equilibrium : too few items or two buyers have identical singleton demand sets*
end if
for (each buyer $i \in N$ in increasing order by $|D_i|$) **do**
  if $(|D_i \setminus \mathcal{A}| \geq 1)$ **then**
    $j \leftarrow \text{argmin}_{\ell \in D_i \setminus \mathcal{A}}\, \ell$ *// Give buyer i an item from its demand set if possible*
  else
    $j \leftarrow \text{argmin}_{\ell \in M \setminus \mathcal{A}}\, \ell$ *// Otherwise, i gets an arbitrary unallocated item*
  end if
  $\mathbf{x}_i \leftarrow \{j\}$
  $p_j \leftarrow 1$
  $\mathcal{A} \leftarrow \mathcal{A} \cup \{j\}$
end for
$L \leftarrow \arg\max_{i \in N} |D_i|$ *// The buyer with the largest demand gets the remaining items*
$\mathbf{x}_L \leftarrow \mathbf{x}_L \cup (M \setminus \mathcal{A})$
$k \leftarrow |\mathbf{x}_L \cap D_L|$
$\epsilon \leftarrow 0$
if $(k < |x_L|)$ **then**
  $\epsilon \leftarrow 1/m^2$ *// Set a low price for the items that L got outside of its demand set*
end if
for $(j \in \mathbf{x}_L)$ **do**
  if $(j \in D_L)$ **then**
    $p_j \leftarrow (1 - \epsilon)/k$
  else
    $p_j \leftarrow \epsilon/(|\mathbf{x}_L| - k)$
  end if
end for
return $(\mathbf{x}, \mathbf{p})$

Algorithm 2. COMPUTE-EQUILIBRIUM($\mathcal{M}$)

Example 2. Consider a market with buyers $N = \{1, \ldots, 6\}$, items $M = \{1, \ldots, 8\}$, and demands: $D_1 = \{1\}$, $D_2 = \{2\}$, $D_3 = \{2, 3\}$, $D_4 = \{2, 3\}$, $D_5 = \{4, 5, 6\}$, $D_6 = \{6, 7, 8\}$. Algorithm 2 sorts the buyers in increasing order of the sizes of their demand sets, breaking ties lexicographically. The order is: $(1, 2, 3, 4, 5, 6)$.

- *Step 1* : Buyer 1 gets item 1 at price 1: $\mathbf{x}_1 = \{1\}$, $p_1 = 1$.
- *Step 2* : Buyer 2 gets item 2 at price 1 : $\mathbf{x}_2 = \{2\}$, $p_2 = 1$.
- *Step 3* : There is one unallocated item left from buyer 3's demand set, and so 3 gets it: $\mathbf{x}_3 = \{3\}$ and $p_3 = 1$.
- *Step 4* : Buyer 4's demand set has been completely allocated, thus 4 gets the free item (outside of its demand) with smallest index: $\mathbf{x}_4 = \{4\}$ and $p_4 = 1$.
- *Step 5* : There are two items (5 and 6) left unallocated in buyer 5's demand set. Thus : $\mathbf{x}_5 = \{5\}$ and $p_5 = 1$.
- *Step 6* : Buyer 6 gets the leftover: $\mathbf{x}_6 = \{6, 7, 8\}$ at $p_6 = p_7 = p_8 = 1/3$.

The characterization obtained through Algorithm 2 raises several important questions. For example, not only do fair division procedures typically guarantee fairness (according to a given solution concept), but also they improve some measure of efficiency when possible.

The utilitarian *social welfare* of an allocation $\mathbf{x}$ is defined as the sum of the buyers' utilities: $\text{SW}(\mathbf{x}) = \sum_{i=1}^{n} u_i(\mathbf{x}_i)$. Note that social welfare normalization is not required for any of our next results.

As the following example illustrates, the allocation computed by Algorithm 2 can be the worst possible among all the market equilibria.

Example 3. Given $n \in \mathbb{N}$, let $N = \{1, \ldots, n\}$ be the set of buyers, $M = \{1, \ldots, 2n\}$ the set of items, and the demand sets given by: $D_i = \{2i - 1, 2i\}$, for each $i \in N$. Algorithm 2 computes the allocation: $\mathbf{x}_1 = \{1\}$, $\mathbf{x}_2 = \{3\}$, ..., $\mathbf{x}_{n-1} = \{2n - 3\}$, $\mathbf{x}_n = \{2, 4, \ldots, 2n - 2, 2n - 1, 2n\}$, with a social welfare of $\text{SW}(\mathbf{x}) = 1$. The optimal allocation supported in a competitive equilibrium is: $\mathbf{x}_i^* = \{2i - 1, 2i\}$, for each $i \in N$, with a social welfare of $\text{SW}(\mathbf{x}^*) = n$.

These observations give rise to the question: *Is there an efficient algorithm for computing a competitive equilibrium from equal incomes with optimal social welfare (among all equilibria) for perfect complements with indivisible goods?*

It is important to note that the allocation that maximizes social welfare among all possible allocations cannot always be supported in a competitive equilibrium. We illustrate this phenomenon in Example 4.

Example 4. Consider a market with buyers: $N = \{1, 2\}$ and items: $M = \{1, 2, 3\}$, where the demand sets are: $D_1 = D_2 = \{1, 2\}$. Concretely, let these demands be induced by the valuations: $v_{1,1} = v_{1,2} = 1$, $v_{1,3} = 0$ and $v_{2,1} = v_{2,2} = 1$, $v_{2,3} = 0$. The optimal social welfare is 1 and can be achieved by giving one of the buyers its entire demand set and the other buyer the remaining item; for example, let $\mathbf{x}_1^* = \{1, 2\}$ and $\mathbf{x}_2^* = \{3\}$, with $p_1 = p_2 = 1/2$ and $p_3 = 1$. Clearly no such allocation can be supported in an equilibrium, because whenever a buyer gets their full demand, the other buyer does not get its own demand but can afford it (their initial budgets are equal). Thus every competitive equilibrium has a social welfare of zero, such as $\mathbf{x}_1 = \{1\}$, $\mathbf{x}_2 = \{2, 3\}$, with $p_1 = 1$, $p_2 = 1$, $p_3 = 0$.

The next result implies that equilibria with optimal social welfare cannot be computed efficiently in the worst case.

Theorem 5. *Given a market $\mathcal{M} = (N, M, \mathbf{v})$ with Leontief valuations, indivisible goods, and an integer $K \in \mathbb{N}$, it is NP-complete to decide if $\mathcal{M}$ has a competitive equilibrium from equal incomes with social welfare at least K.*

We obtain a $1/n$-approximation of the optimal welfare in polynomial time (Algorithm 3) and leave open the question of determining the tight bound.

```
input: Market M with Leontief valuations
output: Equilibrium allocation and prices (x, p), or NULL if none exist.
if ((m < n) or (∃ i, j ∈ N with D_i = D_j and |D_i| = 1) then
    return NULL // Too few items or two buyers have identical singleton demands
end if
P ← ∅ // Buyers eligible for getting their full demand
for (i ∈ N) do
    if (∄ k ∈ N \ {i} such that (D_k ⊆ D_i and m − |D_i| ≥ n − 1)) then
        P ← P ∪ {i}
    end if
end for
if (P = ∅) then
    return Compute-Equilibrium(M) // No good equilibrium: use basic algorithm
end if
k ← arg max_{i∈P} v_i // The buyer in P with highest valuation for its demand gets it
x_k ← D_k
M' ← (N \ {k}, M, (D_i)_{i≠k}) // Reduced market, with buyers N \ {k}
(y, q) ← Compute-Equilibrium(M', D_k) // Call basic algorithm on market M' with
pre-allocated items D_k
x ← (x_k, y) // Final allocation
for (j ∈ D_k) do
    p_j ← 1/|D_k| // Price buyer k's items uniformly
end for
for (j ∈ M \ D_k) do
    p_j ← q_j // For the other buyers, use the prices from the reduced market
end for
return (x, p)
```

Algorithm 3. COMPUTE-EQUILIBRIUM-$1/$N-SW($\mathcal{M}$)

Theorem 6. *There is a polynomial-time algorithm that computes a competitive equilibrium from equal incomes with a social welfare of at least $1/n$ of the optimum welfare attainable in any equilibrium.*

Proof. We claim that Algorithm 3 computes an equilibrium with social welfare at least $1/n$ of the optimal; note that this holds for weighted valuations (i.e. not necessarily normalized). Given a market $\mathcal{M}$, Algorithm 3 computes a set $\mathcal{P}$ of *eligible* buyers, i.e. buyers k for which the following conditions hold:

(1) No buyer i's demand is completely contained in the demand of buyer k.
(2) $m - |D_k| \geq n - 1$

Both conditions are necessary in any equilibrium that gives buyer k its demand set. If Condition 1 is violated, there exists buyer $i \neq k$ with $D_i \subseteq D_k$; i can afford its demand but does not get it, which cannot happen in an equilibrium. If Condition 2 is violated, by allocating k all of its demand, too many items are used up and it's no longer possible to extract all the money.

Algorithm 3 gives the eligible buyer $k \in \mathcal{P}$ with a maximal valuation for its demand all the required items. The remaining buyers and items are allocated using Algorithm 2. We claim that the tuple $(\mathbf{x}, \mathbf{p})$ computed is an equilibrium:

(a) BUDGETS EXHAUSTED: By combining: (i) $\mathbf{x}_k \neq \emptyset$, (ii) $m - |D_k| \geq n - 1$, and (iii) the fact that *Compute-Equilibrium* finds an equilibrium in the reduced market, all buyers get a non-empty bundle at a price of 1.
(b) ITEMS SOLD: Clearly all the items are allocated.
(c) OPTIMALITY FOR EACH BUYER: If the algorithm exits before the set $\mathcal{P}$ is constructed, then no equilibrium exists. Otherwise, buyer k gets its demand.

Let $i \neq k$ be a buyer that does not get its demand; then $|D_i| \geq 2$. If i is not the last buyer allocated, then $|\mathbf{x}_i| = 1$ and there are two subcases:

- $\mathbf{x}_i \subset D_i$: By construction, $\mathbf{p}(\mathbf{x}_i) = 1$, so $\mathbf{p}(D_i) = \mathbf{p}(\mathbf{x}_i) + \mathbf{p}(D_i \setminus \mathbf{x}_i) > 1$ since all the items are priced strictly positively.
- $\mathbf{x}_i \cap D_i = \emptyset$: Then by the time buyer i was allocated, the items in D_i have been exhausted. Since buyer k gets its full demand before everyone else, it cannot be that $D_i \subseteq D_k$ (this holds by choice of the set $\mathcal{P}$, of eligible buyers). Thus there is at least one item from D_i given to a buyer other than k at a price of 1, which combined with: $|D_i| \geq 2$ gives: $\mathbf{p}(D_i) > 1$.

Otherwise, i is the last buyer allocated; note that x_i is always computed in the procedure *Compute-Equilibrium*. Again there are two cases:

- $\mathbf{x}_i \cap D_i = \emptyset$: Then D_i was exhausted before computing $\mathbf{x}_i$. Since it cannot be that $D_i \subseteq D_k$, there exists item $j \in D_i \setminus D_k$ given to another buyer at a price of 1. Using again the fact that $|D_i| \geq 2$ and that all prices are positive, we get: $\mathbf{p}(D_i) > 1$, thus i cannot afford its demand.
- $\mathbf{x}_i \cap D_i \neq \emptyset$: Let $\ell = |\mathbf{x}_i \cap D_i|$. Then i pays $1 - \epsilon$ for ℓ items from D_i, and ϵ for some other items in $\mathbf{x}_i \setminus D_i$. We show that i cannot use ϵ money to purchase the items missing from its allocation. Let $j \in D_i \setminus \mathbf{x}_i$ be such an item and $S' = (\mathbf{x}_i \cap D_i) \cup \{j\}$. Then item j was allocated to a previous buyer, priced at least $1/|D_k|$. Since $\epsilon < 1/|D_k|$, i would have to pay at least $1 - \epsilon + \frac{1}{|D_k|} > 1$ for S'. From $S' \subseteq D_i$, we get: $\mathbf{p}(D_i) \geq \mathbf{p}(S') > 1$, and so i cannot afford D_i.

Thus the market equilibrium conditions are met. Clearly the best equilibrium cannot have a social welfare higher than $n \cdot v_k$, which gives a $1/n$-approximation.

4 Perfect Substitutes

We begin by introducing the utility function in a market with perfect substitutes, represented through additive valuations.

Definition 3 (Additive Utility for Indivisible Goods). *Given a market $\mathcal{M} = (N, M, \mathbf{v})$ with additive utilities and indivisible goods, the utility of a buyer i for a bundle $\mathbf{x} = \langle x_1, \ldots, x_m \rangle \in \{0,1\}^m$ is:*

$$u_i(\mathbf{x}) = \sum_{j=1}^{m} v_{i,j} \cdot x_{i,j} \tag{2}$$

where $v_{i,j}$ are constants and represent the value of buyer i for consuming one unit of good j, while $x_{i,j} = 1$ if buyer i gets good j, and $x_{i,j} = 0$, otherwise.

Next we investigate the computation of competitive equilibria from equal incomes with indivisible goods and additive utilities. Note that if a market $\mathcal{M}$ has a competitive equilibrium at some allocation and prices $(\mathbf{x}, \mathbf{p})$, then $\mathcal{M}$ is guaranteed to have an equilibrium at the same allocation $\mathbf{x}$ where all the prices are rational numbers, $(\mathbf{x}, \mathbf{p}^*)$; this aspect appears implicitly in some of the proofs.

Theorem 7. *Given a market $\mathcal{M} = (N, M, \mathbf{v})$ with additive valuations, indivisible goods, and tuple $(\mathbf{x}, \mathbf{p})$, where $\mathbf{x}$ is an allocation and $\mathbf{p}$ is a price vector, it is coNP-complete to determine whether $(\mathbf{x}, \mathbf{p})$ is a competitive equilibrium for $\mathcal{M}$.*

Theorem 8. *Given a market with indivisible goods and additive valuations, $\mathcal{M} = (N, M, \mathbf{v})$, it is NP-hard to decide if $\mathcal{M}$ has a competitive equilibrium.*

The next question, of computing an equilibrium allocation given a market $\mathcal{M}$ and a price vector $\mathbf{p}$ was raised by Bouveret and Lemaître [3] and studied as well by Aziz in a recent note [2].

Theorem 9. *Given a market $\mathcal{M} = (N, M, \mathbf{v})$ with indivisible goods, additive valuations, and price vector $\mathbf{p}$, it is coNP-hard to decide if there is an allocation $\mathbf{x}$ such that $(\mathbf{x}, \mathbf{p})$ is a market equilibrium.*

Theorem 10. *Given a market $\mathcal{M} = (N, M, \mathbf{v})$ with indivisible goods, additive valuations, and allocation $\mathbf{x}$, it is coNP-hard to decide if there is a price vector $\mathbf{p}$ such that $(\mathbf{x}, \mathbf{p})$ is a market equilibrium.*

Table 1. Summary of the computational results. The market instance is denoted by $\mathcal{M} = \langle N, M, \mathbf{v} \rangle$, where N is a set of buyers, M a set of indivisible items, and $\mathbf{v}$ the values of the buyers for items; $\mathbf{x}$ is an allocation of items to buyers and $\mathbf{p}$ a price vector.

Input \ Valuations	**Perfect Complements**	**Perfect Substitutes**
Market $\mathcal{M}$	$\mathcal{P}$	$\mathcal{NP}$-hard
Market $\mathcal{M}$, allocation $\mathbf{x}$	$\mathcal{P}$	co-$\mathcal{NP}$-hard
Market $\mathcal{M}$, prices $\mathbf{p}$	$\mathcal{NP}$-complete	co-$\mathcal{NP}$-hard
Market $\mathcal{M}$, allocation $\mathbf{x}$, prices $\mathbf{p}$	$\mathcal{P}$	co-$\mathcal{NP}$-complete

Our findings on the complexity of computing a competitive equilibrium from equal incomes for indivisible goods are summarized in Table 1.

Acknowledgements. We would like to thank the anonymous reviewers for useful feedback that helped improve the paper. Simina Brânzei and Peter Bro Miltersen acknowledge support from the Danish National Research Foundation and The National Science Foundation of China (under the grant 61361136003) for the Sino-Danish Center for the Theory of Interactive Computation, and the Center for Research in Foundations of Electronic Markets (CFEM), supported by the Danish Strategic Research Council.

References

1. Arrow, K.J., Debreu, G.: Existence of an equilibrium for a competitive economy. Econometrica **22**(3), 265–290 (1954)
2. Aziz, H.: Competitive equilibrium with equal incomes for allocation of indivisible objects (2015). http://arxiv.org/pdf/1501.06627v1.pdf
3. Bouveret, S., Lemaître, M.: Characterizing conflicts in fair division of indivisible goods using a scale of criteria. In: AAMAS, pp. 1321–1328 (2014)
4. Brainard, W., Scarf, H.E.: How to compute equilibrium prices in 1891. Cowles Foundation Discussion Paper (2000)
5. Budish, E.: The combinatorial assignment problem: approximate competitive equilibrium from equal incomes. J. Polit. Econ. **119**(6), 1061–1103 (2011)
6. Deng, X., Papadimitriou, C., Safra, S.: On the complexity of price equilibria. J. Comput. Syst. Sci. **67**(2), 311–324 (2003)
7. Eisenberg, E., Gale, D.: Consensus of subjective probabilities: the pari-mutuel method. Ann. Math. Stat. **30**, 165–168 (1959)
8. Foley, D.K.: Resource allocation and the public sector. Yale Econ. Essays **7**(1), 45–98 (1967)
9. Hylland, A., Zeckhauser, R.: The efficient allocation of individuals to positions. J. Polit. Econ. **87**(2), 293–314 (1979)
10. Moulin, H.: Fair division and collective welfare. MIT press, Cambridge (2004)
11. Nisan, N., Roughgarden, T., Tardos, E., Vazirani, V.: Algorithmic Game Theory. Cambridge University Press, Cambridge (2007)
12. Othman, A., Papadimitriou, C.H., Rubinstein, A.: The complexity of fairness through equilibrium. In: Proceedings of ACM-EC, pp. 209–226 (2014)
13. Thomson, W., Varian, H.: Theories of justice based on symmetry. In: Hurwicz, L., Schmeidler, D. (eds.) Essays in Memory of Elisha Pazner. Social Goals and Social Organizations. Cambridge University Press, Cambridge (1985)
14. Varian, H.: Equity, envy, and efficiency. JET **9**(1), 63–91 (1974)
15. Walras, L.: Elements d'economie politique pure, ou theorie de la richesse sociale (in French) (1874)

Equilibrium Computation

When Can Limited Randomness Be Used in Repeated Games?

Pavel Hubáček[1](✉), Moni Naor[1], and Jonathan Ullman[2]

[1] Weizmann Institute of Science, Rehovot, Israel
{pavel.hubacek,moni.naor}@weizmann.ac.il

[2] Columbia University Department of Computer Science, New York, USA
jullman@cs.columbia.edu

Abstract. The central result of classical game theory states that every finite normal form game has a Nash equilibrium, provided that players are allowed to use randomized (mixed) strategies. However, in practice, humans are known to be bad at generating random-like sequences, and true random bits may be unavailable. Even if the players have access to enough random bits for a single instance of the game their randomness might be insufficient if the game is played many times.

In this work, we ask whether randomness is necessary for equilibria to exist in finitely repeated games. We show that for a large class of games containing arbitrary two-player zero-sum games, approximate Nash equilibria of the n-stage repeated version of the game exist if and only if both players have $\Omega(n)$ random bits. In contrast, we show that there exists a class of games for which no equilibrium exists in pure strategies, yet the n-stage repeated version of the game has an exact Nash equilibrium in which each player uses only a constant number of random bits.

When the players are assumed to be computationally bounded, if cryptographic pseudorandom generators (or, equivalently, one-way functions) exist, then the players can base their strategies on "random-like" sequences derived from only a small number of truly random bits. We show that, in contrast, in repeated two-player zero-sum games, if pseudorandom generators *do not* exist, then $\Omega(n)$ random bits remain necessary for equilibria to exist.

Keywords: Limited randomness · Repeated games · Nash equilibrium

P. Hubáček—Supported by the I-CORE Program of the Planning and Budgeting Committee and The Israel Science Foundation (grant No. 4/11).

M. Naor—Incumbent of the Judith Kleeman Professorial Chair. Research supported in part by grants from the Israel Science Foundation, BSF and Israeli Ministry of Science and Technology and from the I-CORE Program of the Planning and Budgeting Committee and the Israel Science Foundation (grant No. 4/11).

J. Ullman—Supported by a Junior Fellowship from the Simons Society of Fellows. Part of this work was done while the author was at postdoctoral fellow at the Center for Research on Computation and Society at Harvard University.

M. Hoefer (Ed.): SAGT 2015, LNCS 9347, pp. 259–271, 2015.
DOI: 10.1007/978-3-662-48433-3_20

1 Introduction

The signature result of classical game theory states that a Nash equilibrium exists in every finite normal form game, provided that players are allowed to play randomized (mixed) strategies. It is easy to see that, in some games (e.g. Rock-Paper-Scissors), randomization is necessary for equilibrium to exist. However, the assumption that players are able to randomize their strategies in an arbitrary manner is quite strong, as sources of true randomness may be unavailable and humans are known to be bad at generating random-like sequences.

Motivated by these considerations, Budinich and Fortnow [3] investigated the question of whether Nash equilibria exist when players only have access to *limited randomness*. Specifically, they looked at the "repeated matching pennies game." Matching pennies is a very simple, two-player, two-action, zero-sum game in which the unique equilibrium is for each player to flip a fair coin and play an action uniformly at random. If the game is repeated for n stages, then the unique Nash equilibrium is for each player to play an independent, uniformly random action in each of the n stages. Budinich and Fortnow considered the case where the players only have access to $\ll n$ bits of randomness, which are insufficient to play the unique equilibrium of the game, and showed that there does not even exist an *approximate* equilibrium (where the approximation depends on the deficiency in randomness). That is, if the players cannot choose independent, uniformly random actions in each of the n stages, then no approximate equilibrium exists.

In this work, we further investigate the need for randomness in repeated games by asking whether the same results hold for *arbitrary* games. That is, we start with an arbitrary multi-player game such that Nash equilibria exist only if players can use β bits of randomness. Then we consider the n-stage repetition of that game. Do equilibria exist in the n-stage game if players only have access to $\ll \beta n$ bits of randomness? First, we show that the answer is essentially *no* for arbitrary zero-sum games, significantly generalizing the results of Budinich and Fortnow. On the other hand, we show that the answer is *yes* for a large class of general games.

These results hold when both players are assumed to be computationally unbounded. As noted by Budinich and Fortnow, if we assume that the players are required to run in polynomial time, and cryptographic pseudorandom generators (or, equivalently, one-way functions) exist, then a player equipped with only $\ll n$ truly random bits can generate n *pseudorandom bits* that appear truly random to a polynomial time adversary. Thus, in the computationally bounded regime, if pseudorandom generators exist, then linear randomness is not necessary. We show that, in contrast, in arbitrary repeated two-player zero-sum games, if pseudorandom generators *do not* exist, then linear randomness remains necessary even when the players are polynomial time.

Our Results. Suppose we have an arbitrary finite strategic game among k players. We consider the n-stage repetition of this game in which in each of the n consecutive stages, each of the k players simultaneously chooses an action (which may depend on the history of the previous stages). We assume that in the stage game, $\beta > 0$ bits of randomness for each player are necessary and sufficient for an equilibrium to exist. We ask whether or not the existence of approximate equilibria in the n-stage game requires $\Omega(n)$ bits of randomness per player.

The Case of Computationally Unbounded Players. Our first set of results concerns computationally unbounded players, which is the standard model in classical game theory. In this setting, our first result shows that linear randomness is necessary for a large class of games including every two-player zero-sum game.

Theorem 1 (informal). *For any k-player strategic game in which every Nash equilibrium achieves the minmax payoff profile, in any Nash equilibrium of its repeated version the players' strategies use randomness at least linear in the number of stages.*

An important subset of strategic games where any Nash equilibrium achieves the minmax payoff profile is the class of two-player zero-sum games where von Neumann's minmax theorem implies that the only equilibria are minimax solutions of the game. Hence, in any finitely repeated two-player zero-sum game the players must use randomness at least linear in the number of stages.

Second, we show that the above results cannot be extended to arbitrary games. That is, there exists a class of strategic games that, in their repeated version, admit "randomness efficient" Nash equilibria.

Theorem 2 (informal). *For any k-player strategic game in which for every player there exists a Nash equilibrium that achieves strictly higher expectation than the minmax strategy, there exists a Nash equilibrium of its repeated version where the players use total randomness independent of the number of stages.*

As we shall see, this result is related to the "finite horizon Nash folk theorem," which roughly states that in finitely repeated games every payoff profile in the stage game that dominates the minimax payoff profile can be achieved as a payoff profile of some Nash equilibrium of the repeated game.

The Case of Computationally Efficient Players. For repeated two-player zero-sum games we study the existence of Nash equilibria with limited randomness when the players are computationally bounded. Under the assumption that one-way functions do not exist (see the above discussion), we show that it is possible to *efficiently exploit* any opponent (i.e., gain a non-negligible advantage over the value of the stage game) that uses low randomness in every repeated two-player zero-sum game. Hence, in repeated two-player zero-sum games there are no computational Nash equilibria in which one of the players uses randomness sub-linear in the number of the stages.

Theorem 4 (informal). *In any repeated two-player zero-sum game, if one-way functions do not exist, then for any strategy of the column player using sub-linear randomness, there is a computationally efficient strategy for the row player that* *achieves an average payoff non-negligibly higher than his minimax payoff in the stage game.*

The proof of this result employs the algorithm of Naor and Rothblum [13] for learning adaptively changing distributions. The main idea is to adaptively reconstruct the small randomness used by the opponent in order to render his strategy effectively deterministic and then improve the expectation by playing the best response.

Strong Exploitation of Low-Randomness Players. In the classical setting, i.e., without restrictions on the computational power of the players, it was shown by Neyman and Okada [14] that in every repeated two-player zero-sum game it is possible to extract utility proportional to the randomness deficiency of the opponent. On the other hand, our result in the setting with computationally efficient players guarantees only a non-negligible advantage in the presence of a low-randomness opponent. This leaves open an intriguing question of how much utility can one *efficiently* extract from a low-randomness opponent in a repeated two-player zero-sum game (see the full version [9] for additional discussion).

Other Related Work. In one of the first works to consider the relation between the randomness available to players and the existence of equilibria Halpern and Pass [6] introduced a computational framework of machine games that explicitly incorporates the cost of computation into the utility functions of the players and specifically the possibility of randomness being expensive. They demonstrated this approach on the game of Rock-Paper-Scissors, and showed that in machine games where randomization is costly then Nash equilibria do not necessarily exist. However, in machine games where randomization is free then Nash equilibria always exist.

Based on derandomization techniques, Kalyanaraman and Umans [12] proposed randomness efficient algorithms both for finding equilibria and for playing strategic games. In the context of finitely repeated two-player zero-sum games where one of the players (referred to as the learner) is uninformed of the payoff matrix, they gave an adaptive on-line algorithm for the learner that can reuse randomness over the stages of the repeated game.

Halprin and Naor [7] suggested the possibility of using randomness generated by human players in repeated games for generation of pseudorandom sequences.

2 Notation and Background

Standard Game Theoretic Notation. Formal definitions of main concepts from game theory that we use in this work are provided in the full version [9].

A *strategic game* $G = \langle N, (A_i), (u_i) \rangle$ consists of *(i)* a finite set of players N, *(ii)* for each player $i \in N$ a set of actions A_i, and *(iii)* for each player $i \in N$ a utility function $u_i : A \to \mathbb{R}$ assigning each action profile $a \in A = \times_{j \in N} A_j$ a real-valued payoff $u_i(a)$. In the special case when G is a two-player zero-sum game, i.e., when $N = \{1, 2\}$ and $u_1(a) = -u_2(a)$ for all $a \in A_1 \times A_2$, we use the notation $\langle (A_1, A_2), u \rangle$. We refer to player 1 as the row player (also known as Rowena) and to player 2 as the column player (also known as Colin).[1]

We denote by S_i the set of mixed strategies of player i, i.e., the set $\Delta(A_i)$ of all probability distributions on the action space of player i. For a strategy profile $\sigma \in S = \times_{j \in N} S_j$ we denote by σ_{-i} the profile of strategies of all the players in $N \setminus \{i\}$. A *Nash equilibrium* of a strategic game $\langle N, (A_i), (u_i) \rangle$ is a strategy profile σ such that for every player $i \in N$ we have $\mathbf{E}[u(\sigma_i, \sigma_{-i})] \geq \mathbf{E}[(\sigma'_i, \sigma_{-i})]$ for all $\sigma'_i \in S_i$.

The *minmax payoff of player* i, denoted v_i, is the lowest payoff that the other players *can force upon player* i, i.e., $v_i = \min_{\sigma_{-i} \in S_{-i}} \max_{\sigma_i \in S_i} \mathbf{E}[u_i(\sigma_i, \sigma_{-i})]$. A *minmax strategy of player* i in G is a strategy $\hat{\sigma}_i \in S_i$ such that $\mathbf{E}[u_i(\hat{\sigma}_i, \sigma_{-i})] \geq v_i$ for all $\sigma_{-i} \in S_{-i}$. An *individually rational payoff profile* of G is a vector $p \in \mathbb{R}^{|N|}$ for which $p_i \geq v_i$ for all $i \in N$. A vector $p \in \mathbb{R}^{|N|}$ is a *feasible payoff profile of* G if there exists a collection $\{\alpha_a\}_{a \in A}$ of nonnegative rational numbers such that $\sum_{a \in A} \alpha_a = 1$ and $p_i = \sum_{a \in A} \alpha_a u_i(a)$ for all $i \in N$.

The *n-stage repeated game of* G is an extensive form game with perfect information and simultaneous moves $G^n = \langle N, H, P, (u^*_i) \rangle$ in which: *(i)* $H = \{\emptyset\} \cup \{\bigcup_{t=1}^{n} A^t\}$, where $\emptyset$ is the initial history and A^t is the set of sequences of action profiles in G of length t, *(ii)* $P(h) = N$ for each non-terminal history $h \in H$, and *(iii)* $u^*_i(a^1, \ldots, a^n) = \frac{1}{n} \sum_{t=1}^{n} u_i(a^t)$ for every terminal history $(a^1, \ldots, a^n) \in A^n$.

A *behavioral strategy of player* i is a collection $(\sigma_i(h))_{h \in H \setminus A^n}$ of independent probability measures (one for each non-terminal history), where each $\sigma_i(h)$ is a probability measure over A_i. A *Nash equilibrium* of G^n is a profile σ of behavioral strategies with the property that for every player $i \in N$ and every behavioral strategy σ'_i, we have $\mathbf{E}[u^*(\sigma_i, \sigma_{-i})] \geq \mathbf{E}[u^*(\sigma'_i, \sigma_{-i})]$.

Pseudorandom Generators and One-Way Functions. The notion of cryptographic pseudorandom generators was introduced by Blum and Micali [2], who defined them as algorithms that produce sequences of bits unpredictable in polynomial time, i.e., no efficient next-bit-test is able to predict the next output of the pseudorandom generator given the sequence of bits generated so far. As Yao [15] showed, this is equivalent to a generator whose output is indistinguishable from a truly random string to any polynomial time observer. One of the central questions in cryptography is to understand the assumptions that are sufficient and necessary for implementing a particular cryptographic task. Impagliazzo and Luby [11] (see also Impagliazzo [10]) showed that one-way functions are essential for many cryptographic primitives (e.g., private-key encryption, secure authentication, coin-flipping over telephone). Håstad, Impagliazzo, Levin and Luby [8] showed that pseudorandom generators exist if and only if one-way functions exist. Therefore the existence of one-way functions is the major open problem of cryptography. For an in depth discussion see Goldreich [5].

[1] We have adopted Colin and Rowena from Aumann and Hart [1].

Standard Cryptographic Notation. A function $\mu : \mathbb{N} \to \mathbb{R}^+$ is *negligible* if for all $c \in \mathbb{N}$ there exists $n_c \in \mathbb{N}$ such that for all $n \geq n_c$, $\mu(n) \leq n^{-c}$. A function $\mu : \mathbb{N} \to \mathbb{R}^+$ is *noticeable* if there exists $c \in \mathbb{N}$ and $n_c \in \mathbb{N}$ such that for all $n \geq n_c$, $\mu(n) \geq n^{-c}$.

The *statistical distance* between two distributions X and Y over $\{0,1\}^\ell$, denoted $\mathrm{SD}(X,Y)$, is defined as: $\mathrm{SD}(X,Y) = \frac{1}{2}\sum_{\alpha\in\{0,1\}^\ell} |\Pr[X=\alpha] - \Pr[Y=\alpha]|$. Given a probability distribution $\rho \in \Delta(A)$, the *Shannon entropy of* ρ is defined as $H(\rho) := \mathbf{E}_{a\leftarrow\rho}\left(\log_2\left(\frac{1}{\Pr(\rho=a)}\right)\right)$.

Definition 1 (almost one-way function). *A function f is an* almost one-way function *if it is computable in polynomial time, and for infinitely many input lengths, for any PPTM $\mathcal{M}$, the probability that $\mathcal{M}$ inverts f on a random input is negligible. Namely, for any polynomial p, there exist infinitely many choices of $n \in \mathbb{N}$ such that* $\Pr_{x\sim U_{k(n)},\mathcal{M}}[\mathcal{M}(f(x)) \in f^{-1}(x)] < \frac{1}{p(n)}$.

3 Low-Entropy Nash Equilibria of Repeated Games

In this section we show that, in the setting with players that have unbounded computational power, there are two classes of k-player strategic games at the opposite sides of the spectrum with respect to the amount of randomness necessary for equilibria of their repeated versions. To measure the randomness of a player's strategy we consider the maximal total Shannon entropy of his strategies used along any terminal history.

Definition 2 (Shannon entropy of a strategy in repeated game). *Let $G = \langle N, (A_i), (u_i)\rangle$ be a finite strategic game and let σ_i be a strategy of player i in G^n. For any terminal history $a = (a^1, \ldots, a^n) \in A^n$, let $(\sigma_i(\emptyset), \sigma_i(a^1), \sigma_i(a^1, a^2), \ldots, \sigma_i(a^1, \ldots, a^{n-1}))$ be the n-tuple of strategies of player i in σ_i at all the non-terminal subhistories of a. We define the* Shannon entropy of σ_i, *denoted as $H(\sigma_i)$, as* $H(\sigma_i) := \max_{a\in A^n}\left\{H(\sigma_i(\emptyset)) + \sum_{j=1}^{n-1} H(\sigma_i(a^1,\ldots,a^j))\right\}$.

This is a worst case notion, in that it measures the entropy of the strategy of player i irrespective of the strategies of the other players. For some of our results we consider its alternative variant of *effective Shannon entropy of a strategy σ_i in a strategy profile σ*, i.e., the maximal total entropy of σ_i along terminal histories that are sampled in σ with non-zero probability.

For the restricted class of games in which any Nash equilibrium payoff profile is exactly the minmax payoff profile (e.g. any two-player zero-sum game), we relate Nash equilibria of the strategic game to the structure of Nash equilibria in its repeated version (the proof is provided in the full version [9]).

Proposition 1. *Let $G = \langle N, (A_i), (u_i)\rangle$ be a strategic game such that any Nash equilibrium payoff profile is equal to the minmax payoff profile. For all $n \in \mathbb{N}$, if σ is a Nash equilibrium of $G^n = \langle N, H, P, (u_i^*)\rangle$, the n-stage repeated game of G, then for every non-terminal history $h \in H$ sampled with non-zero probability by σ the strategy profile $\sigma(h)$ is a Nash equilibrium of G.*

For strategic games from this class, Proposition 1 immediately gives a linear lower bound on entropy needed to play Nash equilibria in their repeated games.

Theorem 1. *Let G be a strategic game such that any Nash equilibrium payoff profile is equal to the minmax payoff profile. For all $n \in \mathbb{N}$ and every player $i \in N$, if in any Nash equilibrium of G the strategy of player i is of entropy at least β_i then in any Nash equilibrium of the n-stage repeated game of G the strategy of player i is of entropy at least $n\beta_i$.*

Proof. Assume to the contrary that there exists a Nash equilibrium σ of the n-stage repeated game of G with strategy of entropy strictly smaller than $n\beta_i$ for player i. By Proposition 1, $\sigma(h)$ is a Nash equilibrium of G for all h sampled by σ with non-zero probability. Hence, there must exist a history $h^* \in H$ sampled with non-zero probability in σ such that $\sigma(h^*)$ is a Nash equilibrium of G and the entropy $H(\sigma_i(h^*))$ of $\sigma_i(h^*)$ is strictly smaller than β_i, a contradiction. □

Repeated Non-Zero-Sum Game Requiring a Lot of Randomness. Theorem 1 applies not only to two-player zero-sum games but also to some non-zero-sum games. Consider the strategic game G given by the following payoff matrix:

	Left (L)	Heads (H)	Tails (T)	Right (R)
Up (U)	$0, -1$	$0, -1$	$0, -1$	$0, 0$
Heads (H)	$0, -1$	$1, -1$	$-1, 1$	$-1, 0$
Tails (T)	$0, -1$	$-1, 1$	$1, -1$	$-1, 0$
Down (D)	$0, 0$	$-1, 1$	$-1, 1$	$1, 0$

There are three mixed Nash equilibria in G: $(\frac{1}{2}H+\frac{1}{2}T, \frac{1}{2}H+\frac{1}{2}T)$, $(\frac{1}{2}U+\frac{1}{2}D, \frac{1}{2}H+\frac{1}{2}R)$, and $(\frac{1}{2}U + \frac{1}{2}D, \frac{1}{2}T + \frac{1}{2}R)$; all the three Nash equilibria achieve the same payoff profile $(0, 0)$ and require each player to use one random bit. Notice that the row player can get utility 0 irrespective of the strategy of the column player by selecting his action "Up", and similarly the column player can ensure utility 0 by playing "Right". Hence, the minmax payoff profile is $(0, 0)$. Since none of the three Nash equilibria of G improves over the minmax payoff profile, we get by Theorem 1 that each player must use strategy of entropy at least n in any Nash equilibrium of the n-stage repeated game of G.

Repeated Non-Zero-Sum Game Where Low Randomness Suffices. On the other hand, there are strategic games for which Theorem 1 does not apply, and the players may use in the n-stage repeated game equilibrium strategies of entropy proportional only to the entropy needed in the single-shot game. Consider the strategic game G given by the following payoff matrix:

	Cooperate (C)	Heads (H)	Tails (T)	Punish (P)
Cooperate (C)	3, 3	−3, 6	−3, 6	−3, −3
Heads (H)	6, −3	1, −1	−1, 1	−3, −3
Tails (T)	6, −3	−1, 1	1, −1	−3, −3
Punish (P)	−3, −3	−3, −3	−3, −3	−4, −4

The strategy profile $\sigma = (\frac{1}{2}H + \frac{1}{2}T, \frac{1}{2}H + \frac{1}{2}T)$ is the unique Nash equilibrium of G that achieves payoff profile $(0, 0)$. The minmax payoff profile is $(-3, -3)$, since any player can get utility at least -3 by playing C. We show that the n-stage repeated game of G admits a Nash equilibrium that requires only a single random coin, i.e., the same amount of randomness as the Nash equilibrium σ of the stage game G. Consider the strategy profile in which both players play C in the first $n-1$ rounds and in the last round each player plays H and T with equal probability, and if any player deviates from playing C in one of the first $n-1$ rounds then the opponent plays P throughout all the remaining stages. To see that this strategy profile is a Nash equilibrium of the n-stage repeated game of G note that any deviation from playing C in the first $n-1$ rounds can increase the utility of any player by at most 3 (by playing either H or T instead of C), however the subsequent punishment induces a loss of at least -3 which renders any deviation unprofitable.

The randomness efficient Nash equilibrium from the above example resembles the structure of Nash equilibria constructed in the proof of the *finite horizon Nash folk theorem.* This theorem characterizes the payoff profiles achievable by Nash equilibria of the repeated game. In particular, it shows that in strategic games G such that for very player i there exists a Nash equilibrium σ_i strictly improving over his minmax payoff any feasible and individually rational payoff profile can be approximated by a Nash equilibrium of sufficiently long G^n.

The main idea behind the proof of the folk theorem is that for every player i the gap between the payoff in the Nash equilibrium σ_i and the minmax payoff v_i can be used to punish the player in case he deviates from the strategy that approximates any feasible and individually rational payoff profile. In particular, in any such Nash equilibrium the players use a fixed number of rounds (independent of the number of stages n) before the last round in which they play according to some (possibly mixed) Nash equilibria of the stage game and in the preceding rounds they play pure strategies so that the overall payoff approximates the feasible payoff profile. Hence, the amount of randomness on all the equilibrium paths is independent of the number of stages in any such Nash equilibrium of the repeated game.

Theorem 2. *Let G be a strategic game such that for every player i there exists a Nash equilibrium σ_i of G in which the payoff of player i exceeds his minmax payoff v_i and there exists a feasible and individually rational payoff profile in G. Let β_i be such that in any Nash equilibrium of G the strategy of player i is of entropy at most β_i. There exists $c \in \mathbb{N}$ such that for all sufficiently large*

$n \in \mathbb{N}$ and every player $i \in N$ there exists a Nash equilibrium of G^n, the n-stage repeated game of G, in which the strategy of player i is of effective entropy *at most $c\beta_i$.*

Proof. Let $p \in \mathbb{R}^{|N|}$ be the feasible and individually rational payoff profile of G. There exist coefficients $\{\alpha_a\}_{a\in A} \subset \mathbb{Q}$ such that $\sum_{a\in A} \alpha_a = 1$ and for all $i \in N$, $p_i = \sum_{a\in A} \alpha_a u_i(a)$. Let K be the smallest integer such that each α_a can be written as α'_a/K for $\alpha'_a \in \mathbb{N}$. For some $\ell \in \mathbb{N}$, we divide the stages in G^n into two parts of length ℓK and $m = n - \ell K$. Let s be a strategy profile in G^n that schedules the first ℓK stages such that each action profile a for which $\alpha_a \neq 0$ is played by the players in exactly $\ell\alpha'_a$ number of stages. In the remaining m stages the players cycle between the Nash equilibria $\{\sigma_i\}_{i\in N}$, i.e., for all $j \in \{0, \ldots, m-1\}$ at the stage $n - m + 1 + j$ the players play the Nash equilibrium $\sigma_{j'}$, where $j' = 1 + (j \mod |N|)$. In case any player i deviates from s in one of the first ℓK rounds, the remaining players play the strategy that forces the minmax level v_i on player i.

Note that if the number m of the last stages is such that for all action profiles $a \in A$ with $\alpha_a \neq 0$ and for every player i: $\frac{m}{|N|}\left(\sum_{j\in N} \mathbf{E}[u_i(\sigma_j)] - |N|v_i\right) \geq \max_{a'_i\in A_i} u_i(a'_i, a_{-i}) - u_i(a)$, then no player has a profitable deviation and σ is a Nash equilibrium of G^n. The number m of last stages can be bounded by some constant c selected independently of n. Since the number of stages in which the players play according to some Nash equilibrium of G is at most c (the players take pure actions in all the first $n-c$ stages), for any player i the effective entropy of s_i in s is at most $c\beta_i$. □

4 Low-Entropy Computational Nash Equilibria of Repeated Two-Player Zero-Sum Games

In this section we study randomness in equilibria of repeated two-player zero-sum games with computationally efficient players. The solution concept we consider in this setting is *computational Nash equilibrium* (introduced in the work of Dodis, Halevi and Rabin [4]) that assumes that the players are restricted to computationally efficient strategies and indifferent to negligible improvements in their utilities, i.e., a computational Nash equilibrium is analogous to the concept of ε-Nash equilibrium with a negligible ε, where the player's strategies, as well as any deviations, must be computationally efficient.

To capture the requirement of computational efficiency, the players' strategies must be implemented by families of polynomial-size circuits. For a two-player zero-sum game G, we denote by *repeated game of* G the infinite collection $\{G^n\}_{n\in\mathbb{N}}$ of all the n-stage repeated games of G. A family of polynomial size circuits $\{C_n\}_{n\in\mathbb{N}}$ implements the strategy of the row player in the repeated game of G as follows. In G^n, the n-stage repeated game of G, the circuit C_n takes as input a string corresponding to a non-terminal history h in G^n and $s(n)$ random bits; it outputs an action to be taken at history h. If the strategy of player $i \in \{1, 2\}$ is implemented by family $\{C^i_n\}_{n\in\mathbb{N}}$ then the gameplay in the n-stage repeated game of G is defined in the following way: player i samples a random string $r_i \in \{0,1\}^{s_i(n)}$ and at each stage of G^n takes the action $a = C_n^i(h, r_i) \in A_i$, given that the history of play up to the current stage is h. The utility function u_n^* is for all n defined as in the standard n-stage repeated game of G (i.e., it is the average utility achieved in the stage game over the n stages).

Definition 3 (computational Nash equilibrium of repeated game). *For a two-player zero-sum game $G = \langle (A_1, A_2), u \rangle$,* a computational Nash equilibrium of the repeated game of G *is a strategy profile* $(\{C_n^1\}_{n\in\mathbb{N}}, \{C_n^2\}_{n\in\mathbb{N}})$ *given by polynomial-size circuit families such that for every player $i \in \{1,2\}$ and every strategy $\{\tilde{C}_n^i\}_{n\in\mathbb{N}}$ given by a polynomial-size circuit family it holds for all large enough $n \in \mathbb{N}$ that $\mathbf{E}[u_n^*(C_n^i, C_n^{-i})] \geq \mathbf{E}[u_n^*(\tilde{C}_n^i, C_n^{-i})] + \varepsilon(n)$, where ε is a negligible function.*

We show that if one-way functions do not exist, then in repeated two-player zero-sum games there are no computational Nash equilibria in which the players' strategies use random strings of length sub-linear in the number of the stages.

Our result follows by showing that finding efficiently a best response to the opponent's strategy that uses limited randomness can be seen as a special case of the problem of *learning an adaptively changing distribution* (introduced by Naor and Rothblum [13]). The goal in their framework is for a learner to recover a secret state used to sample a publicly observable distribution, in order to be able to predict the next sample. In particular, this would allow the learner to be competitive to someone who knows the secret state (Naor and Rothblum [13] considered this problem in the context of an adversary trying to impersonate someone in an authentication protocol). In the setting of repeated games, the random string used by the opponent's strategy can be thought of as the secret state. Note that learning it at any non-terminal history would give rise to efficient profitable deviation, since the player could just compute the next move of his opponent and play the best response to it.

Learning Adaptively Changing Distributions. An adaptively changing distribution is given by a pair of algorithms $\mathcal{G}$ and $\mathcal{D}$ for generating an initial state and sampling. The algorithm $\mathcal{G}$ is a randomized function $\mathcal{G} : R \to S_p \times S_{init}$ that outputs an initial public state p_0 and a secret state s_0. The sampling algorithm $\mathcal{D}$ is a randomized function $\mathcal{D} : S_p \times S_s \times R \to S_p \times S_s$ that at each stage takes the current public and secret states, updates its secret sate and outputs a new public state. A learning algorithm $\mathcal{L}$ for $(\mathcal{G}, \mathcal{D})$ is given the initial public state p_0 ($\mathcal{L}$ does not get the initial secret state s_0) and at each round i: i) $\mathcal{L}$ either outputs prediction of the conditional distribution $D_{i+1}^{s_0}(p_0, \ldots, p_i)$ of the public output of $\mathcal{D}$ given the initial secret s_0 and the observed public states $p_0, \ldots, p_i$, or ii) $\mathcal{L}$ proceeds to round $i+1$ after observing a new public state $p_{i+1} \leftarrow D_{i+1}^{s_0}(p_0, \ldots, p_i)$. The goal of the learning algorithm is to output a hypothesis (in a form of a distribution) that is with high probability close in statistical distance to $D_{i+1}^{s_0}(p_0, \ldots, p_i)$. In other words, $\mathcal{L}$ is trying to be competitive to somebody who knows the initial secret state s_0. In the setting where $\mathcal{G}, \mathcal{D}$ are efficiently constructible Naor and Rothblum [13] gave an algorithm $\mathcal{L}$ that learns s_0 in probabilistic polynomial time provided that one-way functions do not exist. Moreover, their algorithm outputs a hypothesis after seeing a number of samples proportional to the entropy of the initial secret state.

Theorem 3 (Naor and Rothblum [13]**).** *Almost one-way functions exist if and only if there exists an adaptively changing distribution* $(\mathcal{G}, \mathcal{D})$ *and polynomials* $\varepsilon(n), \delta(\epsilon)$ *such that it is hard to* $(\delta(n), \epsilon(n))$*-learn the adaptively changing distribution* $(\mathcal{G}, \mathcal{D})$ *with* $O\left(\delta^{-2}(n)\varepsilon^{-4}(n) \log |S_{init}|\right)$ *samples.*

The strategy of the column player (Colin) with limited randomness gives rise to a natural adaptively changing distribution and we show that the algorithm of Naor and Rothblum [13] can be used to construct a computationally efficient strategy for the row player (Rowena) that achieves utility noticeably larger than the value of the stage game. Hence, if one-way functions do not exist, then in repeated two-player strategic games there are no computational Nash equilibria with strategies that use sub-linear randomness in the number of the stages.

Theorem 4. *Let* $G = \langle (A_1, A_2), u\rangle$ *be a two-player zero-sum strategic game with no weakly dominant pure strategies and value* v*. If almost one-way functions do not exist then for any strategy* $\{C_n\}_{n\in\mathbb{N}}$ *of Colin in the repeated game of* G *that uses* $o(n)$ *random bits, there exists a polynomial time strategy of Rowena with expected average utility* $v+\delta(n)$ *against* $\{C_n\}_{n\in\mathbb{N}}$ *for some noticeable function* δ*.*

Proof. Let $\{C_n\}_{n\in\mathbb{N}}$ be an arbitrary strategy of Colin that takes $s(n) \in o(n)$ random bits. Let μ be the minmax strategy of Rowena in G. We define the following adaptively changing distribution $(\mathcal{G}, \mathcal{D})$. The generating algorithm $\mathcal{G}$ on input 1^n outputs a random string of length $s(n)$ as the initial secret state s_0 and the initial history $\emptyset$ of the n-stage repeated game of G as the initial public state p_0. The sampling algorithm $\mathcal{D}$ outputs the new secret state s_{i+1} identical to the secret state s_i that it received as an input (i.e., the secret state remains fixed as the $s(n)$ random coins s_0) and updates the input public state p_i in the following way. The sampling algorithm parses p_i as a history of length i in the n-stage repeated game of G and computes Colin's action $c_i = C_n(p_i, s_i)$ at p_i using randomness s_i. $\mathcal{D}$ additionally samples Rowena's action $r_i \leftarrow \mu$ according to her minmax strategy and then outputs the history $(p_i, (r_i, c_i))$ of length $i + 1$ as the new public state p_{i+1}. Note that after sampling the initial secret state s_0 the only randomness used by $\mathcal{D}$ is to sample the minmax strategy of Rowena.

It follows from Theorem 3 that there exists an efficient learning algorithm $\mathcal{L}$ that after at most $k = k(n) \in O(s(n)\delta^{-2}(n)\epsilon^{-4}(n))$ samples from $\mathcal{D}$ outputs a hypothesis h such that $\Pr[\mathrm{SD}(D_{k+1}^{s_0}, D_{k+1}^{h}) \le \epsilon(n)] \ge 1 - \delta(n)$. Consider the strategy of Rowena that uses $\mathcal{L}$ in order to learn Colin's random coins. In particular, a strategy that at each stage i runs $\mathcal{L}$ on the current history p_{i-1} and if $\mathcal{L}$ outputs some hypothesis h then the strategy plays the best response to Colin's action at stage i sampled according to D_{i+1}^{h}; and otherwise it plays according to Rowena's minmax strategy μ. This strategy can be efficiently implemented and it achieves expectation at least v in the $n - 1$ stages in which Rowena plays according to her minmax strategy. It remains to show that Rowena has a noticeable advantage over the value of the game at the stage in which $\mathcal{L}$ outputs the hypothesis h about s_0 and Rowena selects her strategy as the best response to Colin's action sampled according to D^h_{k+1}.

First, note that since G has no weakly dominant strategies, the best response to any pure action a_2 of Colin achieves a positive advantage over the value of the game. This observation follows from the fact that Rowena's minmax strategy achieves expectation at least v against any action of Colin and from the fact that the minmax strategy must be mixed (as there are no weakly dominant strategies). By moving all the probability in the minmax strategy to the action with highest payoff given that Colin plays a_2, Rowena achieves a value strictly larger than v. Hence, there exists some constant e (depending only on G) such that if D^h_{k+1} is e-close in statistical distance to $D^{s_0}_{k+1}$ then the expectation of the best response against D^h_{k+1} achieves expectation at least $v+c$ for some constant $c > 0$. Moreover, it is good enough if $\mathcal{L}$ outputs such h with probability at least $1-\delta$ for some constant $\delta > 0$. Since ϵ and δ can be constant, for all large enough n the learning algorithm $\mathcal{L}$ outputs the hypothesis after receiving at most $k < n$ samples which allows Rowena to get expectation at least $v + \frac{1}{n}c$. □

It follows from Theorem 4 that if one-way functions do not exist, then there is no computational Nash equilibrium of repeated two-player zero-sum games where one of the players uses randomness sub-linear in the number of stages (the proof is provided in the full version [9]).

References

1. Aumann, R.J., Hart, S.: Long cheap talk. Econometrica **71**(6), 1619–1660 (2003)
2. Blum, M., Micali, S.: How to generate cryptographically strong sequences of pseudo-random bits. SIAM J. Comput. **13**(4), 850–864 (1984)
3. Budinich, M., Fortnow, L.: Repeated matching pennies with limited randomness. In: ACM EC 2011, pp. 111–118 (2011)
4. Dodis, Y., Halevi, S., Rabin, T.: A cryptographic solution to a game theoretic problem. In: Bellare, M. (ed.) CRYPTO 2000. LNCS, vol. 1880, pp. 112–130. Springer, Heidelberg (2000)
5. Goldreich, O.: The Foundations of Cryptography. Basic Techniques, vol. 1. Cambridge University Press, Cambridge (2001)
6. Halpern, J.Y., Pass, R.: Algorithmic rationality: Game theory with costly computation. J. Econ. Theory (2014)
7. Halprin, R., Naor, M.: Games for extracting randomness. ACM Crossroads **17**(2), 44–48 (2010)
8. Håstad, J., Impagliazzo, R., Levin, L.A., Luby, M.: A pseudorandom generator from any one-way function. SIAM J. Comput. **28**(4), 1364–1396 (1999)
9. Hubáček, P., Naor, M., Ullman, J.: When can limited randomness be used in repeated games? CoRR abs/1507.01191 (2015). http://arxiv.org/abs/1507.01191
10. Impagliazzo, R.: Pseudo-random generators for cryptography and for randomized algorithms. Ph.D. thesis, University of California, Berkeley (1992)
11. Impagliazzo, R., Luby, M.: One-way functions are essential for complexity based cryptography (extended abstract). In: FOCS 1989, pp. 230–235 (1989)

12. Kalyanaraman, S., Umans, C.: Algorithms for playing games with limited randomness. In: Arge, L., Hoffmann, M., Welzl, E. (eds.) ESA 2007. LNCS, vol. 4698, pp. 323–334. Springer, Heidelberg (2007)
13. Naor, M., Rothblum, G.N.: Learning to impersonate. In: ICML 2006, pp. 649–656 (2006)
14. Neyman, A., Okada, D.: Repeated games with bounded entropy. Games Econ. Behav. **30**(2), 228–247 (2000)
15. Yao, A.C.: Theory and applications of trapdoor functions (extended abstract). In: FOCS 1982, pp. 80–91 (1982)

Settling Some Open Problems on 2-Player Symmetric Nash Equilibria

Ruta Mehta, Vijay V. Vazirani, and Sadra Yazdanbod(✉)

College of Computing, Georgia Tech, Atlanta, USA
{rmehta,vazirani,syazdanb}@cc.gatech.edu

Abstract. Over the years, researchers have studied the complexity of several decision versions of Nash equilibrium in (symmetric) two-player games (bimatrix games). To the best of our knowledge, the last remaining open problem of this sort is the following; it was stated by Papadimitriou in 2007: find a non-symmetric Nash equilibrium (NE) in a symmetric game. We show that this problem is NP-complete and the problem of counting the number of non-symmetric NE in a symmetric game is #P-complete.

In 2005, Kannan and Theobald defined the *rank of a bimatrix game* represented by matrices (A, B) to be rank$(A + B)$ and asked whether a NE can be computed in rank 1 games in polynomial time. Observe that the rank 0 case is precisely the zero sum case, for which a polynomial time algorithm follows from von Neumann's reduction of such games to linear programming. In 2011, Adsul et al. obtained an algorithm for rank 1 games; however, it does not guarantee symmetric NE in symmetric rank 1 game. We resolve this problem.

1 Introduction

One of the major achievements of complexity theory in recent years is obtaining a fairly complete understanding of the complexity of computing a Nash equilibrium (NE) in a two-player game in various situations; such a game can be represented by two payoff matrices (A, B), and therefore is also known as *bimatrix game*. Of the few remaining open questions, we settle two in this paper regarding symmetric bimatrix games. We note that symmetry arises naturally in numerous strategic situations. In fact, while providing game theory with its central solution concept, Nash [12] felt compelled to also define the notion of a symmetric game and prove, in a separate theorem, that such (finite) games always admit a *symmetric equilibrium, i.e.*, where all players play the same strategy. Examples of well-known bimatrix games that are symmetric are Prisoners' Dilemma and Rock-Paper-Scissors. With the growth of the Internet, on which typically users are indistinguishable, the relevance of symmetric games has further increased.

In a *symmetric game* all players participate under identical circumstances, i.e., strategy sets and payoffs. Thus the payoff of a player i depends only on the strategy, s, played by her and the multiset of strategies, S, played by the others, without reference to their identities; moreover, if any other player j were to play

M. Hoefer (Ed.): SAGT 2015, LNCS 9347, pp. 272–284, 2015.
DOI: 10.1007/978-3-662-48433-3_21

s and the remaining players S, the payoff to j would be identical to that of i. In case of a bimatrix game (A, B) such a symmetry translates to $B = A^T$.

We first provide a brief summary of the known results. The seminal works of Daskalakis, Goldberg and Papadimitriou [6], and Chen, Deng and Teng [2] proved that finding a NE in a bimatrix game, or a symmetric NE in a symmetric bimatrix game, is PPAD-complete [5,13]. For special symmetric games like anonymous games, Daskalakis and Papadimitriou gave an efficient algorithm to compute the approximate Nash equilibrium [7]. Before the resolution of this long-standing question, researchers studied the complexity of computing a NE with desired special properties. For numerous properties, these problems turned out to be NP-hard, even for the case of symmetric games [4,8].

In 2005, Kannan and Theobald [9] defined the *rank of a bimatrix game* (A, B) to be rank$(A + B)$ and asked whether a NE can be computed in rank 1 games in polynomial time. They also gave an example of a bimatrix rank 1 game that has disconnected NE, thereby providing evidence that the problem would be a difficult one[1]. Observe that the rank 0 case is precisely the zero sum case, for which a polynomial time algorithm follows from von Neumann's reduction of such games to linear programming. In 2011, Adsul et al. [1] answered this question in the affirmative; this appears to be the first efficient algorithm for a problem having disconnected solutions. More recently, Mehta [11] showed that for games of rank 3 or more, and for symmetric games of rank 6 or more, the problem is PPAD-complete.

We now list the open problems we are aware of. In 2007, Papadimitriou [14] asked for the complexity of finding a non-symmetric NE in a symmetric game. One motivation for this problem may be the following. In some situations, symmetric equilibrium may imply both the player exhausting the same resources, for instance if they access the same web site, which may be undesirable and non-symmetric equilibrium is called for.

Mehta [11] left open the problem of determining the complexity of the following problems: finding a NE in a rank 2 game, and finding a symmetric NE in a symmetric game of rank $1, 2, 3, 4$, or 5.

In this paper, we show that the problem of finding asymmetric NE in a symmetric game (Papadimitriou's problem) is NP-complete. We further show that the problem of counting the number of non-symmetric NE in a symmetric game is #P-complete. We also give a polynomial time algorithm for finding a symmetric NE in a symmetric game of rank 1. In the full version of the paper, we give some reasons to believe that finding a symmetric NE in a symmetric game of rank 2 or more should not be in P.

Next, we note that given a symmetric bimatrix rank 1 game, the algorithm of Adsul et al. [1] is not guaranteed to produce a symmetric Nash equilibrium, as required in the definition of a symmetric game. Furthermore, the symmetric NE of a symmetric rank 1 game can also be disconnected, thereby making it a difficult problem from the viewpoint of obtaining a polynomial time algorithm.

[1] von Stengel [16] went further to give a symmetric bimatrix rank 1 game that has exponentially many disconnected symmetric Nash equilibria.

In Sect. 1.1 we give the new ideas that are needed, in addition to those of [1], to solve this problem. For an example of a well-known game having disconnected symmetric equilibria, consider Battle of Sexes, with appropriate payoffs so that the game is symmetric.

Recently, McLennan and Tourky [10] gave the notion of imitation games, which simplified the existing proofs of NP-completeness of [4] considerably for the case of symmetric games and led to even more such results. In the full version of the paper we study further properties of imitation games.

1.1 New Techniques

First Problem: Next we give an overview of the approach to show NP (#P) hardness for computing (counting) non-symmetric NE in symmetric games. A quick look at the set of Nash equilibrium problems proven NP-hard suggests Non-Unique NE [8], i.e., whether the given bimatrix game has two or more NE, as the most suitable problem to reduce from. Furthermore, an obvious approach is to use the standard reduction from a bimatrix game (A, B) (where both $A > 0$ and $B > 0$ are $m \times n$ matrices) to a symmetric game (M, M^T), where

$$M = \begin{bmatrix} 0 & A \\ B^T & 0 \end{bmatrix}$$

For any non-zero vector $\boldsymbol{z} \geq 0$, let $\eta(z)$ denote the *normalized vector*, i.e., its components are non-negative and add to 1. It is easy to see that the Nash equilibria $(\boldsymbol{a}, \boldsymbol{b})$ of (A, B) are in one-to-one correspondence with the symmetric NE, $(\eta({}^{a}/_{v}, {}^{b}/_{w}), \eta({}^{a}/_{v}, {}^{b}/_{w}))$ of M, where $v = \boldsymbol{a}^T B\boldsymbol{b}$ and $w = \boldsymbol{a}^T A\boldsymbol{b}$. Furthermore, from a non-symmetric NE $((\boldsymbol{a}, \boldsymbol{b}), (\boldsymbol{a}', \boldsymbol{b}'))$ of game (M, M^T), one can obtain two potential equilibria for (A, B), namely $(\eta(\boldsymbol{a}'), \eta(\boldsymbol{b}))$ and $(\eta(\boldsymbol{a}), \eta(\boldsymbol{b}'))$; indeed, one can readily confirm that they satisfy all complementarity conditions. However, there is a snag, namely all four of vectors $\{\boldsymbol{a}, \boldsymbol{b}, \boldsymbol{a}', \boldsymbol{b}'\}$ may not be nonzero, or one set of vectors may be a scaled up version of the other, thereby not yielding 2 NE for the game (A, B). In fact every NE $(\boldsymbol{a}, \boldsymbol{b})$ of (A, B) yields a non-symmetric NE $((\boldsymbol{a}, \boldsymbol{0}), (\boldsymbol{0}, \boldsymbol{b}))$ for the symmetric game (M, M^T), and such a non-symmetric NE yields only one NE for (A, B).

Let us say that a NE has *full support* if both players play all their strategies. We first seek a small dimensional symmetric game which has a unique NE and moreover it has full support. Recently, [3] showed that all symmetric 2×2 games always have pure NE, leading us to consider 3×3 games. As shown in Lemma 1, the game (D, D^T), for the matrix D specified below, has the right properties.

Next, let us define a "blown up" version of matrix D. For every $k \in \mathbb{R}$, let $K_{n \times m}(k)$ be a matrix with n rows and m columns with all entries equal to k, and define K to be the following $(1 + m + n) \times (1 + m + n)$ matrix:

$$D = \begin{bmatrix} 0 & 4 & 0 \\ 2 & 0 & 4 \\ 3 & 2 & 0 \end{bmatrix} \qquad K = \begin{bmatrix} 0 & K_{1\times m}(4) & K_{1\times n}(0) \\ K_{m\times 1}(2) & K_{m\times m}(0) & K_{m\times n}(4) \\ K_{n\times 1}(3) & K_{n\times m}(2) & K_{n\times n}(0) \end{bmatrix} \qquad G = K + \begin{bmatrix} 0 & 0 & 0 \\ 0 & 0 & A \\ 0 & B^T & 0 \end{bmatrix}$$

Let $\boldsymbol{a}, \boldsymbol{a}' \in \mathbb{R}^m$, $\boldsymbol{b}, \boldsymbol{b}' \in \mathbb{R}^n$ and $c, c' \in \mathbb{R}$. Now let $\boldsymbol{x}$ and $\boldsymbol{y}$ be the following $(1+m+n)$-dimensional vectors, $\boldsymbol{x} = (c, \boldsymbol{a}, \boldsymbol{b})$ and $\boldsymbol{y} = (c', \boldsymbol{a}', \boldsymbol{b}')$. Define the *collapse* of a $(1+m+n)$-dimensional vector say $\boldsymbol{x} = (c, \boldsymbol{a}, \boldsymbol{b})$ to be the 3-dimensional vector whose first component is c, the second is the sum of components of $\boldsymbol{a}$ and the third is the sum of components of $\boldsymbol{b}$; we will denote this by $\text{cl}(\boldsymbol{x})$. Now it is easy to see that if $(\boldsymbol{x}, \boldsymbol{y})$ is a NE of (K, K^T) then $(\text{cl}(\boldsymbol{x}), \text{cl}(\boldsymbol{y}))$ must be a NE of (D, D^T). Therefore, the NE of (K, K^T) must inherit the properties of the NE of D, and hence all four vectors $\{\boldsymbol{a}, \boldsymbol{b}, \boldsymbol{a}', \boldsymbol{b}'\}$ must be nonzero!

The next key idea is to insert the given game (A, B) in this setup in such a way that the composite game has not only the property established above but also captures certain essential features of the game (A, B). For this we will first make the assumption that w.l.o.g. the entries of A and B are positive and $<< 1$, and we construct the matrix G given above. We further define certain 3×3 matrices, $D_{\epsilon_1, \epsilon_2}$, by perturbing D appropriately (see Sect. 3).

We then show that the Nash equilibria of the symmetric game (G, G^T) have an "image" on the Nash equilibria of the perturbed 3×3 matrix, e.g., we show that payoff of the first player, assuming $(\boldsymbol{x}, \boldsymbol{y})$ is played on the symmetric game (G, G^T), is the same as the payoff of first player if $(\text{cl}(\boldsymbol{x}), \text{cl}(\boldsymbol{y}))$ is played on $(D_{\epsilon_1, \epsilon_2}, D_{\epsilon'_1, \epsilon'_2})$, for a suitable choice of $\epsilon_1, \epsilon_2, \epsilon'_1, \epsilon'_2$. Eventually this leads to showing that (G, G^T) has a non-symmetric NE iff (A, B) has at least two NE, and moreover, the non-symmetric NE of (G, G^T) are in a one-to-one correspondence with ordered pairs of NE of (A, B). These give the NP-hardness and #P-completeness results, respectively.

Second Problem: An obvious approach to designing an algorithm for finding symmetric NE in rank-1 symmetric games is to impose symmetry in the approach of Adsul et al. However, this approach fails and a new approach is called for. In order to describe the salient features of the latter, it is important to recall their approach and show where it fails.

Their approach was to start with the standard quadratic program (QP) that captures all Nash equilibria of a given bimatrix game as optimal solutions. Since $\text{rank}(A+B) = 1$, $A+B = \boldsymbol{c}\boldsymbol{d}^T$, for a suitable choice of vectors $\boldsymbol{c}, \boldsymbol{d}$. After making this substitution, the objective function of the QP becomes the product of two linear forms. [1] replaces one of the linear forms by a parameter λ, thereby getting a parameterized linear program LP(λ). They show that the optimal solutions of this linear program, over all choices of $\lambda \in \mathbb{R}$, are precisely all NE of a certain space of rank 1 games, i.e., $(A, \boldsymbol{u}\boldsymbol{d}^T - A)$, for all choices of $\boldsymbol{u} \in \mathbb{R}^m$; we will denote the bimatrix game $(A, \boldsymbol{u}\boldsymbol{d}^T - A)$ by $(A, \boldsymbol{u}, \boldsymbol{d})$. They further show that the union of all the polyhedra defined by the constraints of LP(λ), over all λ, is yet another polyhedron. The one-skeleton of the latter polyhedron contains a path whose points are in one-to-one correspondence with the optimal solutions of LP(λ), $\forall\lambda$. Additionally, λ is monotonic on this path. Therefore, they are able find a NE of game (A, B) via a binary search for the "correct" value of λ.

Adapting this approach to symmetric rank one games will involve the following. We are given game (A, A^T), where $A + A^T = \boldsymbol{c}\boldsymbol{d}^T$, and we seek a symmetric NE $(\boldsymbol{x}, \boldsymbol{x})$. Clearly, we must start with the standard QP that captures symmetric equilibria of symmetric bimatrix games. The optimal solutions of the analogous parameterized linear program are not even NE of games in the corresponding space of rank 1 games, i.e., $(A, \boldsymbol{u}, \boldsymbol{c})$, for all choices of $\boldsymbol{u} \in \mathbb{R}^m$. The reason is that this is not a space of symmetric games.

We rectify this situation by moving to a space of symmetric bimatrix games, but of rank 2. This is made possible by the observation that matrix A can be written as the sum of a skew-symmetric matrix K and the rank one matrix $\frac{1}{2}\boldsymbol{c}\boldsymbol{d}^T$, using the fact that $\boldsymbol{c}\boldsymbol{d}^T$ is a symmetric matrix. Now, replacing the vector $\boldsymbol{d}$ by $\boldsymbol{u}$, for all choices of $\boldsymbol{u} \in \mathbb{R}^m$, we get the space of rank 2 symmetric games

$$\left(\left(K + \frac{1}{2}\boldsymbol{c}\boldsymbol{u}^T\right), \left(K + \frac{1}{2}\boldsymbol{u}\boldsymbol{c}^T\right)^T\right).$$

The new LP(λ) will capture all symmetric NE of this space of symmetric games.

At this stage we introduce another idea, thereby achieving a substantial simplification. We bypass the polyhedra mentioned above completely and reduce the problem of finding the "correct" λ to a one-dimensional fixed-point computation in which every fixed point is guaranteed to be rational. Such a fixed point can be found efficiently by a binary search and yields the "correct" λ, which in turn yields the desired symmetric NE.

In what follows we discuss details of main results, however due to space constraint proofs of some of the lemmas and theorems are deferred to the full version.

2 Preliminaries

A bimatrix game is a two player game, each player having finitely many pure strategies (moves). Let S_i, $i = 1, 2$ be the set of strategies for player i, and let $m \stackrel{\text{def}}{=} |S_1|$ and $n \stackrel{\text{def}}{=} |S_2|$. Then such a game can be represented by two payoff matrices A and B, each of $m \times n$ dimension. If the first player plays strategy i and the second plays j, then the payoff of the first player is A_{ij} and that of the second player is B_{ij}. Note that the rows of these matrices correspond to the strategies of the first player and the columns to the strategies of second player.

Players may randomize among their strategies; a randomized play is called a *mixed strategy*. The set of mixed strategies for the first player is $X = \{\boldsymbol{x} = (x_1, \ldots, x_m) \mid \boldsymbol{x} \geq 0, \sum_{i=1}^m x_i = 1\}$, and for the second player is $Y = \{\boldsymbol{y} = (y_1, \ldots, y_n) \mid \boldsymbol{y} \geq 0, \sum_{j=1}^n y_j = 1\}$. By playing $(\boldsymbol{x}, \boldsymbol{y}) \in X \times Y$ we mean strategies are picked independently at random as per $\boldsymbol{x}$ by the first-player and as per $\boldsymbol{y}$ by the second-player. Therefore the expected payoffs of the first-player and second-player are, respectively $\sum_{i,j} A_{ij}x_iy_j = \boldsymbol{x}^T A\boldsymbol{y}$ and $\sum_{i,j} B_{ij}x_iy_j = \boldsymbol{x}^T B\boldsymbol{y}$.

Definition 1. *(Nash Equilibrium [15]) A strategy profile is said to be a Nash equilibrium strategy profile (NESP) if no player achieves a better payoff by a unilateral deviation [12]. Formally, $(\boldsymbol{x}, \boldsymbol{y}) \in X \times Y$ is a NESP iff $\forall \boldsymbol{x}' \in X$, $\boldsymbol{x}^T A\boldsymbol{y} \geq \boldsymbol{x}'^T A\boldsymbol{y}$ and $\forall \boldsymbol{y}' \in Y$, $\boldsymbol{x}^T B\boldsymbol{y} \geq \boldsymbol{x}^T B\boldsymbol{y}'$.*

Given strategy $\boldsymbol{y}$ for the second-player, the first-player gets $(A\boldsymbol{y})_k$ from her k^{th} strategy. Clearly, her best strategies are $\arg\max_k (A\boldsymbol{y})_k$, and a mixed strategy fetches the maximum payoff only if she randomizes among her best strategies. Similarly, given $\boldsymbol{x}$ for the first-player, the second-player gets $(\boldsymbol{x}^T B)_k$ from k^{th} strategy, and same conclusion applies. These can be equivalently stated as the following complementarity type conditions,

$$\begin{array}{l} \forall i \in S_1,\ x_i > 0 \ \Rightarrow (A\boldsymbol{y})_i = \max_{k \in S_1} (A\boldsymbol{y})_k \\ \forall j \in S_2,\ y_j > 0 \ \Rightarrow (\boldsymbol{x}^T B)_j = \max_{k \in S_2} (\boldsymbol{x}^T B)_k \end{array}$$

It is easy to get the following from the above discussion: $(\boldsymbol{x}, \boldsymbol{y}) \in X \times Y$ is a NE if and only if the following holds, where π_1 and π_2 are scalars.

$$\begin{array}{ll} \forall i \in S_1, (A\boldsymbol{y})_i \leq \pi_1; & x_i((A\boldsymbol{y})_i - \pi_1) = 0 \\ \forall j \in S_2, (\boldsymbol{x}^T B)_j \leq \pi_2; & y_j((\boldsymbol{x}^T B)_j - \pi_2) = 0 \end{array} \tag{1}$$

Game (A, B) is said to be symmetric if $B = A^T$. In a symmetric game the strategy sets of both the players are identical, i.e., $m = n$, $S_1 = S_2$ and $X = Y$. Therefore, we will use n, S and X for both. A Nash equilibrium profile $(\boldsymbol{x}, \boldsymbol{y}) \in X \times X$ is called *symmetric* if $\boldsymbol{x} = \boldsymbol{y}$. Note that at a symmetric strategy profile $(\boldsymbol{x}, \boldsymbol{x})$ both the players get payoff $\boldsymbol{x}^T A \boldsymbol{x}$. Using (1) it follows that $\boldsymbol{x} \in X$ is a symmetric NE of game (A, A^T), with payoff π to both players, iff,

$$\forall i \in S, (A\boldsymbol{x})_i \leq \pi; \quad x_i((A\boldsymbol{x})_i - \pi) = 0 \tag{2}$$

3 NP-Hardness of Non-Symmetric NE in a Symmetric Game

As discussed in the introduction, existence of symmetric NE in a symmetric game is guaranteed [12], however, a symmetric game may not have a non-symmetric equilibrium. In this section, we show that checking existence of non-symmetric NE in general symmetric game is NP-complete and counting such NE is #P-complete. For the NP-completeness result, we will reduce the problem of checking *non-uniqueness of Nash equilibria in bimatrix games*, which is known to be NP-complete [8], to checking if *symmetric game has a non-symmetric equilibrium.* Refer to the second part of Sect. 1.1 for an overview of the reduction. The reduction is strong enough to also give #P-hardness result since counting equilibria in bimatrix games is known to be #P-hard [4].

We will use the definitions and notation established in Sect. 1.1. Given $0 \leq \epsilon_1, \epsilon_2 << 1$, define $D_{\epsilon_1, \epsilon_2}$ as below (similar to D defined in Sect. 1.1):

$$D_{\epsilon_1,\epsilon_2} = \begin{bmatrix} 0 & 4 & 0 \\ 2 & 0 & 4+\epsilon_1 \\ 3 & 2+\epsilon_2 & 0 \end{bmatrix} \tag{3}$$

See the full version of the paper for the proof of the following lemma.

Lemma 1. *Consider the bimatrix game* $(D_{\epsilon_1,\epsilon_2}, D_{\epsilon'_1,\epsilon'_2}{}^T)$ *where* $0 \le \epsilon_i, \epsilon'_i << 1, i = 1,2$. *The game has a unique NE which has full support, and if* $D_{\epsilon_1,\epsilon_2} = D_{\epsilon'_1,\epsilon'_2}$ *then it is a symmetric NE.*

As observed in Sect. 1.1, the well known reduction from a bimatrix game (A,B) to symmetric game $G = \begin{bmatrix} 0 & A \\ B^T & 0 \end{bmatrix}$ is not useful for our purpose. This is because, non-symmetric NE of (G, G^T) can be of the form $((\boldsymbol{a}, \boldsymbol{0}), (\boldsymbol{0}, \boldsymbol{b}))$, and therefore fails to produce more than one NE of game (A,B). Next we show how to circumvent this issue by constructing a suitable matrix G using the game of Lemma 1 such that no component in non-symmetric NE is zero, and it relates to a unique pair of NE of game (A,B). This one-to-one correspondence gives #P-hardness result as well.

Recall the following, where $K_{c \times d}(k)$ is a $c \times d$ dimensional matrix with all entries set to $k \in \mathbb{R}$.

$$K = \begin{bmatrix} 0 & K_{1\times m}(4) & K_{1\times n}(0) \\ K_{m\times 1}(2) & K_{m\times m}(0) & K_{m\times n}(4) \\ K_{n\times 1}(3) & K_{n\times m}(2) & K_{n\times n}(0) \end{bmatrix} \qquad G = K + \begin{bmatrix} 0 & 0 & 0 \\ 0 & 0 & A \\ 0 & B^T & 0 \end{bmatrix}$$

Before we go into proving our claims, we define a few terms and functions, to be used in the rest of the section. For any non-zero, non-negative vector $\boldsymbol{z}$, of any dimension, let $\eta(\boldsymbol{z})$ denote the *normalized vector*, i.e., its components are non-negative and add to 1. For a matrix M and two non-zero vectors $\boldsymbol{z}_1, \boldsymbol{z}_2 \ge 0$ of appropriate dimensions, we define,

$$P(M; \boldsymbol{z}_1, \boldsymbol{z}_2) \stackrel{\text{def}}{=} \eta(\boldsymbol{z}_1)^T M \eta(\boldsymbol{z}_2)$$

i.e., the payoffs obtained by player with payoff matrix M at strategy profile $(\eta(\boldsymbol{z}_1), \eta(\boldsymbol{z}_2))$. Given strategy $(\boldsymbol{x}, \boldsymbol{y})$ of game (G, G^T), where $\boldsymbol{x} = (c, \boldsymbol{a}, \boldsymbol{b})$ and $\boldsymbol{y} = (c', \boldsymbol{a}', \boldsymbol{b}')$, or given NE $(\boldsymbol{a}, \boldsymbol{b}')$ and $(\boldsymbol{a}', \boldsymbol{b})$ for game (A,B), define

$$\epsilon_1 \stackrel{\text{def}}{=} P(A; \boldsymbol{a}, \boldsymbol{b}'), \quad \epsilon_2 \stackrel{\text{def}}{=} P(B; \boldsymbol{a}', \boldsymbol{b}), \quad \epsilon'_1 \stackrel{\text{def}}{=} P(A; \boldsymbol{a}', \boldsymbol{b}) \quad \epsilon'_2 \stackrel{\text{def}}{=} P(B; \boldsymbol{a}, \boldsymbol{b}') \quad (4)$$

For $\boldsymbol{x} = (c, \boldsymbol{a}, \boldsymbol{b})$, recall that $\text{cl}(\boldsymbol{x}) \stackrel{\text{def}}{=} (c, \sum_i a_i, \sum_j b_j)$. Next we show a connection between payoffs in game (G, G^T) and in game $(D_{\epsilon_1,\epsilon_2}, D^T_{\epsilon'_1,\epsilon'_2})$.

Lemma 2. $P(G; \boldsymbol{x}, \boldsymbol{y}) = P(D_{\epsilon_1,\epsilon_2}; \text{cl}(\boldsymbol{x}), \text{cl}(\boldsymbol{y}))$, *and* $P(G^T; \boldsymbol{x}, \boldsymbol{y}) = P(D^T_{\epsilon'_1,\epsilon'_2}; \text{cl}(\boldsymbol{x}), \text{cl}(\boldsymbol{y}))$.

Proof. We will prove the first part, and the second part follows similarly. Let $\boldsymbol{v} = (v_1, v_2, v_3) = \text{cl}(\boldsymbol{x})$ and $\boldsymbol{w} = (w_1, w_2, w_3) = \text{cl}(\boldsymbol{y})$. Then,

$$P(G; \boldsymbol{x}, \boldsymbol{y}) = \boldsymbol{x}^T G \boldsymbol{y} = \boldsymbol{x}^T K \boldsymbol{y} + \boldsymbol{a}^T A \boldsymbol{b}' + \boldsymbol{a}'^T B \boldsymbol{b}$$

where, $\boldsymbol{x}^T K \boldsymbol{y} = 2v_2w_1 + 3v_3w_1 + 4v_1w_2 + 2v_3w_2 + 4v_2w_3$. Note that $\boldsymbol{a} = (\sum_{i \le m} a_i) * \eta(\boldsymbol{a}) = v_2\eta(\boldsymbol{a})$, and similarly $\boldsymbol{b} = v_3\eta(\boldsymbol{b})$, $\boldsymbol{a}' = w_2\eta(\boldsymbol{a}')$ and

$\boldsymbol{b}' = w_3\eta(\boldsymbol{b}')$. Thus, $\boldsymbol{a}^T A \boldsymbol{b}' + \boldsymbol{a}'^T B \boldsymbol{b} = \eta(\boldsymbol{a})^T A \eta(\boldsymbol{b}') v_2 w_3 + \eta(\boldsymbol{a}')^T B \eta(\boldsymbol{b}) v_3 w_2$ and hence $P(G; \boldsymbol{x}, \boldsymbol{y})$ is equal to

$$2v_2w_1 + 3v_3w_1 + 4v_1w_2 + 2v_3w_2 + 4v_2w_3 + \eta(\boldsymbol{a})^T A \eta(\boldsymbol{b}') v_2 w_3 + \eta(\boldsymbol{a}')^T B \eta(\boldsymbol{b}) v_3 w_2.$$

On the other hand we have $P(D_{\epsilon_1,\epsilon_2}; \mathrm{cl}(\boldsymbol{x}), \mathrm{cl}(\boldsymbol{y})) = 2v_2w_1 + 3v_3w_1 + 4v_1w_2 + 2v_3w_2 + 4v_2w_3 + \epsilon_1 v_2 w_3 + \epsilon_2 v_3 w_2$. Since, $\epsilon_1 = P(A; \boldsymbol{a}, \boldsymbol{b}') = \eta(\boldsymbol{a})^T A \eta(\boldsymbol{b}')$ and $\epsilon_2 = P(B; \boldsymbol{a}', \boldsymbol{b}) = \eta(\boldsymbol{a}')^T B \eta(\boldsymbol{b})$ the lemma follows. □

The above lemma implies equivalence between payoffs in games (G, G^T) and $(D_{\epsilon_1,\epsilon_2}, D_{\epsilon_1',\epsilon_2'}{}^T)$, when the strategies are mapped appropriately. Using this, next we establish relation between their NE.

Lemma 3. *If $(\boldsymbol{x}, \boldsymbol{y})$ is a NE for the game (G, G^T) then $(\mathrm{cl}(\boldsymbol{x}), \mathrm{cl}(\boldsymbol{y}))$ is a NE of game $(D_{\epsilon_1,\epsilon_2}, D_{\epsilon_1',\epsilon_2'}{}^T)$, where ϵs are defined as per (4).*

See the full version of the paper for the proof of Lemma 3. The next corollary follows using Lemmas 1 and 3.

Corollary 1. *If $(\boldsymbol{x}, \boldsymbol{y})$ is NE for the game (G, G^T), where $\boldsymbol{x} = (c, \boldsymbol{a}, \boldsymbol{b})$ and $\boldsymbol{y} = (c', \boldsymbol{a}', \boldsymbol{b}')$, then vectors $\boldsymbol{a}, \boldsymbol{a}', \boldsymbol{b}, \boldsymbol{b}'$ are non-zero.*

As was our goal, the above corollary establishes non-zeroness of sub-components of a NE $(\boldsymbol{x}, \boldsymbol{y})$ of the symmetric game (G, G^T) that we constructed from (A, B). Using this property we will show how non-symmetric NE of game (G, G^T) give two distinct NE of game (A, B) and vice-versa.

Lemma 4. *If $(\boldsymbol{x}, \boldsymbol{y})$ is a NE for the game (G, G^T), where $\boldsymbol{x} = (c, \boldsymbol{a}, \boldsymbol{b})$ and $\boldsymbol{y} = (c', \boldsymbol{a}', \boldsymbol{b}')$, then $(\eta(\boldsymbol{a}), \eta(\boldsymbol{b}'))$ and $(\eta(\boldsymbol{a}'), \eta(\boldsymbol{b}))$ both are NE for the game (A, B).*

Proof. We will show that $(\eta(\boldsymbol{a}), \eta(\boldsymbol{b}'))$ is NE for the game (A, B), and the proof for $(\eta(\boldsymbol{a}'), \eta(\boldsymbol{b}))$ is analogous. By contradiction, wlog suppose the first player can change $\eta(\boldsymbol{a})$ to $\eta(\boldsymbol{a}'')$ and get a better payoff, where $\sum_{1 \le i \le m} a_i = \sum_{1 \le i \le m} a_i''$. Then, we have $\mathrm{cl}(\boldsymbol{x}) = \mathrm{cl}(\boldsymbol{x}')$ and $\eta(\boldsymbol{a})^T A \eta(\boldsymbol{b}') < \eta(\boldsymbol{a}'')^T A \eta(\boldsymbol{b}')$.

Let $\boldsymbol{x}' = (c, \boldsymbol{a}'', \boldsymbol{b})$. In the proof of Lemma 2 we showed that for $\boldsymbol{v} = \mathrm{cl}(\boldsymbol{x})$ and $\boldsymbol{w} = \mathrm{cl}(\boldsymbol{y})$, $P(G; \boldsymbol{x}, \boldsymbol{y}) = 2v_2w_1 + 3v_3w_1 + 4v_1w_2 + 2v_3w_2 + 4v_2w_3 + \eta(\boldsymbol{a})^T A \eta(\boldsymbol{b}') v_2 w_3 + \eta(\boldsymbol{a}')^T B \eta(\boldsymbol{b}) v_3 w_2$. Then,

$$\begin{aligned} P(G; \boldsymbol{x}, \boldsymbol{y}) - P(G, \boldsymbol{x}', \boldsymbol{y}) &= \eta(\boldsymbol{a})^T A \eta(\boldsymbol{b}') v_2 w_3 + \eta(\boldsymbol{a}')^T B \eta(\boldsymbol{b}) v_3 w_2 - \\ &\quad \eta(\boldsymbol{a}'')^T A \eta(\boldsymbol{b}') v_2 w_3 - \eta(\boldsymbol{a}')^T B \eta(\boldsymbol{b}) v_3 w_2 \\ &= \eta(\boldsymbol{a})^T A \eta(\boldsymbol{b}') v_2 w_3 - \eta(\boldsymbol{a}'')^T A \eta(\boldsymbol{b}') v_2 w_3 < 0 \end{aligned}$$

A contradiction to $(\boldsymbol{x}, \boldsymbol{y})$ being a NE of (G, G^T). □

Next we prove the reverse of Lemmas 3 and 4; see the full version of the paper for the proof.

Lemma 5. *If* $(\boldsymbol{a}, \boldsymbol{b}')$ *and* $(\boldsymbol{a}', \boldsymbol{b})$ *are NE of game* (A, B)*, and if for* ϵ*s defined in (4),* $(\boldsymbol{v}, \boldsymbol{w})$ *is a NE of game* $(D_{\epsilon_1,\epsilon_2}, D_{\epsilon'_1,\epsilon'_2})$*, then* $((v_1, v_2 * \boldsymbol{a}, v_3 * \boldsymbol{b}), (w_1, w_2 * \boldsymbol{a}', w_3 * \boldsymbol{b}'))$ *is a NE of game* (G, G^T).

The next theorem follows directly using Lemmas 3, 4 and 5.

Theorem 1. *For* $\boldsymbol{x} = (c, \boldsymbol{a}, \boldsymbol{b})$ *and* $\boldsymbol{y} = (c', \boldsymbol{a}', \boldsymbol{b}')$*,* $(\boldsymbol{x}, \boldsymbol{y})$ *is a NE of game* (G, G^T) *iff* $(\mathrm{cl}(\boldsymbol{x}), \mathrm{cl}(\boldsymbol{y}))$ *is a NE of* $(D_{\epsilon_1,\epsilon_2}, D_{\epsilon'_1,\epsilon'_2})$*, where* ϵ*s are defined as in (4), and* $(\eta(\boldsymbol{a}), \eta(\boldsymbol{b}'))$ *and* $(\eta(\boldsymbol{a}'), \eta(\boldsymbol{b}))$ *are both NE of game* (A, B).

To show NP-hardness of computing non-symmetric NE in symmetric games, we need to establish connection between non-symmetric NE of game (G, G^T) and a pair of distinct NE of game (A, B). Theorem 1 almost does the job except that no such conditions are imposed on the NE of (G, G^T) and of (A, B). Next theorem achieves exactly this (see the full version of the paper for proof).

Theorem 2. *The symmetric game* (G, G^T) *has a non-symmetric NE iff the game* (A, B) *has more than one NE.*

Note that, size of (G, G^T) is $O(size(A, B))$, and hence Theorem 2 implies polynomial-time reduction from the problem of checking if a bimatrix game has more than one NE to checking if a symmetric two-player game has a non-symmetric NE. Since former is NP-complete [8], this shows NP-hardness for the latter. Containment in NP follows since all NE of games (G, G^T) are rational numbers of size polynomial in the size (bit-length) of G [14]. Thus we get the next theorem.

Theorem 3. *Checking existence of a non-symmetric Nash equilibrium in a symmetric game is NP-complete.*

The proof of Theorem 2 does not indicate how the number of equilibria in the games relate to each other. We explore this in the next theorem to show the $\#P$-completeness result (see the full version of the paper for proof).

Theorem 4. *There is a 1-to-1 correspondence between ordered pairs of (distinct) NE of game* (A, B) *and non-symmetric NE of the symmetric game* (G, G^T).

The next theorem follows using the fact that counting the number of NE in a bimatrix game is $\#P$-hard [4], and Theorem 4. Here, containment in $\#P$ follows from the rationality of NE in bimatrix games.

Theorem 5. *Counting the number of non-symmetric equilibria in a symmetric game is* $\#P$*-complete.*

4 Efficient Algorithm for Symmetric Rank-1 Games

In this section we consider computing symmetric NE in symmetric constant rank games. Recall that rank of a two player game (A, B) is defined as $rank(A + B)$. Adsul et al. [1] gave a polynomial time algorithm to compute a Nash equilibrium of a rank-1 bimatrix game. In case of a symmetric game, the Nash equilibrium found by the algorithm need not be symmetric. In what follows, we design a polynomial time algorithm to compute a symmetric Nash equilibrium in symmetric rank-1 games. Our algorithm is an extension of the Adsul et al. approach.

Let (A, A^T) be a symmetric game, where A is an $n \times n$ square matrix. As discussed in Sect. 2 a mixed-strategy $\boldsymbol{x} \in X$ is a symmetric Nash equilibrium of game (A, A^T) if and only if it satisfies (2). Using this, the next lemma follows.

Lemma 6. *If $\boldsymbol{x} \in X$ satisfies first part of (2) then $\boldsymbol{x}^T A\boldsymbol{x} - \pi \leq 0$. Equality holds iff $\boldsymbol{x}$ is a symmetric NE of (A, A^T).*

Using Lemma 6 we get the following quadratic program which exactly captures the symmetric Nash equilibria of game (A, A^T).

$$\begin{array}{ll} \max : \boldsymbol{x}^T A\boldsymbol{x} - \pi & \\ s.t. \quad A\boldsymbol{x} \leq \pi; \quad \boldsymbol{x} \geq \boldsymbol{0}; \quad \sum_{i \in S} x_i = 1 \end{array}$$

Note that, the objective value of the above program is at most zero, and exactly zero at the optimal (Lemma 6). If the rank of game (A, A^T) is one then $A + A^T = \boldsymbol{c} \cdot \boldsymbol{d}^T$, where $\boldsymbol{c}, \boldsymbol{d} \in \mathbb{R}^n$. Note that $\boldsymbol{c} \cdot \boldsymbol{d}^T$ is a symmetric matrix, and therefore, we have $\boldsymbol{c} \cdot \boldsymbol{d}^T = \boldsymbol{d} \cdot \boldsymbol{c}^T$

We will represent matrix A as sum of a skew-symmetric matrix and a rank-1 symmetric matrix. Let K be a matrix such that $k_{ij} = a_{ij} - \frac{c_i d_j}{2}$. This implies $A = K + \frac{1}{2}\boldsymbol{c} \cdot \boldsymbol{d}^T$. Since, $\boldsymbol{c} \cdot \boldsymbol{d}^T = \boldsymbol{d} \cdot \boldsymbol{c}^T$ and $A + A^T = \boldsymbol{c} \cdot \boldsymbol{d}^T$, we get $K + K^T = 0$. Thus, K is skew-symmetric, and therefore $\boldsymbol{z}^T K\boldsymbol{z} = 0$ for any vector $\boldsymbol{z} \in \mathbb{R}^n$. Replacing $A = K + \frac{1}{2}\boldsymbol{c} \cdot \boldsymbol{d}^T$ in the above quadratic program we get,

$$\begin{array}{ll} \max : \frac{1}{2}(\boldsymbol{x}^T \boldsymbol{c})(\boldsymbol{d}^T \boldsymbol{x}) - \pi & \\ s.t. \quad K\boldsymbol{x} + \frac{\boldsymbol{c}}{2}(\boldsymbol{d}^T \boldsymbol{x}) \leq \pi; \quad \boldsymbol{x} \geq \boldsymbol{0}; \quad \sum_i x_i = 1 \end{array}$$

The above formulation is a rank-1 quadratic program, which is NP-hard in general. However, we will show that it can be solved in polynomial time using the Nash equilibrium properties. The feasible region of the above program is linear, while the cost function is quadratic which introduces the difficulty. The idea is to construct an LP-type formulation, using the fact that the quadratic term is a product of two linear terms, while maintaining the fact that optimal value of the new formulation is also zero and it is achieved only when complementarity is satisfied. Towards this, we first replace $\boldsymbol{d}^T \boldsymbol{x}$ by λ in the objective function as well as in the inequality. This gives the following optimization problem where $\boldsymbol{x}$ is a variable vector, and π and λ are scalar variables.

$$\begin{array}{ll} \max : \frac{1}{2}\lambda(\boldsymbol{x}^T \boldsymbol{c}) - \pi & \\ s.t. \quad K\boldsymbol{x} + \frac{\boldsymbol{c}}{2}\lambda \leq \pi; \quad \boldsymbol{x} \geq \boldsymbol{0}; \quad \sum_i x_i = 1 \end{array} \tag{5}$$

Lemma 7. *Let $(\boldsymbol{x}, \lambda, \pi)$ be a feasible point of (5), then $\frac{1}{2}\lambda(\boldsymbol{x}^T \boldsymbol{c}) - \pi \leq 0$. Equality holds iff $x_i(K\boldsymbol{x} + \frac{\boldsymbol{c}}{2}\lambda - \pi)_i = 0$, $\forall i \in [n]$.*

Note that, formulation (5) is independent of vector $\boldsymbol{d}$, an essential for our original game. This seems very counter intuitive at first. However, this very property allows (5) to capture NE of a space of games, as established next. Finally, we will use this rich structure to formulate one-dimensional fixed point to solve our game. The next lemma shows that the solution set (5) is rich enough to contain a point for every value of λ; see the full version of the paper for its proof.

Lemma 8. *Given $\lambda \in \mathbb{R}$, $\exists(\boldsymbol{x}, \pi) \in \mathbb{R}^{n+1}$ such that $(\boldsymbol{x}, \lambda, \pi)$ is a solution of (5), and the objective value is zero at $(\boldsymbol{x}, \lambda, \pi)$.*

Lemmas 7 and 8 imply that the optimal value of (5) is zero, and for every $a \in \mathbb{R}$ there is an optimal solution with $\lambda = a$. If we substitute some value for λ in (5), then it becomes an LP. Therefore, consider it as a parameterized linear program $LP(\lambda)$. The optimal value of $LP(\lambda)$ for any $\lambda \in \mathbb{R}$ is zero (due to Lemma 8). Therefore, solutions of (5) are exactly the solutions of $LP(\lambda)$, $\forall \lambda \in \mathbb{R}$; see the full version of the paper for a brief discussion on its structure.

Result of the next lemma is central to the construction of one-dimensional fixed point formulation for solving our original game (A, A^T).

Lemma 9. *Given a $\lambda \in \mathbb{R}$, if $(\boldsymbol{x}, \pi)$ is a solution of $LP(\lambda)$ then for any $\boldsymbol{v} \in \mathbb{R}^n$ satisfying $\boldsymbol{v}^T\boldsymbol{x} = \lambda$, $\boldsymbol{x}$ is a symmetric NE of game (Z, Z^T), where $Z = K + \frac{1}{2}\boldsymbol{c}\boldsymbol{v}^T$.*

Proof. Let $(\boldsymbol{x}, \pi)$ be a solution of $LP(\lambda)$, then since the feasible region of $LP(\lambda)$ is a subset of the feasible region of (5), vector $(\boldsymbol{x}, \lambda, \pi)$ satisfies $(K\boldsymbol{x} - \frac{\boldsymbol{c}}{2}\lambda)_i \leq \pi$; $\boldsymbol{x} \geq \boldsymbol{0}$; $\sum_i x_i = 1$. This ensures that $\boldsymbol{x}$ is a probability distribution vector. Due to Lemma 8, it also satisfies $x_i(K\boldsymbol{x} + \frac{\boldsymbol{c}}{2}\lambda - \pi)_i = 0,\ \forall i \in [n]$. Setting, $\lambda = \boldsymbol{v}^T\boldsymbol{x}$, these conditions are exactly that of (2) for strategy $\boldsymbol{x}$ and game (Z, Z^T) where $Z = K + \frac{1}{2}\boldsymbol{c}\boldsymbol{v}^T$. □

Remark 1. Note that both the matrices of the games constructed in Lemma 9 change with $\boldsymbol{v}$, and $\boldsymbol{c}\boldsymbol{v}^T$ need not be a symmetric matrix. Therefore, rank$(Z + Z^T) = 2$. In the Adsul et al. approach, the first matrix is same in all the games, and the solutions of $LP(\lambda)$ are NE of a family of rank-1 games, which crucially uses the fact that $\boldsymbol{y}$ need not be same as $\boldsymbol{x}$ (non-symmetric). For this reason, their approach is not immediately applicable for finding symmetric NE.

Lemma 9 implies that if we can find a λ such that the solution $(\boldsymbol{x}, \pi)$ of $LP(\lambda)$ satisfies $\boldsymbol{d}^T\boldsymbol{x} = \lambda$, then $\boldsymbol{x}$ is a symmetric Nash equilibrium of our original rank-1 game (A, A^T). Using this observation, consider a 1-dimensional correspondence $F : [d_{min},\ d_{max}] \to 2^{[d_{min},\ d_{max}]}$, where $d_{min} = \min_{i\in[n]} d_i$ and $d_{max} = \max_{i\in[n]} d_i$.

$$F(\lambda) = \{\boldsymbol{d}^T\boldsymbol{x} \mid \boldsymbol{x} \text{ is a solution of } LP(\lambda)\}$$

By definition we have that $\forall\lambda \in [d_{min},\ d_{max}]$, $F(\lambda)$ is non-empty (Lemma 8) and convex. Now using the Kakutani fixed-point theorem, F has fixed-points. Clearly, every fixed-point of F gives a Nash equilibrium by Lemma 9, and the next theorem follows.

Theorem 6. *The fixed points of F exactly capture the Nash equilibria of the game (A, A^T).*

Since the Nash equilibrium profiles of game (A, A^T) are rational vectors of size polynomial in the size of A [15], the fixed-points of F are also rational numbers of polynomial sized (using Theorem 6). Thus, one can compute an exact fixed point of F in polynomial time using a simple binary search starting with the pivots d_{min} and d_{max}, and the next theorem follows.

Theorem 7. *The problem of computing a symmetric Nash equilibrium in a symmetric rank-1 game is in P.*

References

1. Adsul, B., Garg, J., Mehta, R., Sohoni, M.: Rank-1 bimatrix games: a homeomorphism and a polynomial time algorithm. In: ACM Symposium on the Theory of Computing, pp. 195–204 (2011)
2. Chen, X., Deng, X., Teng, S.-H.: Settling the complexity of computing two-player Nash equilibria. J. ACM **56**(3), 14:1–14:57 (2009)
3. Cheng, S.-F., Reeves, D.M., Vorobeychik, Y., Wellman, M.P.: Notes on equilibria in symmetric games. In: Proceedings of International Workshop On Game Theoretic And Decision Theoretic Agents (GTDT), pp. 71–78 (2004)
4. Conitzer, V., Sandholm, T.: New complexity results about Nash equilibria. Games Econ. Behav. **63**(2), 621–641 (2008)
5. Daskalakis, C.: Survey: Nash equilibria: complexity, symmetries, and approximation. Comput. Sci. Rev. **3**(2), 87–100 (2009)
6. Daskalakis, C., Goldberg, P.W., Papadimitriou, C.H.: The complexity of computing a nash equilibrium. SIAM J. Comput. **39**(1), 195–259 (2009). Special issue for STOC 2006
7. Daskalakis, C., Papadimitriou, C.: Computing equilibria in anonymous games. In: 48th Annual IEEE Symposium on Foundations of Computer Science (FOCS) (2007)
8. Gilboa, I., Zemel, E.: Nash and correlated equilibria: some complexity considerations. Games Econ. Behav. **1**, 80–93 (1989)
9. Kannan, R., Theobald, T.: Games of fixed rank: a hierarchy of bimatrix games. Econ. Theor. **42**(1), 157–174 (2010). Preliminary version appeared in SODA 2007, and available at arXiv:cs/0511021 since 2005
10. McLennan, A., Tourky, R.: Simple complexity from imitation games. Games Econ. Behav. **68**(2), 683–688 (2010)
11. Mehta, R.: Contant rank bimatrix games are PPAD-hard. In: ACM Symposium on the Theory of Computing, pp. 545–554 (2014)
12. Nash, J.: Non-cooperative games. Ann. Math. **54**(2), 289–295 (1951)
13. Papadimitriou, C.H.: On the complexity of the parity argument and other inefficient proofs of existence. JCSS **48**(3), 498–532 (1992)
14. Papadimitriou, C.H.: The complexity of finding Nash equilibriam, Chapter 2. In: Nisan, N., Roughgarden, T., Tardos, E., Vazirani, V. (eds.) Algorithmic Game Theory, pp. 29–50. Cambridge University Press, Cambridge (2007)

15. von Stengel, B.: Equilibrium computation for two-player games in strategic and extensive form Chapter 3. In: Nisan, N., Roughgarden, T., Tardos, E., Vazirani, V. (eds.) Algorithmic Game Theory, pp. 53–78. Cambridge University Press, Cambridge (2007)
16. von Stengel, B.: Rank-1 games with exponentially many Nash equilibria. arXiv preprint (2012). arXiv:1211.2405

Approximating Nash Equilibria in Tree Polymatrix Games

Siddharth Barman[1(✉)], Katrina Ligett[1], and Georgios Piliouras[2]

[1] California Institute of Technology, Pasadena, CA, USA
{barman,katrina}@caltech.edu
[2] Singapore University of Technology and Design, Singapore, Singapore
georgios@sutd.edu.sg

Abstract. We develop a quasi-polynomial time Las Vegas algorithm for approximating Nash equilibria in polymatrix games over trees, under a mild renormalizing assumption. Our result, in particular, leads to an expected polynomial-time algorithm for computing approximate Nash equilibria of tree polymatrix games in which the number of actions per player is a fixed constant. Further, for trees with constant degree, the running time of the algorithm matches the best known upper bound for approximating Nash equilibria in bimatrix games (Lipton, Markakis, and Mehta 2003).

Notably, this work closely complements the hardness result of Rubinstein (2015), which establishes the inapproximability of Nash equilibria in polymatrix games over constant-degree bipartite graphs with two actions per player.

1 Introduction

The complexity of equilibrium computation is a central area of research in algorithmic game theory. Recent years have seen significant progress in this line of work, especially in the context of two-player games [3–7, 13, 16, 17]. Furthermore, the computation of approximate Nash equilibrium in games over networks has emerged as an important research direction [2, 8, 10, 12, 15]. Motivation for studying such multiplayer games stems in part from the prevalence and importance of large networks of interconnected, self-interested agents.

The prototypical family of large network games is that of *polymatrix games*. These games merge two classical concepts, two-player games and networks. In a polymatrix game, each player corresponds to a node in a network, and each edge encodes a two-player game between the two endpoints of the edge. A player's payoff is the sum of her payoffs across the bimatrix games (edges) she participates in. Polymatrix games capture complex settings with arbitrarily many players while keeping the description complexity of the game polynomially small in the number of players. Computation of equilibria for polymatrix games is hence a natural test case, and has emerged at the boundary of computational tractability.

The seminal PPAD hardness reductions for computing ε-Nash equilibria[1] by Daskalakis, Goldberg, and Papadimitriou [5] along with their extensions by Chen,

[1] In an ε-Nash equilibrium, a player can gain at most ε by unilaterally deviating from her current strategy.

M. Hoefer (Ed.): SAGT 2015, LNCS 9347, pp. 285–296, 2015.
DOI: 10.1007/978-3-662-48433-3_22

Deng, and Teng [3] to two-player games were crucially developed within the context of polymatrix games.[2] Recently, Rubinstein [16] strengthened these inapproximability guarantees by establishing that there exists a constant ε such that finding an ε-Nash equilibrium in polymatrix games over bipartite graphs of constant degree is computationally hard. Our positive algorithmic result is inspired by this work and *explores the boundary between tractability and intractability of ε-Nash computation in polymatrix games.*

The study of equilibria in polymatrix games has had a long history [11,18]. To avoid hardness, most algorithmic results have focused on structured subclasses of polymatrix games. These include polymatrix generalizations of zero-sum games where exact Nash equilibria can be computed in polynomial time [2,8]. Games on trees is another family of multiplayer games that has received attention [10,12,15]. The proposed algorithm in [10] finds an exact Nash equilibrium in two-action games on paths and runs in polynomial time, but in the case of trees the running time may be exponential even if the degree of the underlying tree is bounded. In contrast, we study computation of approximate Nash equilibrium in trees of arbitrary degree, and develop an algorithm that runs in quasi-polynomial time. Finally, some interesting progress has been made in the case of general polymatrix games as well, where it has been shown that a $(0.5 + \varepsilon)$-Nash equilibrium of a polymatrix game can be computed in time polynomial in the input size and $1/\varepsilon^2$ [9].

Results. We develop a quasi-polynomial time algorithm for approximating Nash equilibrium in polymatrix games over trees under a mild renormalizing assumption on the players' payoffs. Specifically, instead of normalizing the entries of each bimatrix game to lie in [0,1], which results in each player i's payoff depending linearly on its degree, we normalize them to lie in $[0, 1/\text{degree}(i)]$, so that players' total payoffs lie in [0,1]. Our results actually extend even under weaker renormalization conditions; see Sect. 2 for details. We show that, given an n-player, m-action normalized polymatrix game over a tree, we can find an ε-Nash equilibrium of the game in expected time

$$m^{O\left(\frac{\log m(\log m+\log n-\log \varepsilon)}{\varepsilon^4}\right)}.$$

Our approach immediately implies a polynomial time approximation scheme for computing Nash equilibria when the number of actions per player is constant. The case of standard bimatrix games can be trivially captured in our setting via a single-edge polymatrix game. Further, for trees of constant degree our framework yields an algorithm that finds an ε-Nash equilibrium in time $m^{O\left(\frac{\log m+\log n}{\varepsilon^2}\right)}$. Note that in the single edge case (i.e., the case of standard bimatrix games) this running-time bound matches the best known upper bound for approximating Nash equilibria [13].

Techniques. We develop a dynamic program to find an approximate Nash equilibrium of the given tree polymatrix game. The idea is to root the underlying

[2] These hardness result hold for polynomially small ε.

tree and process it in a bottom-up manner. For each node/player p we maintain a set of mixed strategies—i.e., probability distributions over player's actions—that can be extended into a "partial" (approximate) equilibrium of the subtree rooted at the node. That is, for each mixed strategy assigned to p there exist mixed strategies for the descendants of p under which no descendant can benefit more than ε, in expectation, by unilateral deviation. We find such extendable mixed strategies of a player p after processing all of its children; in other words, we start from the leaves of the tree and move towards the root. Note that such an extendable mixed strategy for the root corresponds to an approximate Nash equilibrium of the game. Also, it is worth pointing out that the tree structure enables us to find partial equilibria of disjoint subtrees separately. In particular, the fact that the utilities of players depend only on the actions of its parent and its children implies that disjoint subtrees can be processed separately.

In and of itself, using a dynamic program to find an approximate Nash equilibrium over a tree is a natural idea. In fact, similar approaches have been adopted in prior work; see, e.g., [10]. The key technical contribution in this paper is to show that the update step in the dynamic program can be performed in quasi-polynomial time. To do this, we focus on a specific set of mixed strategies U, which is the set of all uniform distributions with support size polynomial in the approximation parameter ε and logarithmic in the number of players and the number of actions; see Sect. 2 for a formal definition. It was shown in [1] that every multiplayer game admits an ε-Nash equilibrium wherein the mixed strategy of each player is contained in U. Hence, given an n-player game, an exhaustive search over the set U^n is guaranteed to find an approximate Nash equilibrium. But, such a search runs in time exponential in n. We show that for tree polymatrix games, an exponential-time exhaustive search can be bypassed. The idea is to follow the above mentioned dynamic program and consider, for each player p, mixed strategies in the set U that can be extended into partial equilibria of the subtree rooted at p. To perform the update step in the dynamic program we employ a linear program that, interestingly, gives a tight characterization of mixed strategies that can be extended. Together, these ideas lead us to a quasi-polynomial time approximation algorithm.

2 Notation and Preliminaries

We study games with n players and m actions per player.[3] Write $[n]$ and $[m]$ to denote the set of players and the set of actions of each player, respectively. The utilities of the players are normalized between 0 and 1; in particular, for each player p we have utility $u_p : [m]^n \to [0, 1]$. Let Δ^m be the set of probability distributions over $[m]$. In addition, for mixed strategy profile $x = (x_q)_{q \in [n]} \in \Delta^m \times \ldots \times \Delta^m$, we denote the expected utility of player p by $u_p(x)$. Following

[3] We assume that each player has m actions for ease of presentation. The developed result directly extends to the case wherein the number of actions of each player is different.

standard notation, we use x_{-p} to denote the mixed strategy profile of all players besides p.

Definition 1 (ε-Nash equilibirum). *A mixed strategy profile $x = (x_q)_{q \in [n]}$, where each $x_q \in \Delta^m$, is said to be an ε-Nash equilibrium iff for every player $p \in [n]$ and action $a \in [m]$ we have $u_p(x) \geq u_p(a, x_{-p}) - \varepsilon$.*

Here, setting $\varepsilon = 0$ gives us the definition of a Nash equilibrium.

Polymatrix Games. In a polymatrix game, the players correspond to vertices of a graph $G = (V, E)$ and the utility of each player $p \in V$ depends only on her action and the actions of her neighbors. Moreover, the utility of each player is *separable*, i.e., for each edge $(p, q) \in E$ we have a bimatrix game specified by $m \times m$ matrices $A_{p,q}$ and $A_{q,p}$, and the utility of player p, under action profile $(a_q)_{q \in [n]} \in [m]^n$, is specified as follows $u_p(a_1, a_2, \ldots, a_n) := \sum_{q:(p,q) \in E} e_{a_p}^T A_{p,q} e_{a_q}$. Here, $e_k \in \mathbb{R}^m$ denotes the standard basis vector with 1 in the kth component and 0's elsewhere. Along these lines, for a mixed strategy profile $(x_q)_{q \in [n]} \in (\Delta^m)^n$, the expected utility of player p, $u_p(x_1, x_2, \ldots, x_n) := \sum_{q:(p,q) \in E} x_p^T A_{p,q} x_q$.

As mentioned above, the utility of each player is normalized between 0 and 1. A typical way to accomplish this normalization (see, e.g., [9]) is to assume that for each player $p \in [n]$ the associated payoff matrices, $A_{p,q}$s, are entry-wise between 0 and 1, and the utility of player p with degree d (in the graph) is obtained by dividing the sum of the payoffs by d, i.e., $u_p(a_1, a_2, \ldots, a_n) := \frac{1}{d} \sum_{q:(p,q) \in E} e_{a_p}^T A_{p,q} e_{a_q}$. This normalization ensures that the same approximation guarantee is achieved for all players, irrespective of their degrees. If, instead, one assumes that entry-wise the $A_{p,q}$s are between 0 and 1 and simply add the payoffs $e_{a_i}^T A_{p,q} e_{a_j}$, then the approximation guarantee for players with higher degree—since ε is the same of all the players—is stronger. This would lead to an undesirable, nonuniform approximation bound.

The degree-normalized scaling mentioned above is equivalent to the assumption that for player $p \in [n]$, with degree d, the matrices $A_{p,q}$s are contained in $[0, 1/d]^{m \times m}$ and $u_p(a_1, \ldots, a_n) := \sum_{q:(p,q) \in E} e_{a_p}^T A_{p,q} e_{a_q}$. In this paper we in fact consider a more general setup in which, for a player with degree d, entries of $A_{p,q}$s are between 0 and $\max \left\{ \frac{1}{d}, \frac{\varepsilon}{2\sqrt{6d \log m}} \right\}$. Here, again we assume that for each action profile a we have $u_i(a) \in [0, 1]$. Developing a quasi-polynomial time algorithm without an entry-wise assumption (i.e., without requirement (i) in the following definition) remains an interesting direction for future work.

Definition 2 (Normalized Polymatrix Game). *Let $\mathcal{G}$ be an n-player m-action polymatrix game over graph $G = (V, E)$ and with payoff matrices $A_{p,q}$ and $A_{q,p}$, for $(p, q) \in E$. Given parameter ε, we say that $\mathcal{G}$ is normalized iff for each player $p \in [n]$ we have (i) the entries of $A_{p,q}$s are contained in $\left[0, \max \left\{ \frac{1}{d}, \frac{\varepsilon}{2\sqrt{6d \log m}} \right\}\right]$; here d is the degree of player p in G, and (ii) for every action profile $(a_1, \ldots, a_n) \in [m]^n$, the utility $u_p(a_1, \ldots, a_n) := \sum_{q:(p,q) \in E} e_{a_p}^T A_{p,q} e_{a_q}$ is between 0 and 1.*

Given mixed strategies of the neighbors of a player p, say $(x_q)_{q:(p,q)\in E}$, $x_p \in \Delta^m$ is said to be an ε-*best response* of p against $(x_q)_{q:(p,q)\in E}$ if p cannot benefit more than ε in expectation by deviating from x_p, i.e.,

$$\sum_{q:(p,q)\in E} x_p^T A_{p,q} x_q \geq \max_{j\in[m]} \left(\sum_{q:(p,q)\in E} e_j^T A_{p,q}\, x_q \right) - \varepsilon. \tag{1}$$

This paper studies polymatrix games in which the underlying graph G is a tree. Note that a polymatrix game with exactly two players over a single edge $(p,q) \in E$—which is trivially a tree—corresponds to a bimatrix game between players p and q. Hence, computation of an approximate Nash equilibrium in tree polymatrix games is at least as hard as computation of approximate Nash equilibrium in bimatrix games. Therefore, our running-time benchmark for finding an ε-Nash equilibrium is quasi-polynomial: $m^{O\left(\frac{\log m}{\varepsilon^2}\right)}$, which is the best known upper bound for approximating Nash equilibria in bimatrix games [13].

Uniform Probability Distributions. A probability distribution $x \in \Delta^m$ is said to be b uniform if it is a uniform distribution over a size-b multiset of $[m]$. Write $U \subset \Delta^m$ to denote the set of all $\left(\frac{8(\ln m+\ln n-\ln \varepsilon+\ln 8)}{\varepsilon^2}\right)$-uniform probability distributions. Note that

$$|U| = m^{O\left(\frac{\log m+\log n-\log \varepsilon}{\varepsilon^2}\right)} \tag{2}$$

As mentioned above, the work of Babichenko et al. [1] establishes that every n-player m-action game admits an ε-Nash equilibrium $x = (x_q)_{q\in[n]}$ such that $x_q \in U$ for all $q \in [n]$. Hence, an exhaustive search over the set U^n is guaranteed to find an ε-Nash equilibrium. Note that the running time of such a search is $m^{O\left(\frac{n(\log m+\log n-\log \varepsilon)}{\varepsilon^2}\right)}$, which is exponential in n. In contrast to this exponential-time algorithm, we show that for tree polymatrix games an approximate Nash equilibrium can be computed in expected time $m^{O\left(\frac{\log m(\log m+\log n-\log \varepsilon)}{\varepsilon^4}\right)}$, which is quasi-polynomial in n and m.

Next we state McDiarmid's inequality [14]. We use this concentration bound to prove our main result.

McDiarmid Inequality. *Let $Z_1, Z_2, \ldots, Z_d \in \mathcal{Z}$ be independent random variables and $f : \mathcal{Z}^d \to \mathbb{R}$ be a function of $Z_1, Z_2, \ldots, Z_d$. If for all $i \in [d]$ and for all $z_1, z_2, \ldots, z_d, z_i' \in \mathcal{Z}$ the function f satisfies*

$$|f(z_1,\ldots,z_i,\ldots,z_d) - f(z_i,\ldots,z_i',\ldots,z_d)| \leq c_i,$$

then for $\delta > 0$,

$$\Pr(|f - \mathbb{E}[f]| \geq \delta) \leq 2\ exp\left(\frac{-2\delta^2}{\sum_{i=1}^d c_i^2}\right).$$

3 Quasi-Polynomial Time Algorithm

This section develops the dynamic program that finds an approximate Nash equilibrium. We will consider G to be a rooted tree and process it in a bottom-up manner. We start with players all whose descendants are leaves, and then iteratively proceed onto the remaining players.

Write $\mathcal{C}(q)$ and $\mathcal{D}(q)$ to denote the set of children and the set of descendants of player q, respectively. The iterative process maintains a set $U_{p,q}(z)$ for each parent-child pair $(p, q) \in E$ and each $z \in U$.[4] Intuitively, $U_{p,q}(z)$ denotes the set of mixed strategies for player q that can be extended into a "partial" ε-Nash equilibrium of the subtree rooted at q. Here p, the parent of q, is playing mixed strategy z and might not be best responding. Formally, the inductive definition of the sets $U_{p,q}(z)$s is as follows:

- If q is a leaf player (i.e., q corresponds to a leaf in tree G), then $U_{p,q}(z) := \{y \in U \mid y \text{ is an } \varepsilon\text{-best response of } q \text{ against } z\}$.
- Else, if q is a not a leaf player, we define $U_{p,q}(z) := \{y \in U \mid$ there exist mixed strategies $(x_c)_{c \in \mathcal{C}(q)} \in \prod_{c \in \mathcal{C}(q)} U_{q,c}(y)$ such that y is an ε-best response of q against $(x_c)_{c \in \mathcal{C}(q)}$ and $z\}$; here, mixed strategy z is associated with parent player p.

We also define the set U_r for the root r of tree G: $U_r := \{y \in U \mid$ there exist mixed strategies $(x_c)_{c \in \mathcal{C}(r)} \in \prod_{c \in \mathcal{C}(r)} U_{r,c}(y)$ such that y is an ε-best response of r against $(x_c)_{c \in \mathcal{C}(r)}\}$.

If $y \in U_{p,q}(z)$ and q is not a leaf, then, by the above definition, there exist mixed strategy profiles $(x_c)_{c \in \mathcal{C}(q)} \in \prod_{c \in \mathcal{C}(q)} U_{q,c}(y)$ such that y is an ε-best response of q against $(x_c)_{c \in \mathcal{C}(q)}$ and z. We will use $E_{p,q}(z, y)$ to denote such a collection of mixed strategies, $(x_c)_{c \in \mathcal{C}(q)}$.

Along these lines, for the root r of the tree G we define $E_r(y)$, for each $y \in U_r$, to be a collection of mixed strategies $(x_c)_{c \in \mathcal{C}(r)} \in \prod_{c \in \mathcal{C}(r)} U_{r,c}(y)$ such that y is an ε-best response of r against $(x_c)_{c \in \mathcal{C}(r)}$.

Note that mixed strategies in $E_{p,q}(z, y)$ extend y into a "partial" ε-Nash equilibrium of the subtree rooted at q. Specifically, we can inductively use $E_{p,q}(z, y)$, then $E_{q,c}(y, x_c)$, for each $c \in \mathcal{C}(q)$, and so on, to determine mixed strategies $(x_s)_{s \in \mathcal{D}(q)}$ for each descendant $s \in \mathcal{D}(q)$ such that no player in the subtree rooted at q can benefit more than ε, in expectation, by deviating unilaterally. Here we do not assert that the parent player p is at an approximate equilibrium. In addition, note that the utilities of all the players $s \in \mathcal{D}(q) \cup \{q\}$ depend only on the mixed strategies of players in $\mathcal{D}(q) \cup \{p, q\}$ and, hence, these utilities can be determined even if the mixed strategies of players in $[n] \setminus (\mathcal{D}(q) \cup \{p, q\})$ are unspecified. Following the definition of $U_{p,q}(z)$, Algorithm 1 constructs these sets and extensions $E_{p,q}(z, y)$ for all parent-child pairs $(p, q) \in E$ and $z \in U$ in a bottom-up manner. At the end, the algorithm uses the set U_r defined for the root r to find an ε-Nash equilibrium of the game. Overall, the applicability of the sets $U_{p,q}$ and $E_{p,q}$ is established in Lemma 1 below.

[4] Recall that U is the set of all $O\left(\frac{\log m + \log n - \log \varepsilon}{\varepsilon^2}\right)$-uniform probability distributions.

Algorithm 1. Algorithm for finding ε-Nash equilibrium in tree polymatrix games

---

Given: A normalized polymatrix game over tree $G = (V, E)$. **Return:** An ε-Nash equilibrium of the game.

1: Initialize processed set P to be the leaves in G and all $U_{p,q}(z) = \varnothing$
2: **while** $V \setminus P \neq \phi$ **do**
3:    Select $p \in V \setminus P$ such that $\mathcal{C}(p) \subseteq P$
4:    **for all** $q \in \mathcal{C}(p)$ and $z \in U$ **do**
5:       **for all** $y \in U$ **do**
6:          **if** q is a leaf node and y is an ε-best response of q against z **then**
7:             Update $U_{p,q}(z) \leftarrow U_{p,q}(z) \cup \{y\}$
8:          **else if** there exist mixed strategy profiles $(x_c)_{c \in \mathcal{C}(q)} \in \prod_{c \in \mathcal{C}(q)} U_{q,c}(y)$ such
             that y is an ε-best response against $(x_c)_{c \in \mathcal{C}(q)}$ and z **then**
9:             Update $U_{p,q}(z) \leftarrow U_{p,q}(z) \cup \{y\}$ and set $E_{p,q}(z, y) \leftarrow (x_c)_{c \in \mathcal{C}(q)}$ {There
             could be multiple tuples $(x_c)_{c \in \mathcal{C}(q)}$ that satisfy this best-response condi-
             tion. We set $E_{p,q}(z, y)$ to be any one of them.}
10:          **end if**
11:       **end for**
12:    **end for**
13:    $P \leftarrow P \cup \{p\}$
14: **end while**
15: For the root r of the tree G, initialize $U_r = \phi$.
16: **for all** $y \in U$ **do**
17:    **if** there exist mixed strategy profiles $(x_c)_{c \in \mathcal{C}(r)} \in \prod_{c \in \mathcal{C}(r)} U_{r,c}(y)$ such that y is
          an ε-best response against $(x_c)_{c \in \mathcal{C}(r)}$ **then**
18:       Update $U_r \leftarrow U_r \cup \{y\}$ and set $E_r(y) = (x_c)_{c \in \mathcal{C}(r)}$. Use Lemma 1 to find an
          ε-Nash equilibrium of the game
19:    **end if**
20: **end for**

Lemma 1. *Let $\mathcal{G}$ be a polymatrix game over a tree $G = (V, E)$. Given sets $U_{p,q}(z)$—for each parent-child pair $(p, q) \in E$—and mixed strategy collections $E_{p,q}(z, y)$—for $y \in U_{p,q}(z)$—along with a mixed strategy profile $\hat{y} \in U_r$ and associated collection $E_r(\hat{y})$ for the root r, we can find an ε-Nash equilibrium of the game $\mathcal{G}$ in time polynomial in $|U|$.*

Proof. The lemma is implied directly by the underlying definitions. For each parent-child pair $(p, q) \in E$ there exists at least one set $U_{p,q}$ which is nonempty: as mentioned above, every n-player m-action game admits an ε-Nash equilibrium $(\hat{x}_q)_{q \in [n]}$ where each $\hat{x}_q \in U$. Hence, in particular, $\hat{x}_q \in U_{p,q}(\hat{x}_p)$. Moreover, we have $\hat{x}_r \in U_r$.

In fact to find an ε-Nash equilibrium we can start with the given mixed strategy profile $x_r = \hat{y}$ then, for each $c \in \mathcal{C}(r)$, pick the corresponding mixed strategy x_c in $E_r(x_r)$. The definition of $E_r(x_r)$ implies that x_c can be extended to obtain an ε-Nash equilibrium of the subtree rooted at c. We can in fact find such an ε-Nash equilibrium by proceeding inductively down the tree; in particular, by setting x_s for $s \in \mathcal{C}(c)$ to be the strategy associated with s in $E_{r,c}(x_r, x_c)$).

The definitions of $E_{p,q}$s ensure that this inductive process will run to completion and find an ε-Nash equilibrium of the subtree rooted at c. By repeating the process for each $c \in \mathcal{C}(r)$ we will find a mixed strategy x_p for each player $p \in [n]$. Furthermore, the definitions of the underlying sets also imply that the found mixed strategy profile $(x_p)_{p\in[n]}$ is an ε-Nash equilibrium. □

Algorithm 1 tests whether $y \in U_{p,q}(z)$ (i.e., tests whether there exist mixed strategy profiles $(x_c)_{c\in\mathcal{C}(q)} \in \prod_{c\in\mathcal{C}(q)} U_{q,c}(y)$ such that y is an ε-best response of q against $(x_c)_{c\in\mathcal{C}(q)}$ and z) in Step 8, and the same idea is employed in Step 17. In particular, if the number of children of q is $\Omega\left(\frac{\log m}{\varepsilon^2}\right)$ then Algorithm 1 uses the following linear-programming relaxation LP(p,q,z,y) to perform this test. The other case, wherein $|\mathcal{C}(q)| = o\left(\frac{\log m}{\varepsilon^2}\right)$, is addressed directly via exhaustive search, see proof of Theorem 1 for details.

$$\begin{aligned}
\max_{\alpha_x,\sigma_c} \quad & 0 \\
\text{subject to} \quad & \sum_{x\in U_{q,c}(y)} \alpha_x = 1 \qquad \forall c \in \mathcal{C}(q) \\
& \sigma_c = \sum_{x\in U_{q,c}(y)} \alpha_x\, x \qquad \forall c \in \mathcal{C}(q) \\
& y^T A_{q,p} z + \sum_{c\in\mathcal{C}(q)} y^T A_{q,c}\,\sigma_c \geq e_j^T A_{q,p} z + \sum_{c\in\mathcal{C}(q)} e_j^T A_{q,c}\,\sigma_c - \frac{\varepsilon}{2} \qquad \forall j \in [m] \qquad (3) \\
& \alpha_x \geq 0 \qquad \forall x \in \cup_{c\in\mathcal{C}(q)} U_{q,c}(y) \\
& \sigma_c \in \Delta^m \qquad \forall c \in \mathcal{C}(q).
\end{aligned}$$

Formally, Lemma 2 below establishes that the feasibility of the linear program LP(p,q,z,y) implies the required containment $y \in U_{p,q}(z)$, when $|\mathcal{C}(q)| = \Omega\left(\frac{\log m}{\varepsilon^2}\right)$. Note that LP$(p,q,z,y)$ is parameterized by players p and q along with mixed strategies z and y. In addition, inequality (3) in LP(p,q,z,y) enforces that y is an $\varepsilon/2$-best response against σ_cs and z. Also, if for some player $c \in \mathcal{C}(q)$ the set $U_{q,c}(y)$ is empty, then LP(p,q,z,y) is trivially infeasible.

Lemma 2. *Let player p be the parent of player q in a normalized polymatrix game over rooted tree $G = (V,E)$. Also, let the number of children of q, $|\mathcal{C}(q)| = \Omega\left(\frac{\log m}{\varepsilon^2}\right)$. Then, the feasibility of the linear program LP(p,q,z,y), for mixed strategies $z,y \in U$, implies that $y \in U_{p,q}(z)$. Moreover, using a feasible solution of LP(p,q,z,y) we can find mixed strategy profiles $E_{p,q}(z,y)$ via a sampling algorithm whose expected running time is polynomial in $|U|$.*

Proof. Scalars $(\alpha_x)_{x\in U_{q,c}(y)}$ are nonnegative and sum up to one; hence, they induce a probability distribution over $U_{q,c}(y)$, for $c \in \mathcal{C}(q)$. Write α^c to denote this distribution. Also, let χ_c be the random variable that is equal to mixed strategy $x \in U_{q,c}(y)$ with probability α_x, i.e., χ_c is drawn from α^c. Note that $\mathbb{E}_{\alpha^c}[\chi_c] = \sigma_c$.

Let d denote the number of children of q, $d := |\mathcal{C}(q)|$. For fixed $j \in [m]$, we consider function $f_j(\chi_1, \ldots, \chi_d) := \sum_{c \in \mathcal{C}(q)} e_j^T A_{q,c} \chi_c$. The expected value of the function satisfies $\mathbb{E}_{\alpha^1,\ldots,\alpha^d}[f_j] = \sum_{c \in \mathcal{C}(q)} e_j^T A_{q,c}\, \sigma_c$.

Given that the underlying game is normalized (see Definition 2) and $d = \Omega\left(\frac{\log m}{\varepsilon^2}\right)$, each entry of $A_{q,c}$ is between 0 and $\frac{\varepsilon}{2\sqrt{6d \log m}}$.

This entry-wise bound implies that for any $c \in \mathcal{C}(q)$ and $\chi_1, .., \chi_c, .., \chi_d, \chi_c' \in U$ the following Lipscihtz condition holds for f_j:

$$|f_j(\chi_1, \ldots, \chi_c, \ldots, \chi_d) - f_j(\chi_1, \ldots, \chi_c', \ldots, \chi_d)| \leq \frac{\varepsilon}{2\sqrt{6d \log m}}.$$

Using McDiarmid's inequality (see Sect. 2) we get that

$$\Pr_{\alpha^1,\ldots,\alpha^d}(|f_j - \mathbb{E}[f_j]| \geq \varepsilon/4) \leq \frac{2}{m^3}.$$

Say, $\mathcal{E}$ denotes the event that for all $j \in [m]$, we have $|f_j - \mathbb{E}[f_j]| \leq \varepsilon/4$. Using the union bound we get that $\Pr_{\alpha^1,\ldots,\alpha^d}(\mathcal{E}) \geq 1 - 2/m^2$. Therefore, the probabilistic method guarantees the existence of mixed strategies $x_c \in U_{q,c}(y)$, for $c \in \mathcal{C}(q)$, that satisfy $\mathcal{E}$. Note that to obtain such a collection of mixed strategies the expected number of times that we need to sample—the product distribution $\prod_{c \in \mathcal{C}(q)} \alpha^c$—is at most two.

Say that mixed strategies $x_c \in U_{q,c}(y)$, for $c \in \mathcal{C}(q)$, satisfy event $\mathcal{E}$. Next we will show that y is an ε-best response of q against $(x_c)_{c \in \mathcal{C}(q)}$, and z. Overall, this implies that $y \in U_{p,q}(z)$, and we can set $E_{p,q}(z, y) = (x_c)_{c \in \mathcal{C}(q)}$.

Mixed strategies x_cs satisfy $|f_j(x_1, \ldots, x_d) - \mathbb{E}[f_j]| \leq \varepsilon/4$ for all $j \in [m]$. That is,

$$\left| \sum_{c \in \mathcal{C}(d)} e_j^T A_{q,c}\, x_c - \sum_{c \in \mathcal{C}(q)} e_j^T A_{q,c}\, \sigma_c \right| \leq \frac{\varepsilon}{4} \qquad \forall j \in [m]. \tag{4}$$

Using inequality (4) for each j in the support of distribution $y \in \Delta^m$, we have

$$\left| \sum_{c \in \mathcal{C}(d)} y^T A_{q,c}\, x_c - \sum_{c \in \mathcal{C}(q)} y^T A_{q,c}\, \sigma_c \right| \leq \frac{\varepsilon}{4}. \tag{5}$$

Note that y satisfies inequality (3) in the linear program, i.e., y is an $\varepsilon/2$-best response against σ_cs and z. Using inequalities (4) and (5) to bound the change in the left-hand-side of (3) and the right-hand-side of (3) respectively, we get that y is an ε-best response against x_cs and z:

$$y^T A_{q,p} z + \sum_{c \in \mathcal{C}(q)} y^T A_{q,c}\, x_c \geq e_j^T A_{q,p} z + \sum_{c \in \mathcal{C}(q)} e_j^T A_{q,c}\, x_c - \varepsilon \qquad \forall j \in [m]$$

Therefore, if $\text{LP}(p, q, z, y)$ is feasible then $y \in U_{p,q}(z)$. Also, note that the size of $\text{LP}(p, q, z, y)$ is at most $O(nm|U|)$, therefore we can solve the linear program in time polynomial in $|U|$. As mentioned above, given a feasible solution of $\text{LP}(p, q, z, y)$, to obtain $E_{p,q}(y, z)$ (i.e., a collection of mixed strategies $(x_c)_{c \in \mathcal{C}(q)}$ that satisfy $\mathcal{E}$) the expected number of times that we need to sample is at most two. This establishes the running time bound stated in the lemma, and we get the desired claims. □

Next we prove the main result.

Theorem 1. *Given an n-player m-action normalized polymatrix game over a tree, Algorithm 1 determines an ε-Nash equilibrium of the game in expected time*

$$m^{O\left(\frac{\log m(\log m + \log n - \log \varepsilon)}{\varepsilon^4}\right)}.$$

Proof. Let $G = (V, E)$ be the underlying tree of the given normalized polymatrix game. First, we will prove that Algorithm 1 necessarily finds a mixed strategy in U_r, for the root r of G, in the specified amount of time. Hence, via Lemma 1, we get that Algorithm 1 successfully finds an ε-Nash equilibrium of the game.

As mentioned above, it was established in [1] that every n-player m-action game admits an $\varepsilon/2$-Nash equilibrium $(\hat{x}_q)_{q \in [n]}$ where each $\hat{x}_q \in U$.[5] Hence, for each parent-child pair $(p, q) \in E$ there exists at least one set $U_{p,q}$ which is nonempty; in particular, $\hat{x}_q \in U_{p,q}(\hat{x}_p)$. Moreover, for $z = \hat{x}_p$ and $y = \hat{x}_q$ the relaxation $\text{LP}(p, q, z, y)$ is guaranteed to be feasible. Therefore, contingent on the fact that the "if" condition in Step 8 and 17 is performed correctly, we get that Algorithm 1 is guaranteed to move up the tree with non-empty $U_{p,q}$s and, finally, find a mixed strategy in U_r.

Specifically, the correctness of the "if" condition (which we establish below) ensures that for an $\varepsilon/2$-Nash equilibrium $(\hat{x}_p)_{p \in [n]}$ and the sets $U_{p,q}$s populated by the algorithm we have $\hat{x}_q \in U_{p,q}(\hat{x}_p)$ for every parent-child pair $(p, q) \in E$. This follows via an inductive argument over levels of the tree: if q is a leaf node then $\hat{x}_q$ is an ε best response against $\hat{x}_p$ and we get the desired containment $\hat{x}_q \in U_{p,q}(\hat{x}_p)$. Furthermore, using the induction hypothesis that $\hat{x}_c \in U_{q,c}(\hat{x}_q)$ for all $c \in \mathcal{C}(q)$, we get that the "if" condition in Step 8 will be satisfied for $\hat{x}_q$ and $\hat{x}_p$, i.e., the algorithm will include $\hat{x}_q$ in $U_{p,q}(\hat{x}_p)$ and the inductive claim holds. In particular, this observation implies that the algorithm will never encounter the situation wherein the set $U_{p,q}(x)$ is remain empty for all $x \in U$ after the for loops, i.e., the algorithm will always run to completion. It is also relevant to note that the algorithm can set $E_{p,q}(\hat{x}_q, \hat{x}_p)$ to be any tuple $(x_c)_{c \in \mathcal{C}(q)}$ that satisfies the best response condition for $\hat{x}_q$ and $\hat{x}_p$. That is, it is not necessary that the algorithm sets $E_{p,q}(\hat{x}_q, \hat{x}_p) = (\hat{x}_c)_{c \in \mathcal{C}(q)}$. But still, the above mentioned argument goes though and we get that the algorithm always runs to completion.

The "if" condition in Step 8 and 17 is performed $O(n|U|^2)$ times. Next we show that the "if" condition is verified correctly in expected time $|U|^{O\left(\frac{\log m}{\varepsilon^2}\right)}$. This overall establishes the stated claims.

[5] The change from ε-Nash equilibrium to $\varepsilon/2$-Nash equilibrium can be easily addressed by adjusting the size of U.

If the number of children of a player q is $o\left(\frac{\log m}{\varepsilon^2}\right)$ then we can go over the entire set $\prod_{c\in\mathcal{C}(q)} U_{q,c}(y)$ in time $|U|^{o\left(\frac{\log m}{\varepsilon^2}\right)}$ and determine whether the "if" condition in Step 8 is satisfied. The same argument works in Step 17, if the number of children of the root r is $o\left(\frac{\log m}{\varepsilon^2}\right)$.

For the remainder of the proof we consider the other case wherein the number of children of q (or the root r) is $\Omega\left(\frac{\log m}{\varepsilon^2}\right)$. In this case we verify the "if" condition in Step 8 (and Step 17) by solving the linear-programming relaxation $\text{LP}(p, q, z, y)$ and employing Lemma 2. Note the size of $\text{LP}(p, q, z, y)$ is $O(n|U|)$ and hence (again, via Lemma 2) in expected time polynomial in $|U|$ we can test if $y \in U_{p,q}(z)$ and find $E_{p,q}(z, y)$. Recall that this test is guaranteed to succeed for the $\varepsilon/2$-Nash equilibrium $(\hat{x}_q)_{q\in[n]}$, since the corresponding $\text{LP}(p, q, z, y)$s will be feasible. Hence, we get that Algorithm 1 proceeds up the tree with $\hat{x}_q$s, and eventually after processing the root r finds an ε-Nash equilibrium of the game.

Steps 8 and 17 are executed $O(n|U|^2)$ times, and the expected running time of these steps is $|U|^{O\left(\frac{\log m}{\varepsilon^2}\right)}$. These observations establish the time complexity of the algorithm and complete the proof. $\square$

Acknowledgement. This work was supported by NSF grants CNS-0846025, CCF-1101470, CNS-1254169, SUTD grant SRG ESD 2015 097, along with a Microsoft Research Faculty Fellowship, a Google Faculty Research Award, a Linde/ SISL Post-doctoral Fellowship and a CMI Wally Baer and Jeri Weiss postdoctoral fellowship. Katrina Ligett gratefully acknowledges the support of the Charles Lee Powell Foundation. The bulk of the work was conducted while Georgios Piliouras was a postdoctoral scholar at Caltech.

References

1. Babichenko, Y., Barman, S., Peretz, R.: Simple approximate equilibria in large games. In: Proceedings of the Fifteenth ACM Conference on Economics and Computation, pp. 753–770. ACM (2014)
2. Cai, Y., Daskalakis, C.: On minmax theorems for multiplayer games. In: Proceedings of the Twenty-Second Annual ACM-SIAM Symposium on Discrete Algorithms, pp. 217–234. SIAM (2011)
3. Chen, X., Deng, X., Teng, S.H.: Settling the complexity of computing two-player nash equilibria. J. ACM (JACM) **56**(3), 14 (2009)
4. Daskalakis, C.: On the complexity of approximating a Nash equilibrium. ACM Trans. Algorithms (TALG) **9**(3), 23 (2013)
5. Daskalakis, C., Goldberg, P.W., Papadimitriou, C.H.: The complexity of computing a nash equilibrium. SIAM J. Comput. **39**(1), 195–259 (2009)
6. Daskalakis, C., Mehta, A., Papadimitriou, C.: A note on approximate Nash equilibria. In: Spirakis, P.G., Mavronicolas, M., Kontogiannis, S.C. (eds.) WINE 2006. LNCS, vol. 4286, pp. 297–306. Springer, Heidelberg (2006)
7. Daskalakis, C., Mehta, A., Papadimitriou, C.: Progress in approximate Nash equilibria. In: Proceedings of the 8th ACM Conference on Electronic Commerce. pp. 355–358. ACM (2007)

8. Daskalakis, C., Papadimitriou, C.H.: On a network generalization of the minmax theorem. In: Albers, S., Marchetti-Spaccamela, A., Matias, Y., Nikoletseas, S., Thomas, W. (eds.) ICALP 2009, Part II. LNCS, vol. 5556, pp. 423–434. Springer, Heidelberg (2009)
9. Deligkas, A., Fearnley, J., Savani, R., Spirakis, P.: Computing approximate nash equilibria in polymatrix games. In: Liu, T.-Y., Qi, Q., Ye, Y. (eds.) WINE 2014. LNCS, vol. 8877, pp. 58–71. Springer, Heidelberg (2014)
10. Elkind, E., Goldberg, L.A., Goldberg, P.: Nash equilibria in graphical games on trees revisited. In: Proceedings of the 7th ACM Conference on Electronic Commerce, pp. 100–109. ACM (2006)
11. Howson Jr., J.T.: Equilibria of polymatrix games. Manage. Sci. **18**(5–part–1), 312–318 (1972)
12. Kearns, M., Littman, M.L., Singh, S.: Graphical models for game theory. In: Proceedings of the Seventeenth Conference on Uncertainty in Artificial Intelligence, pp. 253–260. Morgan Kaufmann Publishers Inc. (2001)
13. Lipton, R.J., Markakis, E., Mehta, A.: Playing large games using simple strategies. In: Proceedings of the 4th ACM Conference on Electronic Commerce, pp. 36–41. ACM (2003)
14. McDiarmid, C.: On the method of bounded differences. Surv. Comb. **141**(1), 148–188 (1989)
15. Ortiz, L.E., Kearns, M.: Nash propagation for loopy graphical games. In: Advances in Neural Information Processing Systems. pp. 793–800 (2002)
16. Rubinstein, A.: Inapproximability of nash equilibrium. In: Proceedings of the Forty-Seventh Annual ACM Symposium on Theory of Computing (STOC) (2015)
17. Tsaknakis, H., Spirakis, P.G.: An optimization approach for approximate Nash equilibria. In: Deng, X., Graham, F.C. (eds.) WINE 2007. LNCS, vol. 4858, pp. 42–56. Springer, Heidelberg (2007)
18. Yanovskaya, E.B.: Equilibrium points in polymatrix games. Lithuanian Mathematical Journal (1968) (in Russian)

Abstracts and Brief Announcements

Commitment in First-Price Auctions

Yunjian Xu[1](✉) and Katrina Ligett[2]

[1] Engineering Systems and Design, Singapore University of Technology and Design, Singapore, Singapore
xuyunijian@gmail.com
[2] Computing and Mathematical Sciences and Division of the Humanities and Social Sciences, California Institute of Technology, Pasadena, USA

We study a variation of the single-item sealed-bid first-price auction where one bidder (the leader) is given the option to publicly pre-commit to a distribution from which her bid will be drawn. We formulate the auction as a two-stage Stackelberg game: in the first round, one bidder (the leader) is allowed to publicly commit to a mixed strategy; in the second round, the other bidders submit their bids simultaneously. Given the publicly known commitment of the leader, the other bidders simultaneously play (possibly randomized) actions in the second stage.

Intuitively, the leader may have incentive to commit to a distribution with support below what she would have played in a simultaneous first price auction, because in doing so, she induces her opponents to lower their bids as well. Our results provide support for this intuition. For example, we see that even a very simple form of commitment, in which the leader announces to bid zero (effectively, to exit the auction) with some positive probability p and to bid some other announced real number with probability $1-p$, can strictly (and significantly) benefit both bidders in a simple two-bidder setting.

We also completely characterize the optimal commitment strategy for the leader in terms of the bidder valuations, for arbitrary numbers of bidders. The characterized optimal commitment, together with best responses on the part of the followers, forms a subgame perfect equilibrium (SPE). We find that if the leader has the highest or the second highest valuation, then the two bidders with highest valuations strictly benefit from the presence of a committing bidder. This result establishes the leader's optimal commitment as a *coordinative* mechanism that allows the top two valued bidders to collude without money transfer. Somewhat surprisingly, compared with the simultaneous first-price auction, the leader's optimal commitment yields the top two bidders the same net utility benefit. This observation could eliminate possible conflicts on who should commit and who should follow.

A leader's optimal commitment may result in *inefficient* outcomes, i.e., the highest-valued bidder does not always win the item. Because the leader's optimal commitment benefits the bidders but hurts the welfare, it must decrease the auctioneer's revenue. Indeed, the auctioneer's revenue strictly decreases whenever the leader (and the highest valued follower) strictly benefits from the commitment.

Y. Xu—This research was supported in part by NSF grants CCF-0910940 and CNS-1254169, the Charles Lee Powell Foundation, and a Microsoft Research Faculty Fellowship.

M. Hoefer (Ed.): SAGT 2015, LNCS 9347, p. 299, 2015.
DOI: 10.1007/978-3-662-48433-3_23

Brief Announcement: Effect of Strategic Grading and Early Offers in Matching Markets

Hedyeh Beyhaghi[1](✉), Nishanth Dikkala[2], and Éva Tardos[1]

[1] Department of Computer Science, Cornell University, 336 Gates Hall, Ithaca, NY 14853, USA
{hedyeh,eva}@cs.cornell.edu

[2] Department of Electrical Engineering and Computer Science, Massachusetts Institute of Technology, Cambridge, USA
nishanthd@csail.mit.edu

In this paper we consider the effect of strategic behavior in matching markets as school graduates get assigned to jobs (or to further education) reacting to multiple incentives:

- Companies want to hire the best students,
- Students want to take the best jobs,
- Schools want to help their graduating students take great jobs.

Strategic suppression of grades, as well as early offers and contracts, are well-known phenomena in this matching process. To help the placement of their students, schools often like suppressing grades, especially grades of their top performing students. An important reason for suppressing grades is the desire to have better placements for all students in the school, not only the top performing students. Suppressing grades is an explicit policy of many high schools and universities, but the same effect can also be achieved by allowing grade inflation, or by not having clear grading guidelines. When a large fraction of the class receives a grade of A, the expressiveness of grades suffer, effectively creating the same effect as suppressing grades by other schools. Similarly, randomness in how grades are assigned also decreases the information content of the transcripts.

Students and companies or schools of further education also behave strategically, acting to get better jobs or to improve the quality of students they can hire or attract. One tool in this area is making early offers, referred to as the unraveling of the matching market. Companies at times make offers to students quite a bit before they graduate, based on transcripts with significant amount of course work, and hence important information is still missing. Students often accept these early offers, or even apply for them.

Effect of Strategic Grading. In the first part of the paper, we study the loss in social welfare resulting from strategic grading by schools using a model introduced by Ostrovsky and Schwarz [in *American Economic Journal* 2010]. To do

A full version of the paper can be found at http://arxiv.org/abs/1507.02718.
Supported in part by NSF grants CCF-0910940 and CCR-1215994, ONR grant N00014-08-1-0031, and a Google Research Grant.

M. Hoefer (Ed.): SAGT 2015, LNCS 9347, pp. 300–302, 2015.
DOI: 10.1007/978-3-662-48433-3_24

so, we need to model the way that placement of a student with ability a in a job with quality q, will contribute to welfare. We assume that the resulting welfare is a monotone increasing function of both a and q, and the effects of these two contributing factors are separable. Concretely, we assume that the resulting welfare is expressed as $f(q)g(a)$ with both f and g nondecreasing, and g also concave. We think of grades as a form of signaling, and identify all grades with the expected ability of the group of students who receive that grade, and assume that all employers are risk-neutral and aim to hire students with higher expected ability.

We consider the decrease of the quality of the matching resulting from the fully strategic grading used by a school, i.e., the price of anarchy of the information disclosure game of Ostrovsky and Schwarz. For this part of the paper we use the continuous model of Ostrovsky and Schwarz, assuming that there are infinitely many students and each school is infinitesimally small. Their main result is that (under mild assumption, such as the equilibrium being continuous), at the unique Nash equilibrium of their grading game, schools disclose the right amount of information, so that students and employers will not find it profitable to contract early. We give a bound on the loss of quality focusing on the case when welfare is measured as aq.

Theorem 1. *If the students aggregate ability distribution is uniform, and f and g are the identity, the loss of efficiency in connected equilibria of the strategic grading game is bounded by* 1.36*, and give an example with loss of 1.07.*

Effect of Unraveling of the Matching Market. While Ostrovsky and Schwarz show that early contracting is not advantageous under an equilibrium grading policy of their information disclosure game, we observe that early contracting is increasingly common. Schools do aim to optimize the placements of their students, but we believe that they do not fully optimize grading. The pervasiveness of early contracting does suggest that the information released in grades is not at the equilibrium of the disclosure game. Exact optimization of grades at the full generality proposed by the model is not feasible, or even advisable as grades play many roles, including motivating the students.

To study the effect of unraveling we consider a two stage game where in stage one some student-employer pairs can agree on early contracts. In the second stage grades are released based on each school's grading policy, and the remaining students and jobs are matched based on their grades. In this two stage game, prospective employers have to make decisions about early offers without any grade information about the students, solely based on the school that the student attends, where we assume that the distribution of students and grading policy in each school is public knowledge.

Theorem 2. *Suppose functions f and g are increasing and g is concave. Assuming overall student abilities are uniformly distributed, the efficiency loss of matching is at most a factor of 2 under any grading strategy, and this bound is tight.*

We also consider the quality of the matching resulting from early contracting in isolation, assuming fully informative grading, and again focusing on the case when welfare is measured as aq. When all schools have identical (and uniform) distribution of student abilities we improve the above bound of 2 to 4/3.

Theorem 3. *The loss of efficiency for the two-stage assignment game in the case that all of the schools have uniform distribution of students on the same interval of abilities is at most* $\frac{4}{3}$*, and give an example with loss of 1.11.*

Brief Announcement: New Mechanisms for Pairwise Kidney Exchange

Hossein Efsandiari[1(✉)] and Guy Kortsarz[2]

[1] University of Maryland, College Park, MD, USA
hossein@cs.umd.edu
[2] Rutgers University, Camden, NJ, USA
guyk@camden.rutgers.edu

Abstract. In this paper, we consider the *pairwise kidney exchange game*. Ashlagi et al. [1] present a 2-approximation randomized truthful mechanism for this problem. We note that the variance of the utility of an agent in this mechanism may be as large as $\Omega(n^2)$, which is not desirable in a real application. Here, we resolve this issue by providing a 2-approximation randomized truthful mechanism in which the variance of the utility of each agent is at most $2 + \epsilon$. Later, we derandomize our mechanism and provide a *deterministic* mechanism such that, if an agent deviates from the mechanism, she does not gain more than $2\lceil \log_2 m \rceil$.

1 Introduction

Kidney transplant is the only treatment for several types of kidney diseases. Since people have two kidneys and can survive with only one kidney, they can potentially donate one of their kidneys. It may be the case that a patient finds a family member or a friend willing to donate her kidney. Nevertheless, at times the kidney's donor is not compatible with the patient. Consider two incompatible patient-donor pairs. If the donor of the first pair is compatible with the patient of the second pair and vise-versa, we can efficiently serve both patients without affecting the donors.

To make the pool of donor-patient pairs larger, hospitals combine their lists of pairs to one big pool, trying to increase the number of treated patients by exchanging pairs from different hospitals. This process is managed by some national supervisor. A centralized mechanism can look at all of the hospitals together and increase the total number of kidney exchanges. The problem is that each hospital is interested in is teasinge the numbe of its own served patients. Thus, the hospital may not report some patient-donors pairs, namely, the hospital may report a partial list. This partial list is then matched by the national supervisors. Undisclosed set of pairs are matched by the hospitals locally, without the knowledge of the supervisor. This may have a negative effect on the number of served patients.

The full version of this paper is available at http://arxiv.org/abs/1507.02746.
H. Efsandiari and G. Kortsarz—Supported by NSF grant number 1218620.

M. Hoefer (Ed.): SAGT 2015, LNCS 9347, pp. 303–304, 2015.
DOI: 10.1007/978-3-662-48433-3_25

Notations and Definitions. In a kidney exchange game we have a graph G, and each agent owns a disjoint set of vertices of G. A mechanism for this game receives the reported set of vertices from every agent. After the vertices are reported, the mechanism chooses a matching on the induced subgraph of the reported vertices. After this global run, each agent matches her unmatched vertices, including her undisclosed vertices, privately.

In this game, the utility of each agent is the expected number of her matched vertices and the *social welfare* of a mechanism is the size of the output matching. A kidney exchange mechanism is *truthful* if no agent gains more by reporting a partial subset of her vertex. A kidney exchange mechanism F is α-approximation if for every graph G the number of matched vertices in the maximum matching of G is at most α times the expected number of matched vertices by F in G.

Related Work. Ashlagi et al. [1] provide a randomized 2-approximation truthful mechanism for the multi-agent kidney exchange game. Moreover, they show that there is no truthful mechanism with an approximation ratio better than 8/7. They also introduce a deterministic 2-approximation truthful mechanism for two player kidney exchange game. However, they conjectured that there is no deterministic constant-approximation truthful mechanism for the multi-agent kidney exchange game, even with three agents.

Our Results. In this paper, first, we show that the variance of the utility of an agent in the mechanism proposed by Ashlagi et al. may be as large as $\Omega(n^2)$, where n is the number of vertices. The variance of the utility can be interpreted as the risk of the agent caused by the randomness in the mechanism. Indeed, in a real application agents prefer to take less risk for the same expected utility. In this paper, we provide a tool to lower the variance of the utility of each agent in a kidney exchange mechanism, while keeping the expected utility of each agent the same. We used this tool to provide a 2-approximation randomized truthful mechanism in which the variance of the utility of each agent is at most $2 + \epsilon$.

Theorem 1. *There exists a truthful 2-approximation mechanism for multi-agent kidney exchange such that the variance of the utility of each agent is at most $2 + \epsilon$, where ϵ is an arbitrary small constant.*

Interestingly, we could apply our technique to design a 2-approximation *deterministic* mechanism such that if an agent deviates from the mechanism, she does not gain more than $2\lceil \log_2 m \rceil$. We call such a mechanism *almost truthful.* Indeed, in a practical scenario an almost truthful mechanism is likely to imply a truthful mechanism. To the best of our knowledge this is the first deterministic mechanism for the multi-agent kidney exchange game.

Theorem 2. *There exists an almost truthful deterministic 2-approximation mechanism for multi-agent kidney exchange.*

Reference

1. Ashlagi, I., Fischer, F., Kash, I.A., Procaccia, A.D.: Mix and match: A strategyproof mechanism for multi-hospital kidney exchange. Games and Economic Behavior (2013)

Brief Announcement: On Effective Affirmative Action in School Choice

Yun Liu(✉)

Department of Economics, Copenhagen Business School,
Porcelænshaven 16A, 2000 Frederiksberg, Denmark
yliueco@gmail.com

The purpose of affirmative action in school choice is to create a more equal and diverse social environment, i.e., granting students from disadvantaged social groups preferential treatments in school admission decisions to maintain racial, ethnic or socioeconomic balance. Recent evidences from both academia and practice, however, indicate that implementing affirmative action policies in school choice problems may induce substantial welfare loss on the purported beneficiaries (i.e., *minority* students). Using the *minority reserve* policy in the *student optimal stable mechanism* as an example, this paper addresses the following two questions: what are the causes of such perverse consequence, and when we can effectively implement affirmative action policies without unsatisfied outcomes.

The minimal requirement of an effective affirmative action is that it should not make at least one minority student strictly worse off, while leaves all the rest minority students weakly worse off. We first show that a variant of the Ergin-acyclicity structure, *type-specific acyclicity*, is necessary and sufficient to guarantee this minimal effectiveness criterion in a stable mechanism. Next, we introduce a more demanding effectiveness criterion which requires implementing a (stronger) affirmative action does not harm any minority students. We show that a stable mechanism makes no minority students strictly worse off if and only if the priority structure is *strongly type-specific acyclic*. These two findings clearly reveal the source of perverse affirmation actions in school choice, which also implies that such adverse effects are not coincidences but rather a fundamental property concealed in the priority structures.

We then responses to the second question such that when we can effectively implement affirmative action policies without unsatisfied outcomes. We show that priority structures in practice are very unlikely to be neither type-specific acyclic nor strongly type-specific acyclic. More specifically, if there is more than one minority student has lower priority than any majority students in two (or more) schools in any given priority structures, then the priority structure is not type-specific acyclic. Strongly type-specific acyclicity condition is even more confined which requires no minority student has lower priority than any majority

A draft full version is available at http://ssrn.com/abstract=2617877.

Y. Liu—Financial support from the Center for research in the Foundations of Electronic Markets (CFEM), supported by the Danish Council for Strategic Research, is gratefully acknowledged.

M. Hoefer (Ed.): SAGT 2015, LNCS 9347, pp. 305–306, 2015.
DOI: 10.1007/978-3-662-48433-3_26

students in all schools. This result suggests that even if helping disadvantaged social groups is deemed desirable for the society, caution should be exercised when applying affirmative action to rebalance education opportunities among different social groups.

Brief Announcement: Resource Allocation Games with Multiple Resource Classes

Roy B. Ofer and Tami Tamir(✉)

School of Computer Science, The Interdisciplinary Center, Herzliya, Israel
tami@idc.ac.il

1 Introduction

Media streaming is among the most popular services provided over the Internet. The lack of a central authority that controls the users, motivates the analysis of Media on Demand (MoD) services using game theoretic concepts. We define and study the corresponding resource-allocation game, where users correspond to self-interested players who choose a MoD server with the objective of minimizing their individual cost. Each user requires a certain media-file which determines the user's class. A server provides both broadcasting and storage needs. Accordingly, the user's cost function encompasses both negative and positive, class-dependent, congestion effects.

An instance of the multi-class resource allocation game is defined by a tuple $G = \langle I, M, A, U \rangle$, where I is the set of players, M is the set of servers and A is the set of classes. Let $n = |I|$ and $m = |M|$. Each player belongs to a single class from A, thus, $I = I_1 \cup I_2 \cdots \cup I_{|A|}$, where all players from I_k belong to class k. For $i \in I$, let $a_i \in A$ denote the class to which player i belongs. The parameter $U \in \mathbb{R}^+$ is the class activation-cost, which is assumed to be uniform for all classes.

An *allocation* of players to servers is a function $f : I \rightarrow M$. For a given allocation, the *load* on a server j, denoted by L_j, is the number of players assigned to j, and $L_{j,k}$ denotes the number of players from I_k assigned to j.

The cost of a player i in an allocation f consists of two components: the load on the server the player is assigned to (as in job scheduling games [3]), and the player's share in the class activation-cost (as in cost-sharing games [1]). Formally, $c_f(i) = L_{f(i)} + \frac{U}{L_{f(i),a_i}}$. Note that the class activation-cost is shared evenly among the players from this class serviced by a server. Our model generalizes the one studied in [2], where all players belong to the same class.

In MoD systems, the bandwidth required for transmitting a certain media-file corresponds to one unit of load. The storage cost of a media-file on a server is shared by the users requiring its transmission that are serviced by the server.

2 Our Results and Techniques

We provide answers to the basic questions regarding resource allocation games with multiple resource classes. Namely, equilibrium existence, convergence, calculation and efficiency. We prove that a *Pure Nash Equilibrium* (PNE) exists

M. Hoefer (Ed.): SAGT 2015, LNCS 9347, pp. 307–308, 2015.
DOI: 10.1007/978-3-662-48433-3_27

for any instance by presenting an exact potential function for the game. By analyzing this function we show:

Theorem 1. *For every instance G, better-response dynamics converges to a PNE within $O(n^4)$ steps.*

The equilibrium inefficiency is analyzed with respect to the objective of minimizing the maximal cost among the players. That is, given an allocation f, the social cost of f is given by $c_{max}(f) = \max_{i \in I} c_f(i)$.

We provide several lower bounds on the social cost of an optimal solution, and then combine them to present the following tight bound on the Price of Anarchy (PoA).

Theorem 2. *For the family $\mathcal{G}$ of resource allocation games with multiple resource classes, $PoA(\mathcal{G}) = m$.*

We show that for any number of servers, there exists a game for which the Price of Stability (PoS) is $2 - \frac{1}{m}$. This upper bound is almost matched. Our main result is a polynomial time algorithm that constructs a PNE whose social cost is at most twice the optimum. For two servers, we present a simpler algorithm and our analysis is tight.

Theorem 3. *For the family $\mathcal{G}$ of resource allocation games with multiple resource classes, $2 - \frac{1}{m} \leq PoS(\mathcal{G}) \leq m$. For two servers, $PoS(\mathcal{G}) = 3/2$.*

Our algorithms for finding a stable assignment with low social cost are based on two new methods:

1. While all the players create the same unit-load on the servers, our algorithms group the players into sets, based on their classes. An initial assignment is found by considering these sets as an instance of a multiple-knapsack packing problem with arbitrary-size elements. This method enables analysis of the assignment using known packing techniques and their properties.
2. The stabilization phase that follows the initial assignment consists of iterations in which the algorithm may reassign complete sets of players, or perform a *supervised* sequence of improving steps. The sequence is initiated by one player i, and is then limited to players of i's class who may benefit from following i by performing exactly the same migration. Analyzing the configuration after each improving step is complex; however, it is possible to analyze the effect of each supervised sequence of improving steps on the potential function and to bound the cost of an assignment derived by this method.

References

1. Anshelevich, E., Dasgupta, A., Kleinberg, J.M., Tardos, É., Wexler, T., Roughgarden, T.: The price of stability for network design with fair cost allocation. SIAM J. Comput. **38**(4), 1602–1623 (2008)
2. Feldman, M., Tamir, T.: Conflicting congestion effects in resource allocation games. J. Oper. Res. **60**(3), 529–540 (2012)
3. Vöcking, B.: In: Nisan, N., Roughgarden, T., Tardos, E., Vazirani, V. (eds.) Algorithmic Game Theory. ch. 20: Selfish load balancing. Cambridge University Press, NY, (2007)

Brief Announcement: On the Fair Subset Sum Problem

Gaia Nicosia[1], Andrea Pacifici[2], and Ulrich Pferschy[3](✉)

[1] Università degli Studi Roma Tre, Roma, Italy
nicosia@ing.uniroma3.it
[2] Università degli Studi di Roma Tor Vergata, Roma, Italy
andrea.pacifici@uniroma2.it
[3] University of Graz, Graz, Austria
pferschy@uni-graz.at

The Addressed Problem. This work considers a fair allocation problem, called the *Fair Subset Sum problem* (FSSP), where a common and bounded resource is to be shared among two agents A and B. Each agent is willing to select a set of items from an available ground set of weighted items. We address two types of items structure. In the *Separate Items* case, each agent owns a set of items having nonnegative weights and each agent can only use its own items. In the *Shared Items* case the agents select the items from the same ground set. The total utility of a solution x is denoted by $U(x)$ and is equal to the sum of weights over all selected items. A solution x is feasible if $U(x) \leq 1$. We denote the set of feasible solutions by X. Obviously, the computation of the system optimum solution x^* amounts to the solution of a classical *subset sum problem* (SSP).

A desirable feasible solution divides the utilities of the selected objects in a *fair* way. The trade-off between fairness and efficiency, quantified by the so-called *Price of Fairness* [1], is the central theme in this paper. We use two different notions of fairness focusing on the individual utilities obtained by each agent. *Maximin fairness*: Based on the principle of *Rawlsian justice*, a maximin fair solution is such that the least happy agent gains as much as possible. *Proportional fairness* [4]: A solution is *proportional fair*, if any other solution does not give a relative improvement for one agent which is larger than the relative loss inflicted on the other agent. Recently, we also extended our work to a normalized variant of maximin fairness known as *Kalai-Smorodinski* fairness.

The contribution of this work regards the assessment of the quality of fair solutions compared to the global system optimum solution for a general multi-agent allocation problem. A special focus is put on FSSP for which we perform a thorough analysis in terms of the largest items weight. In particular, we measure the Price of Fairness (PoF) as defined in [1] and introduced in [2]. By definition PoF $\in [0, 1]$ and PoF ≈ 1 indicates that the fair solution may be arbitrarily far from system optimum, while PoF ≈ 0 means that fair solutions are always close to optimum. We denote the Prices of Fairness corresponding to maximin and proportional fair solutions as PoF_{MM} and PoF_{PF}.

It is easy to provide worst case instances of FSSP with PoF $= 1$, corresponding to pathological instances in which items weights are either very large or very

M. Hoefer (Ed.): SAGT 2015, LNCS 9347, pp. 309–311, 2015.
DOI: 10.1007/978-3-662-48433-3_28

small. To avoid similar unrealistic settings, as for many bin-packing heuristics, the Price of Fairness is studied as a function of an upper bound $\alpha \leq 1$ on the size of the maximum item weight. Formally, we extend the definition of PoF from [1] as follows: Let $\mathcal{I}_\alpha$ denote the set of all instances of our FSSP where all items weights are not larger than α, let U_I^* be the maximum total utility of instance I, and let f_I be a fair solution of instance I. Then we can define the Price of Fairness depending on α as follows: $\text{PoF}(\alpha) = \sup_{I \in \mathcal{I}_\alpha} \frac{U_I^* - U(f_I)}{U_I^*}$. Obviously, PoF = PoF(1).

Summary of Results. First, we provide some basic, yet very general results for proportional fair solutions valid for any k-agent problem. In particular, we show that if there exists a proportional fair solution, then such a solution is unique and maximizes the product of agents utilities. Similar results were derived in different contexts (e.g. for convex and compact utility sets) but here we provide simple but fairly general proofs. Moreover, we present a general upper bound on the Price of Fairness for any proportional fair solution. Additionally, for two agents it is possible to show that the global utility of a proportional fair solution, if it exists, is never smaller than that of a maximin fair solution. (This is not true anymore as soon as the number of agents becomes three.) When the problem is symmetric, we give a full characterization of proportional fair solutions by showing that if a fair solution exists then it also system optimal and all agents get the same utility value.

Table 1. Separate items.

α	PoF_{MM}	PoF_{PF}
1	1	
[2/3, 1]	$2 - 1/\alpha$	1/2
[1/2, 2/3]	1/2	
(0, 1/2]	$\left[\frac{1}{\lceil\frac{1}{\alpha}\rceil}, \alpha\right]$	$\left[\frac{1}{\lceil\frac{1}{\alpha}\rceil}, \alpha\right]$

Table 2. Shared items.

α	PoF_{MM}
1	1
[2/3, 1]	$2\alpha - 1$
[1/3, 2/3]	1/3
(0, 1/3]	$\left[\frac{1}{1+2\lceil\frac{1}{2\alpha}\rceil}, \alpha\right]$

The main contribution of this work is an almost complete description of PoF(α) for FSSP for all values of α. A summary of the obtained expressions (resp. intervals) for PoF is given in Table 1 for separate items and Table 2 for shared items. Note that for the case of shared items we prove that, if a proportional fair solution exist, it is the system optimum and coincides also with the maximin fair solution. Conversely, whenever a proportional fair solution does not exists, it is possible to provide almost tight bounds on PoF_{MM}. A different variant of a game-theoretic setting of SSP with two agents was recently considered in [3].

References

1. Bertsimas, D., Farias, V., Trichakis, N.: The price of fairness. Oper. Res. **59**(1), 17–31 (2011)

2. Caragiannis, I., Kaklamanis, C., Kanellopoulos, P., Kyropoulou, M.: The efficiency of fair division. In: Leonardi, S. (ed.) WINE 2009. LNCS, vol. 5929, pp. 475–482. Springer, Heidelberg (2009)
3. Darmann, A., Nicosia, G., Pferschy, U., Schauer, J.: The subset sum game. Eur. J. Oper. Res. **233**(3), 539–549 (2014)
4. Kelly, F.P., Maulloo, A.K., Tan, D.K.H.: Rate control in communication networks: shadow prices, proportional fairness, and stability. J. Oper. Res. Soc. **49**, 237–252 (1998)

Brief Announcement: Computation of Fisher-Gale Equilibrium by Auction

Yurii Nesterov and Vladimir Shikhman(✉)

Center for Operations Research and Econometrics (CORE),
Catholic University of Louvain (UCL), 34 voie du Roman Pays,
1348 Louvain-la-Neuve, Belgium
{yurii.nesterov,vladimir.shikhman}@uclouvain.be

Abstract. We study the Fisher model of a competitive market from the algorithmic perspective. For that, the related convex optimization problem due to Gale and Eisenberg, [3], is used. The latter problem is known to yield a Fisher equilibrium under some structural assumptions on consumers' utilities, e.g. homogeneity of degree 1, homotheticity etc. We just assume the concavity of consumers' utility functions. For this case we suggest a novel concept of Fisher-Gale equilibrium by introducing consumers' utility prices. We develop a subgradient-type algorithm from Convex Analysis to compute a Fisher-Gale equilibrium by auction. In worst case, the number of price updates needed to achieve the ε-tolerance is proportional to $\frac{1}{\varepsilon^2}$.

Keywords: Fisher equilibrium · Computation of equilibrium · Price adjustment · Convex optimization · Subgradient methods · Auction

The concept of Fisher equilibrium for a competitive market dates back to 1891, see e.g. [1]. Due to Fisher's model, consumers buy goods by spending given wealths in order to maximize their utility functions. There are fixed amounts of supplied goods available at the market. Fisher equilibrium comprises of optimal consumption bundles and equilibrium prices which clear the market of goods. Aiming at the efficient computation of a Fisher equilibrium, a related convex optimization problem has been proposed in [3]. This so-called Gale's problem consists of maximizing an aggregated logarithmic utility function subject to market feasibility constraints. The feasibility constraints ensure that the aggregated consumption does not exceed the fixed amounts of supplied goods. The solutions of Gale's problem give equilibrium allocations for the Fisher market. Moreover, the Lagrange or dual multipliers for its feasibility constraints yield equilibrium prices. It is crucial to point out that the solutions of Gale's problem provide Fisher equilibrium mainly if the wealths are fully spent within the budget constraints. To guarantee the latter fact some structural assumptions on the consumers' utility functions have been made in the literature.

Our goal is to examine the applicability of the Gale's approach by departing from the structural assumptions on the consumers' utilities. We just assume the

M. Hoefer (Ed.): SAGT 2015, LNCS 9347, pp. 312–313, 2015.
DOI: 10.1007/978-3-662-48433-3_29

concavity of consumers' utility functions. In this case, we cannot guarantee the full spending of wealths within the budget constraints. This is the main reason why under our concavity assumption the concepts of Fisher and Gale equilibrium may come apart. To deal with this difficulty, we generalize both concepts of Fisher and Gale equilibrium by introducing the so-called utility prices attributed to consumers. Prices of utility allow to dynamically transfer the utility of a consumption bundle to a common numéraire. Using this transferable utility (cf. [2]), we introduce a novel concept of Fisher-Gale equilibrium. Here, consumers maximize their revenues as the differences of transferred utilities and expenditures expressed in a numéraire. It turns out that Fisher and Gale equilibria can be viewed as Fisher-Gale equilibrium. In particular, for Fisher equilibrium the utility prices are inverse Lagrange multipliers associated to budget constraints. For Gale equilibrium, the utility prices appear as ratios of wealths to achieved utilities. The latter gives rise to the efficient computation of a Fisher-Gale equilibrium by following the Gale's approach.

In this paper we develop a subgradient-type algorithm to compute a Fisher-Gale equilibrium by Gale's approach. Its convergence properties are crucially based on Convex Analysis. The price adjustment corresponds to the quasi-monotone subgradent method for nonsmooth convex minimization, recently suggested in [4]. As objective function for the latter method we take the total logarithmic revenue of the market. Equilibrium prices can be then characterized as its minimizers. We refer to [5] for the similar approach using the total excessive revenue of the market. In order to decentralize prices, the auction design is used: *consumers settle and update their individual prices, and producers sell at the highest offer price.* Our price adjustment is based on a tâtonnement procedure, i.e. the prices change proportionally to consumers' individual excess supplies. While our algorithm proceeds, the market clearance is achieved on average. The latter means that during the price adjustment supply meets demand statistically. Altogether, the sequence of highest offer prices, historical averages of consumption bundles and historical averages of utility prices generated by our algorithm, converges to a Fisher-Gale equilibrium. In worst case, the number of price updates needed to achieve the ε-tolerance is proportional to $\frac{1}{\varepsilon^2}$. Note that this rate of convergence is optimal for nonsmooth convex minimization.

References

1. Brainard, W.C., Scarf, H.: How to compute equilibrium prices in 1891. Am. J. Econ. Sociol. **64**, 57–83 (2005)
2. Cherchye, L., Demuynck, T., De Rock, B.: Is utility transferable? a revealed preference analysis. Theoret. Econ. **10**, 57–83 (2015)
3. Eisenberg, E., Gale, D.: Consensus of subjective probabilities: the pari-mutuel method. Ann. Math. Stat. **30**, 165–168 (1959)
4. Nesterov, Y., Shikhman, V.: Quasi-monotone subgradient methods for nonsmooth convex minimization. J. Optim. Theory Appl. **165**, 917–940 (2015)
5. Nesterov, Y., Shikhman, V.: Excessive revenue model of competitive markets. In: Proceedings of the IMU-AMS Special Session on Nonlinear Analysis and Optimization (to appear)

Author Index

Printed in the United States
By Bookmasters